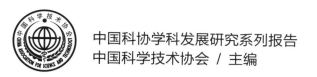

中国科协学科发展研究系列报告
中国科学技术协会 / 主编

2022—2023

学科发展报告
综合卷

中国科协科学技术创新部　组编

中国科学技术出版社
·北　京·

图书在版编目（CIP）数据

2022—2023 学科发展报告综合卷 / 中国科学技术协会主编；中国科协科学技术创新部组编 . -- 北京：中国科学技术出版社，2024.6

（中国科协学科发展研究系列报告）

ISBN 978-7-5236-0694-0

Ⅰ. ① 2… Ⅱ. ①中… ②中… Ⅲ. ①科学技术 – 技术发展 – 研究报告 – 中国 –2022–2023 Ⅳ. ① N12

中国国家版本馆 CIP 数据核字（2024）第 090144 号

策　　划　刘兴平　秦德继
责任编辑　赵　佳
封面设计　北京潜龙
正文设计　中文天地
责任校对　吕传新
责任印制　徐　飞

出　　版　中国科学技术出版社
发　　行　中国科学技术出版社有限公司
地　　址　北京市海淀区中关村南大街 16 号
邮　　编　100081
发行电话　010-62173865
传　　真　010-62173081
网　　址　http://www.cspbooks.com.cn

开　　本　787mm × 1092mm　1/16
字　　数　545 千字
印　　张　23
版　　次　2024 年 6 月第 1 版
印　　次　2024 年 6 月第 1 次印刷
印　　刷　河北鑫兆源印刷有限公司
书　　号　ISBN 978-7-5236-0694-0 / N · 324
定　　价　150.00 元

2022—2023

学科发展报告综合卷

专　　家（按姓氏笔画排序）

于　剑	万建民	马　峥	王　巍	王国强
王家臣	仇保兴	卞　科	邓中亮	甘晓华
巩馥洲	乔　杰	任胜利	刘　冰	刘培一
江　亿	杜　鹏	李　东	杨学明	杨新泉
肖　川	肖　宏	吴爱祥	陈剑平	赵巍胜
胡洪营	黄　维	梅旭荣	雒建斌	樊代明
潘云涛	魏炳波			

编写组成员（按姓氏笔画排序）

于相龙	马林岭	王　云	王一然	王子莲
王元元	王丽伟	王建华	王建康	王晓欢
王雁玲	王瑞元	王新河	石慧峰	史　琳
冯培松	朱　军	朱林琦	朱斯语	伍军红
刘　挺	刘广立	刘荣志	刘奕群	许大涛
阮竹恩	孙　璐	孙　巍	孙楚原	巫寅虎
李　政	李宏岩	李紫阳	杨　睿	杨胜利
杨晓静	杨韵龙	邱利民	何　颖	邹亚飞

沈　超	宋　浦	张　昂	张　瑶	张　霞
张学莹	张哲源	陆宴辉	陈　曦	陈小兵
陈少军	范开军	林伯阳	罗二仓	庞思平
郑传临	赵　勇[1]	赵　勇[2]	赵　晶	赵扬玉
姜　军	姜开利	姚颖垠	袁军鹏	倪中福
徐　莉	徐婉桢	高　歌	郭子锋	郭文涛
陶　益	黄　雷	曹红玉	董照辉	韩佳慧
程维红	漆洪波	戴东旭	戴其根	魏东斌

注：1 恶性肿瘤学科

2 航空科学技术学科

序

习近平总书记强调，科技创新能够催生新产业、新模式、新动能，是发展新质生产力的核心要素。要求广大科技工作者进一步增强科教兴国强国的抱负，担当起科技创新的重任，加强基础研究和应用基础研究，打好关键核心技术攻坚战，培育发展新质生产力的新动能。当前，新一轮科技革命和产业变革深入发展，全球进入一个创新密集时代。加强基础研究，推动学科发展，从源头和底层解决技术问题，率先在关键性、颠覆性技术方面取得突破，对于掌握未来发展新优势，赢得全球新一轮发展的战略主动权具有重大意义。

中国科协充分发挥全国学会的学术权威性和组织优势，于 2006 年创设学科发展研究项目，瞄准世界科技前沿和共同关切，汇聚高质量学术资源和高水平学科领域专家，深入开展学科研究，总结学科发展规律，明晰学科发展方向。截至 2022 年，累计出版学科发展报告 296 卷，有近千位中国科学院和中国工程院院士、2 万多名专家学者参与学科发展研讨，万余位专家执笔撰写学科发展报告。这些报告从重大成果、学术影响、国际合作、人才建设、发展趋势与存在问题等多方面，对学科发展进行总结分析，内容丰富、信息权威，受到国内外科技界的广泛关注，构建了具有重要学术价值、史料价值的成果资料库，为科研管理、教学科研和企业研发提供了重要参考，也得到政府决策部门的高度重视，为推进科技创新做出了积极贡献。

2022 年，中国科协组织中国电子学会、中国材料研究学会、中国城市科学研究会、中国航空学会、中国化学会、中国环境科学学会、中国生物工程学会、中国物理学会、中国粮油学会、中国农学会、中国作物学会、中国女医师协会、中国数学会、中国通信学会、中国宇航学会、中国植物保护学会、中国兵工学会、中国抗癌协会、中国有色金属学会、中国制冷学会等全国学会，围绕相关领域编纂了 20 卷学科发展报告和 1 卷综合报告。这些报告密切结合国家经济发展需求，聚焦基础学科、新兴学科以及交叉学科，紧盯原创性基础研究，系统、权威、前瞻地总结了相关学科的最新进展、重要成果、创新方法和技术发展。同时，深入分析了学科的发展现状和动态趋势，进行了国际比较，并对学科未来的发展前景进行了展望。

报告付梓之际，衷心感谢参与学科发展研究项目的全国学会以及有关科研、教学单位，感谢所有参与项目研究与编写出版的专家学者。真诚地希望有更多的科技工作者关注学科发展研究，为不断提升研究质量、推动成果充分利用建言献策。

前　言

2022 年 10 月，党的二十大报告指出我国已经“进入创新型国家行列”，到 2035 年，要“实现高水平科技自立自强，进入创新型国家前列”。近年来，国家不断持续加强学科建设投入，全面谋划科技创新工作，在科技界的共同努力下，我国科技创新取得新的历史性成就。随着各学科研究成果的不断涌现，我国的学科发展环境、学科发展水平也呈现持续向好的良性态势。

为充分把握学科发展规律和发展趋势，引领学科发展方向，助力构筑全面均衡发展的高质量学科体系，中国科协持续深入开展学科发展研究。2022 年，中国科协组织中国数学会、中国物理学会、中国化学会、中国电子学会等 20 个全国学会，分别就数学、高端科学仪器与集成电路先进装备、化学、水环境、制冷及低温工程、人工智能、工业互联网、航空科学技术、航天科学技术、现代先进毁伤技术及效应评估、采矿工程、种子学、材料学、粮油科学与技术、基础农学、植物保护学、作物学、恶性肿瘤、产科学、城市科学共计 20 个学科 / 领域的发展情况进行系统研究，编辑出版了《中国科协学科发展研究系列报告》。

受中国科协委托，重庆同方知网科技发展有限公司和《中国学术期刊（光盘版）》电子杂志社有限公司组织有关单位和相关专家在上述 20 个学科发展报告的基础上，编写了《2022—2023 学科发展报告综合卷》（以下简称《综合卷》）。《综合卷》分为 3 章和附件部分：第一章以科学技术部、教育部、国家统计局、国家自然科学基金委员会、新华社等官方网站的资料文件和 WOS、ESI、OECD、CNKI 等数据库以及第三方权威报告中的数据信息作为客观数据来源依据，结合各学科领域专家的咨询建议，梳理 2022—2023 年我国科学技术领域学科发展的总体情况，并从宏观层面总结我国科技领域总体学科发展态势，评析学科发展存在的问题与挑战，提出促进学科发展的启示与建议；第二章以 20 个学科发展报告为基础，对 20 个学科近年的研究现状、国内外研究进展比较、学科发展趋势与展望等进行综述；第三章为20个学科发展报告主要内容的英文介绍；附件为2022—2023年与学科发展相关的部分资料。《综合卷》第二、第三章学科排序根据承担研究任务的相关全国学会在中国科协的编号顺序排列。

为做好《综合卷》的研究工作，《中国学术期刊（光盘版）》电子杂志社有限公司成立了专家组和编写组，由《中国学术期刊（光盘版）》电子杂志社有限公司、中国科学院文献情报中心、中国科学院科技战略咨询研究院、中国科学技术信息研究所等单位的情报学领域专家以及20个学会选派的学科专家组成。其中：第一章和附件由《中国学术期刊（光盘版）》电子杂志社组织开展研究工作并完成编写任务；第二章和第三章由相关学科对应的20个学会负责组织开展研究工作并完成编写任务。

《综合卷》第一章内容是对我国科技领域学科发展的宏观概述，由于各个数据源的学科分类标准和数据颗粒度不相一致，不同指标的数据难以放在相同的学科分类标准下进行比较，因此相关指标数据均采用原始数据的学科分类予以呈现。另外，由于各个学科成果众多而篇幅有限，所以第一章中的“学科发展动态”一节无法将各学科重大成果一一列举，敬请见谅。《综合卷》的第二章主要在20个学科发展报告的基础上综合而成，仅概括相关学科的重要进展和总体情况，难以完整地反映我国科技领域学科发展的全貌。尤其考虑到编写时间仓促，数据统计口径众多，任务量大，加之调研工作确有一定难度，虽经多方努力，仍难免存在问题或遗憾，敬请广大读者谅解并指正。在此，也谨向所有为《综合卷》编写付出辛勤劳动的专家学者和工作人员表达诚挚的谢意！

本书编写组

2024年3月

目 录

第三章　相关学科进展与趋势（英文）

附件

第一章

学科发展综述

第一节　学科发展概况

学科是具有特定研究对象的知识系统，是一种知识分类体系，一般从破解交叉领域需求的难题及前沿领域分化、发展而形成，学科的形成是科学知识体系化的重要特征和标志。当前，新一轮科技革命和产业变革加速推进，人工智能、量子信息、脑科学等新兴研究领域不断出现和发展，学科呈现高度分化的趋势；但是当今科学已经进入了以多学科交叉融合为主要特征的“大科学”时代，单一学科的思维模式已难以实现科技创新的重大突破以及复杂问题的解决，因此，各学科之间的交叉融合又在日益深入。学科之间高度分化又日益融合的特征正在逐步改变着学科结构，学科体系随着科研和知识体系的动态发展在不断动态调整，但总体来说，当今科学时代的学科体系仍然是包括基础研究、技术研究、应用研究的结构化知识体系。在科学研究中形成的知识体系又反过来成为科学研究和人才培养的框架，是科学技术创新发展的重要基础。学科建设是开展知识活动的组织通过人财物等方面的投入，提升学科知识生产与传播能力的过程。因此，大力推进各学科建设是学科发展的目的与需要，也是推动我国科学技术高质量发展的基石和核心动力，对提升我国自主创新能力、建设创新型国家具有重要意义。

近年来，党中央全面分析国际科技创新竞争态势，深入研判国内外发展形势，针对我国科技事业面临的突出问题和挑战，坚持把科技创新摆在国家发展全局的核心位置，全面谋划科技创新工作，科技创新取得新的历史性成就。2022 年 10 月，党的二十大报告指出我国已经“进入创新型国家行列”，到 2035 年，要“实现高水平科技自立自强，进入创新型国家前列”。这一目标的实现，离不开我国各学科领域研究水平的提升和发展。

我国在 2022—2023 年继续明确学科建设的顶层设计路径，密集出台了一系列促进学科发展的改革措施；持续稳定增加学科发展投入，特别是基础研究财政投入力度，优化支出结构；国家自然科学基金委员会（以下简称基金委）、教育部、各高校等主体联合发力，进一步优化学科布局。在各项举措和投入的指引下，我国各类科研产出成果丰富，根据多个国际数据库统计，我国已经成为国际科研论文产出最多的国家，2022 年 SCI 论文量为 73.56 万篇，占世界份额的 28.9%。并且，科研产出已经从量的积累迈向质的飞跃，根据 ESI 数据库统计，我国在工程学、化学、材料科学等 7 个科技领域高水平论文量位居世界第一。根据世界知识产权组织（WIPO）最新发布的数据，2022 年中国的申请人提交了 158 万件专利申请，远超美国（50 万件）等其他国家，已连续多年成为专利申请数量最多的国家，专利成果转化率进一步提升。与此同时，国家科技战略力量逐渐强化，形成层次清晰、目标明确的各类科技创新要素聚集平台，作为各学科学术成果交流平台的科技期刊也取得显著的进步；大科学装置发展迅速、高端仪器国产化情况进一步改善，科学数据市场化建设加快，对学科发展的支撑作用日益凸显；科技人才队伍不断壮大，高水平领军人才持续涌现；各学科的国内外交流合作规模均在不断扩大，我国不仅国内同行交流频繁而活跃，也在主动融入全球科技创新网络，促进了我国学科发展水平的整体提升和青年人才的培养。本节从学科建设的顶层设计、学科建设投入、学科研究成果、基础支撑条件、学科平台建设、科技人才队伍、国内外交流合作等能够有力反映学科发展概况的资料、数据出发进行分析，力求客观反映我国学科的整体发展现状。

一、2022—2023 年促进学科发展的相关政策

2023 年 2 月 21 日，中共中央总书记习近平在中共中央政治局第三次集体学习时强调切实加强基础研究，夯实科技自立自强根基。要优化基础学科建设布局，支持重点学科、新兴学科、冷门学科和薄弱学科发展，推动学科交叉融合和跨学科研究，构筑全面均衡发展的高质量学科体系①。

新时代下，我国牢牢把握建设世界科技强国的战略目标，始终把基础研究和前沿技术摆在突出位置，针对各学科科技事业的发展，密集出台了一系列重大改革举措，本部分从基础研究、高技术研究、应用研究与成果转化、创新生态环境 4 个方面对 2022—2023 年促进学科发展的相关政策进行总结和梳理。

（一）面向世界科技前沿，加强基础研究

基础研究是整个科学体系的源头，是所有技术问题的总机关。加强基础研究，是实现高水平科技自立自强的必然要求，是建设世界科技强国的必由之路。党的十八大以来，党中央把提升原始创新能力摆在更加突出的位置，成功组织了一批重大基础研究任务、建成一批重大科技基础设施，基础前沿方向重大原创成果持续涌现②，我国基础研究整体实力和学术水平显著增强。支持基础研究发展的政策体系不断完善，制定出台《国务院关于全面加强基础科学研究的若干意见》《加强“从 0 到 1”基础研究工作方案》《新形势下加强基础研究若干重点举措》等文件，加强基础研究工作。

---

① 新华网．习近平在中共中央政治局第三次集体学习时强调　切实加强基础研究　夯实科技自立自强根基［EB/OL］.（2023-02-22）. http://www.news.cn/politics/leaders/2023-02/22/c_1129386597.htm.

② 习近平．加强基础研究　实现高水平科技自立自强［J］. 求是，2023（15）：4-9.

2021年2月26日，国务院新闻办公室举行新闻发布会，会上透露科技部将根据中央要求制定《基础研究十年行动方案（2021—2030）》[①]，方案耗时一年时间完成制定。该方案是未来十年国家层面推动基础研究发展的政策指导性和任务部署性文件，明确我国基础研究发展的总体思路、发展目标和重点任务，提出具体的行动措施。基础研究十年行动方案将围绕重大原始创新和关键核心技术突破，着力打造基础研究体系化力量、建设高水平人才队伍、夯实科研基础能力、营造良好创新生态，增强基础研究对创新发展的源头供给和支撑引领作用。

2021年12月24日，《中华人民共和国科学技术进步法》[②]（以下简称《科学技术进步法》）完成了第二次修订，并自2022年1月1日起施行。修订后的《科学技术进步法》将基础研究单列成章，并提出：国家加强基础研究能力建设，尊重科学发展规律和人才成长规律，强化项目、人才、基地系统布局，为基础研究发展提供良好的物质条件和有力的制度保障。国家加强规划和部署，推动基础研究自由探索和目标导向有机结合，围绕科学技术前沿、经济社会发展、国家安全重大需求和人民生命健康，聚焦重大关键技术问题，加强新兴和战略产业等领域基础研究，提升科学技术的源头供给能力。国家鼓励科学技术研究开发机构、高等学校、企业等发挥自身优势，加强基础研究，推动原始创新。

2022年8月，教育部印发《关于加强高校有组织科研　推动高水平自立自强的若干意见》[③]，明确了加强高校有组织科研的重点举措，提出要加快目标导向的基础研究重大突破。研究设立基础研究和交叉学科专项，启动基础学科研究中心、医药基础研究创新中心建设。持续实施“高等学校基础研究珠峰计划”。

2022年9月，为鼓励企业加大创新投入，支持我国基础研究发展，财政部、税务总局公布了《关于企业投入基础研究税收优惠政策的公告》[④]，对企业投入基础研究相关税收政策作出如下规定：对企业出资给非营利性科研机构、高等学校和政府性自然科学基金用于基础研究的支出，在计算应纳税所得额时可按实际发生额在税前扣除，并可按100%在税前加计扣除；对非营利性科研机构、高等学校接收企业、个人和其他组织机构基础研究资金收入，免征企业所得税。

2023年7月，《求是》2023年第15期发表了习近平总书记文章《加强基础研究　实现高水平科技自立自强》。总书记指出，当前新一轮科技革命和产业变革深入发展，学科交叉融合不断推进，科学研究范式发生深刻变革，科学技术和经济社会发展加速渗透融

---

① 国家发展和改革委员会．“十四五”规划《纲要》名词解释之18：基础研究十年行动方案［EB/OL］．（2021-12-24）．https://www.ndrc.gov.cn/fggz/fzzlgh/gjfzgh/202112/t20211224_1309267.html.

② 全国人民代表大会常务委员会．中华人民共和国科学技术进步法（2021年修订）［Z］．（2021-12-24）．https://www.most.gov.cn/xxgk/xinxifenlei/fdzdgknr/fgzc/flfg/202201/t20220118_179043.html.

③ 教育部．教育部印发《关于加强高校有组织科研　推动高水平自立自强的若干意见》［EB/OL］．（2022-08-29）．http://www.moe.gov.cn/jyb_xwfb/gzdt_gzdt/s5987/202208/t20220829_656091.html.

④ 财政部，税务总局．关于企业投入基础研究税收优惠政策的公告［Z］．（2022-09-30）．https://www.gov.cn/zhengce/zhengceku/2022-10/11/content_5717700.htm.

合，基础研究转化周期明显缩短，国际科技竞争向基础前沿前移。应对国际科技竞争、实现高水平科技自立自强，推动构建新发展格局、实现高质量发展，迫切需要我们加强基础研究，从源头和底层解决关键技术问题。总书记强调，加强基础研究，要从以下 6 个方面着手：第一，强化基础研究前瞻性、战略性、系统性布局；第二，深化基础研究体制机制改革；第三，建设基础研究高水平支撑平台；第四，加强基础研究人才队伍建设；第五，广泛开展基础研究国际合作；第六，塑造有利于基础研究的创新生态。

（二）面向国家重大需求，支持高技术研究

工程科技是推动人类进步的发动机，是产业革命、经济发展、社会进步的有力杠杆。党的二十大报告提出要“以国家战略需求为导向，集聚力量进行原创性引领性科技攻关，坚决打赢关键核心技术攻坚战”。

国务院总理李强 2023 年 5 月 5 日主持召开国务院常务会议，审议通过《关于加快发展先进制造业集群的意见》①。会议指出，发展先进制造业集群，是推动产业迈向中高端、提升产业链供应链韧性和安全水平的重要抓手，有利于形成协同创新、人才集聚、降本增效等规模效应和竞争优势。会议强调，要把发展先进制造业集群摆到更加突出位置，坚持全国一盘棋，引导各地发挥比较优势，在专业化、差异化、特色化上下功夫，做到有所为、有所不为。要统筹推进传统产业改造升级和新兴产业培育壮大，促进技术创新和转化应用，推动高端化、智能化、绿色化转型，壮大优质企业群体，加快建设现代化产业体系。要坚持有效市场和有为政府更好结合，着力营造产业发展的良好生态。2023 年的政府工作报告中提到，今年要重点加快建设现代化产业体系，围绕制造业重点产业链，集中优质资源合力推进关键核心技术攻关等。国务院总理李强 2023 年 8 月 25 日主持召开国务院常务会议，审议通过《医药工业高质量发展行动计划（2023—2025 年）》《医疗装备产业高质量发展行动计划（2023—2025年）》②。会议强调，医药工业和医疗装备产业是卫生健康事业的重要基础，事关人民群众生命健康和高质量发展全局。要着力提高医药工业和医疗装备产业韧性和现代化水平，增强高端药品、关键技术和原辅料等供给能力，加快补齐我国高端医疗装备短板。要着眼医药研发创新难度大、周期长、投入高的特点，给予全链条支持，鼓励和引导龙头医药企业发展壮大，提高产业集中度和市场竞争力。要充分发挥我国中医药独特优势，加大保护力度，维护中医药发展安全。要高度重视国产医疗装备的推广应用，完善相关支持政策，促进国产医疗装备迭代升级。要加大医工交叉复合型人才培养力度，支持高校与企业联合培养一批医疗装备领域领军人才。

教育部印发《关于加强高校有组织科研　推动高水平自立自强的若干意见》，就推动

---

① 中国政府网．李强主持召开国务院常务会议　审议通过关于加快发展先进制造业集群的意见等［EB/OL］．（2023-05-05）. https://www.gov.cn/yaowen/2023-05/05/content 5754266.htm?_esid=3948532.

② 新华社．李强主持召开国务院常务会议　审议通过《医药工业高质量发展行动计划（2023—2025 年）》等［EB/OL］．（2023-08-25）. https://www.gov.cn/yaowen/liebiao/202308/content_6900133.htm.

高校充分发挥新型举国体制优势，加强有组织科研，全面加强创新体系建设，着力提升自主创新能力，更高质量、更大贡献服务国家战略需求作出部署。高校要立足新发展阶段、贯彻新发展理念、构建新发展格局，把服务国家战略需求作为最高追求，坚持战略引领、组织创新、深度融合、系统推进的指导原则，要在继续充分发挥好自由探索基础研究主力军和主阵地作用，持续开展高水平自由探索研究的基础上，加快变革高校科研范式和组织模式，强化有组织科研，更好服务国家安全和经济社会发展面临的现实问题和紧迫需求，为实现高水平科技自立自强、加快建设世界重要人才中心和创新高地提供有力支撑。明确了加强高校有组织科研的重点举措之一是：加快国家战略急需的关键核心技术重大突破。实施“有组织攻关重大项目培育计划”，布局建设集成攻关大平台。实施“千校万企”协同创新伙伴计划。深入实施高等学校人工智能、区块链、碳中和科技创新行动。

在人工智能领域，科技部等六部门2022年8月印发《关于加快场景创新以人工智能高水平应用促进经济高质量发展的指导意见》①，提出要以“企业主导、创新引领、开放融合、协同治理”为基本原则，着力打造人工智能重大场景。围绕高端高效智能经济、安全便捷智能社会建设、高水平科研活动、国家重大活动和重大工程打造重大场景。随后，科技部印发《科技部关于支持建设新一代人工智能示范应用场景的通知》②，提出要首批支持建设10个示范应用场景，分别是：智慧农场、智能港口、智能矿山、智能工厂、智慧家居、智能教育、自动驾驶、智能诊疗、智慧法院和智能供应链。2023年1月，工业和信息化部（以下简称工信部）等十七部门联合印发了《“机器人+”应用行动实施方案》③，提出“到2025年，制造业机器人密度较2020年实现翻番，服务机器人、特种机器人行业应用深度和广度显著提升，机器人促进经济社会高质量发展的能力明显增强”的发展目标。要聚焦制造业、农业、建筑、能源、商贸物流、医疗健康、养老服务、教育、商业社区服务、安全应急和极限环境应用等十大经济发展和社会民生重点领域，深化“机器人+”应用。2022年年底，因ChatGPT的发布，生成式人工智能（Generative Artificial Intelligence，简称GenAI）引起了公众的广泛关注。2023年8月15日，由国家互联网信息办公室、国家发展和改革委员会（以下简称国家发展改革委）等七部门联合发布的《生成式人工智能服务管理暂行办法》④正式实行。该办法从技术发展与治理、服务规范、监督检查与法律

① 科技部. 科技部等六部门关于印发《关于加快场景创新以人工智能高水平应用促进经济高质量发展的指导意见》的通知［EB/OL］.（2022-08-12）. https://www.most.gov.cn/xxgk/xinxifenlei/fdzdgknr/fgzc/gfxwj/gfxwj2022/202208/t20220812_181851.html.

② 科技部. 科技部关于支持建设新一代人工智能示范应用场景的通知［EB/OL］.（2022-08-15）. https://www.most.gov.cn/xxgk/xinxifenlei/fdzdgknr/qtwj/qtwj2022/202208/t20220815_181874.html.

③ 工业和信息化部. 工业和信息化部等十七部门关于印发“机器人+”应用行动实施方案的通知［EB/OL］.（2023-01-19）. https://wap.miit.gov.cn/jgsj/zbys/wjfb/art/2023/art_ceee3ccbae884f458f6c94d19c1c057a.html.

④ 中国政府网. 生成式人工智能服务管理暂行办法［Z］.（2023-07-10）. https://www.gov.cn/zhengce/zhengceku/202307/content_6891752.htm.

责任等方面规定了各行业组织、企业、教育和科研机构等主体利用生成式人工智能技术提供生成式人工智能服务时的相关要求和规范，促进了生成式人工智能的健康发展和规范应用。

在电力领域，2022 年 8 月 29 日，工信部等五部门联合印发《加快电力装备绿色低碳创新发展行动计划》①（以下简称《行动计划》）。《行动计划》提出主要目标是：通过 5~8 年时间，电力装备供给结构显著改善，保障电网输配效率明显提升，高端化智能化绿色化发展及示范应用不断加快，国际竞争力进一步增强，基本满足适应非化石能源高比例、大规模接入的新型电力系统建设需要。煤电机组灵活性改造能力累计超过 2 亿千瓦，可再生能源发电装备供给能力不断提高，风电和太阳能发电装备满足 12 亿千瓦以上装机需求，核电装备满足 7000 万千瓦装机需求。重点任务是：①装备体系绿色升级行动；②电力装备技术创新提升行动；③网络化智能化转型发展行动；④技术基础支撑保障行动；⑤推广应用模式创新行动；⑥电力装备对外合作行动。并且，《行动计划》明确了电力装备绿色低碳发展的重点方向，即推进火电、水电、核电、风电、太阳能、氢能、储能、输电、配电及用电这 10 个领域电力装备绿色低碳发展。

在钢铁工业领域，工业和信息化部、国家发展和改革委员会、生态环境部联合发布《关于促进钢铁工业高质量发展的指导意见》（工信部联原〔2022〕6 号）②。提出坚持创新发展的基本原则：突出创新驱动引领，推进产学研用协同创新，强化高端材料、绿色低碳等工艺技术基础研究和应用研究，强化产业链工艺、装备、技术集成创新，促进产业耦合发展，强化钢铁工业与新技术、新业态融合创新。主要目标包括：力争到 2025 年，钢铁工业基本形成布局结构合理、资源供应稳定、技术装备先进、质量品牌突出、智能化水平高、全球竞争力强、绿色低碳可持续的高质量发展格局。创新能力显著增强。行业研发投入强度力争达到 1.5%，氢冶金、低碳冶金、洁净钢冶炼、薄带铸轧、无头轧制等先进工艺技术取得突破进展。关键工序数控化率达到 80% 左右，生产设备数字化率达到 55%，打造 30 家以上智能工厂。

在信息技术领域，2023 年 1 月，为贯彻落实《中华人民共和国数据安全法》，推动数据安全产业高质量发展，工信部等十六部门印发了《关于促进数据安全产业发展的指导意见》③。提出数据安全产业发展的目标为：到 2025 年，数据安全产业基础能力和综合实力明显增强。产业规模迅速扩大、核心技术创新突破、应用推广成效显著、产业生态完备有序。在

---

① 工业和信息化部. 工业和信息化部等五部门联合印发加快电力装备绿色低碳创新发展行动计划的通知［EB/OL］.（2022-08-29）. https://wap.miit.gov.cn/zwgk/zcwj/wjfb/tz/art/2022/art_4ccbd89465cc4336b88b19a02bbf473b.html.

② 工业和信息化部. 工业和信息化部 国家发展和改革委员会 生态环境部关于促进钢铁工业高质量发展的指导意见［EB/OL］.（2022-05-20）. https://www.ndrc.gov.cn/xwdt/ztzl/cjgyjjpwzz/bmgzqk/202205/t20220520_1324965.html.

③ 工业和信息化部. 工业和信息化部等十六部门关于促进数据安全产业发展的指导意见［EB/OL］.（2023-01-13）. https://wap.miit.gov.cn/jgsj/waj/wjfb/art/2023/art_e92c30f708884a3db7a77e135682ea8b.html.

提升产业创新能力方面的具体任务是：①加强核心技术攻关，推进新型计算模式和网络架构下数据安全基础理论和技术研究，支持后量子密码算法、密态计算等技术在数据安全产业的发展应用；②构建数据安全产品体系，加快发展数据资源管理、资源保护产品，重点提升智能化水平，加强数据质量评估、隐私计算等产品研发；③布局新兴领域融合创新，加快数据安全技术与人工智能、大数据、区块链等新兴技术的交叉融合创新，赋能提升数据安全态势感知、风险研判等能力水平。

（三）面向经济主战场和人民生命健康，实现基础与应用研究融通发展

应用研究主要指直接面向市场应用需求，能够将学术研究成果紧密连接到推动客观世界发展的一端的研究，为基础研究和工程技术提供重要的实践源泉。

2022 年 1 月 1 日起开始施行的《中华人民共和国科学技术进步法》（2021 年修订），专门设立“应用研究与成果转化”章节。其中提到：国家鼓励以应用研究带动基础研究，促进基础研究与应用研究、成果转化融通发展；国家加强面向产业发展需求的共性技术平台和科学技术研究开发机构建设，鼓励地方围绕发展需求建设应用研究科学技术研究开发机构；国家鼓励科学技术研究开发机构、高等学校加强共性基础技术研究，鼓励以企业为主导，开展面向市场和产业化应用的研究开发活动；国家加强科技成果中试、工程化和产业化开发及应用，加快科技成果转化为现实生产力；利用财政性资金设立的科学技术研究开发机构和高等学校，应当积极促进科技成果转化，加强技术转移机构和人才队伍建设，建立和完善促进科技成果转化制度；国家鼓励企业、科学技术研究开发机构、高等学校和其他组织建立优势互补、分工明确、成果共享、风险共担的合作机制，按照市场机制联合组建研究开发平台、技术创新联盟、创新联合体等，协同推进研究开发与科技成果转化，提高科技成果转移转化成效等内容。

教育部印发《关于加强高校有组织科研　推动高水平自立自强的若干意见》提出的加强高校有组织科研的重点举措包括：提升科技成果转移转化能力服务产业转型升级，启动实施“百校千项”高价值专利转化行动，加强国家知识产权试点示范高校建设；启动实施“百校千城”未来产业培育行动，进一步发挥好国家大学科技园国家级创新平台作用，试点未来产业科技园建设；围绕区域协调发展战略，发挥关键省份和节点城市作用，加强教育部创新平台和高水平科研机构建设。

企业是创新的主体，为鼓励企业创新，加速项目成果转化，2022 年 8 月，科技部、财政部联合印发《企业技术创新能力提升行动方案（2022—2023 年）》[①]（以下简称《方案》），从 10 个方面进一步加强企业技术创新。《方案》要求，到 2023 年年底，一批惠企创新政策落地见效，创新要素加速向企业集聚，各类企业依靠科技创新引领高质量发展

① 科技部，财政部. 科技部　财政部关于印发《企业技术创新能力提升行动方案（2022—2023 年）》的通知［Z］.（2022-08-05）. https://www.gov.cn/zhengce/zhengceku/2022-08/15/content_5705464.htm.

取得积极成效，一批骨干企业成为国家战略科技力量，一大批中小企业成为创新重要发源地，形成更加公平公正的创新环境。《方案》制定了支持企业加强技术创新的十大举措：推动惠企创新政策扎实落地、建立企业常态化参与国家科技创新决策的机制、引导企业加强关键核心技术攻关、支持企业前瞻布局基础前沿研究、促进中小企业成长为创新重要发源地、加大科技人才向企业集聚的力度、强化对企业创新的风险投资等金融支持、加快推进科技资源和应用场景向企业开放、加强产学研用和大中小企业融通创新、提高企业创新国际化水平。《方案》提出要在"十四五"国家重点研发计划应用类重点专项及部分科技创新 2030—重大项目中设立科技型中小企业项目。通过国家科技成果转化引导基金等支持科技型中小企业转移转化科技成果，提升技术创新水平。要支持企业与高校、科研院所共建一批新型研发机构等内容。

2022 年 9 月，科技部印发了《"十四五"技术要素市场专项规划》①（以下简称《规划》），围绕建设统一开放、竞争有序、制度完备、治理完善的高标准技术要素市场目标，提出若干重点任务。《规划》的一个重点任务是要健全科技成果产权制度，其中包括深化科技成果使用权、处置权和收益权改革。探索科技成果产权制度改革，开展赋予科研人员职务科技成果所有权或长期使用权试点，形成一批可操作、可复制、有效果的经验，在全社会范围内推广。要建立职务科技成果赋权机制，推动试点单位建立高效畅通的职务科技成果赋权管理制度、工作流程和决策机制，建立赋权科技成果的负面清单，明确转化职务科技成果各方的权利和义务。要优化科技成果转化全过程管理和服务体系，健全充分体现知识价值导向的职务科技成果转化收益分配机制，优化科技成果转化管理体系，畅通科技成果转移转化道路。要建立科技成果转化尽职免责机制，以是否符合中央精神和改革方向、是否有利于科技成果转化作为对科技成果转化活动的定性判断标准，实行审慎包容监管。

在生态环境领域，2022 年 9 月 19 日，科技部、生态环境部、住房和城乡建设部、气象局、林草局联合印发《"十四五"生态环境领域科技创新专项规划》②。指出我国生态环境保护面临的形势与挑战以及国际生态环境科技发展趋势，提出"十四五"我国生态环境科技发展需求包括：①深入打好污染防治攻坚战需要科技创新解决污染治理中难啃的"硬骨头"；②生态环境治理体系与治理能力现代化需要构建服务型科技创新体系，提升环保产业竞争力；③应对气候变化等全球共同挑战需要通过科技创新提出中国方案；④改善生态环境质量、保障公众健康需要依靠科技创新提升生态环境健康风险应对水平。

在能源领域，我国先后出台《中共中央、国务院关于完整准确全面贯彻新发展理念做好碳达峰碳中和工作的意见》、《国务院关于印发 2030 年前碳达峰行动方案的通知》（国

---

① 科技部. 科技部关于印发《"十四五"技术要素市场专项规划》的通知［Z］.（2022-09-30）. https://www.gov.cn/zhengce/zhengceku/2022-11/02/content_5723781.htm.

② 科技部. 科技部 生态环境部 住房和城乡建设部 气象局 林草局关于印发《"十四五"生态环境领域科技创新专项规划》的通知［EB/OL］.（2022-09-19）. https://www.most.gov.cn/xxgk/xinxifenlei/fdzdgknr/fgzc/gfxwj/gfxwj2022/202209/t20220926_182638.html.

发〔2021〕23号)、《国家发展改革委等部门关于促进炼油行业绿色创新高质量发展的指导意见》(发改能源〔2023〕1364号)、《国家发展改革委、国家能源局关于印发〈“十四五”现代能源体系规划〉的通知》(发改能源〔2022〕210号)等文件。提出扎实推进我国炼油行业绿色创新高质量发展的基本原则之一——创新驱动、自立自强。坚持科技是第一生产力、人才是第一资源、创新是第一动力。加强创新链和产业链对接，强化企业科技创新主体地位，发挥各类企业的创新优势和活力，着力提升行业自主创新能力。大力推进绿色低碳技术创新，以创新引领行业绿色发展[①]。

在通信领域，2022年1月28日，工信部发布《工业和信息化部关于大众消费领域北斗推广应用的若干意见》[②]，提出要提升北斗系统用户体验和竞争优势，将大众消费领域打造成为北斗规模化应用的动力引擎。“十四五”末，突破一批关键技术和产品，健全覆盖芯片、模块、终端、软件、应用等上下游各环节的北斗产业生态，培育20家以上专精特新“小巨人”企业及若干家制造业单项冠军企业，树立一批应用典型样板，建设一批融合应用示范工程，形成大众消费领域好用易用的北斗时空服务体系。在提升产业基础能力的意见中，提出要：①突破关键核心技术和产品，针对大众消费领域应用需求，重点突破短报文集成应用、融合卫星/基站/传感器的室内外无缝定位、自适应防欺骗抗干扰等关键技术，加快推进高精度、低功耗、低成本、小型化的北斗芯片及关键元器件研发和产业化，形成北斗与5G、物联网、车联网等新一代信息技术融合的系统解决方案；②构建北斗应用服务基础设施，完善北斗网络辅助公共服务平台建设，扩大平台用户规模，进一步提高北斗定位速度。

在交通运输领域，2022年5月交通运输部发布了《交通运输部促进科技成果转化办法》[③]，主要面向交通运输部所属事业单位，包括部属高校、科研机构和其他具有研发能力的事业单位的成果转化工作，明确转化方式、技术权益、机制建设、收益分配、转化激励、经费投入、绩效评价、人员兼职、离岗创业、法律责任等各环节、全流程制度内容。

(四)强化支撑基础，营造良好创新生态

2020年12月23日，国家发展改革委、中央网信办、工业和信息化部、国家能源局联合发布《关于加快构建全国一体化大数据中心　协同创新体系的指导意见》(发改高技〔2020〕1922号)[④]指出，数据是国家基础战略性资源和重要生产要素。以深化数据要素市

① 国家发展改革委．国家发展改革委等部门关于促进炼油行业绿色创新高质量发展的指导意见［EB/OL］．(2023-10-10)．https://www.ndrc.gov.cn/xxgk/zcfb/tz/202310/t20231025_1361500.html.

② 工业和信息化部．工业和信息化部关于大众消费领域北斗推广应用的若干意见［EB/OL］．(2022-01-28)．https://wap.miit.gov.cn/jgsj/dzs/wjfb/art/2022/art_f205fad69fc44a8da9b296577a8eaf38.html.

③ 交通运输部．交通运输部促进科技成果转化办法［Z］．(2023-02-22)．https://www.gov.cn/zhengce/zhengceku/2022-05/30/content_5693005.htm.

④ 国家发展改革委．关于加快构建全国一体化大数据中心　协同创新体系的指导意见［EB/OL］．(2020-12-23)．https://www.ndrc.gov.cn/xwdt/ztzl/dsxs/zcwj2/202201/t20220112_1311852.html.

场化配置改革为核心，优化数据中心建设布局，推动算力、算法、数据、应用资源集约化和服务化创新，对于深化政企协同、行业协同、区域协同，全面支撑各行业数字化升级和产业数字化转型具有重要意义。提出统筹规划，协同推进的原则：坚持发展与安全并重，统筹数据中心、云服务、数据流通与治理、数据应用、数据安全等关键环节，协同设计大数据中心体系总体架构和发展路径。以市场实际需求决定数据中心和服务资源供给。着眼引领全球云计算、大数据、人工智能、区块链发展的长远目标，适度超前布局，预留发展空间。正确处理政府和市场关系，破除制约大数据中心协同创新体系发展的政策瓶颈，着力营造适应大数据发展的创新生态，发挥企业主体作用，引导市场有序发展。

健康、良好、可持续的创新生态是学科发展的必要条件，这其中离不开高水平的科技人才队伍建设、学术治理环境的改善、全社会科学氛围和公民科学素质的提升。近两年，我国围绕营造良好创新生态，出台了一系列的政策举措。

教育部印发的《关于加强高校有组织科研　推动高水平自立自强的若干意见》提出，推进高水平人才队伍建设打造国家战略人才力量，依托重大科研平台组织实施重大科技任务和重大工程，培养造就一批战略科学家；推进科教融合、产教协同培育高质量创新人才，认定一批国家科教协同创新平台，深入实施基础学科拔尖学生培养计划和国家急需高层次人才培养专项；推进高水平国际合作，布局建设一批一流国际联合实验室等平台；推进科研评价机制改革营造良好创新生态，完善“双一流”建设动态监测系统，引导高校主动对接国家战略布局，提升支撑国家重大科技任务的能力，大力弘扬科学家精神，加强学风作风建设。

2022 年 10 月，中共中央办公厅、国务院办公厅印发《关于加强新时代高技能人才队伍建设的意见》[①]，制定了“力争到 2035 年，技能人才规模持续壮大、素质大幅提高，高技能人才数量、结构与基本实现社会主义现代化的要求相适应”的目标任务，并围绕加大高技能人才培养力度、完善技能导向的使用制度、建立技能人才职业技能等级制度和多元化评价机制、建立高技能人才表彰激励机制等方面做了具体部署。包括创新高技能人才培养模式、健全高技能人才岗位使用机制、拓宽技能人才职业发展通道、加大高技能人才表彰奖励力度等内容。2023 年 8 月，中共中央办公厅、国务院办公厅印发《关于进一步加强青年科技人才培养和使用的若干措施》，提出要引导支持青年科技人才服务高质量发展；支持青年科技人才在国家重大科技任务中“挑大梁”“当主角”；要加大基本科研业务费对职业早期青年科技人才稳定支持力度；要完善自然科学领域博士后培养机制；要更好发挥青年科技人才决策咨询作用；要提升科研单位人才自主评价能力；要减轻青年科技人才非科研负担；要加大力度支持青年科技人才开展国际科技交流合作；要加大青年科技人才生活服务保障力度；要加强对

---

① 中国政府网. 中共中央办公厅　国务院办公厅印发《关于加强新时代高技能人才队伍建设的意见》[EB/OL].（2022-10-07）. https://www.gov.cn/zhengce/2022-10/07/content_5716030.htm.

青年科技人才工作的组织领导；国家科技创新基地要大力培养使用青年科技人才等措施。

在探索分类评价和优化学术治理环境方面，继2021年7月16日国务院办公厅发布《关于完善科技成果评价机制的指导意见》（国办发〔2021〕26号）后，2022年9月，科技部等八部门印发了《关于开展科技人才评价改革试点的工作方案》[①]，针对人才评价“破四唯”后“立新标”不到位、资源配置评价改革不到位、用人单位评价制度建设不到位等突出问题，按照承担国家重大攻关任务的人才评价以及基础研究类、应用研究和技术开发类、社会公益研究类的人才评价，建立分类评价机制，引导各类科技人才人尽其才、才尽其用、用有所成。例如，针对承担国家重大攻关任务的人才评价，突出支撑国家重大战略需求导向，注重个人评价与团队评价相结合，对承担“卡脖子”国家重大攻关任务、国家重大科技基础设施建设任务等并作出重要贡献的科研人员，在岗位聘用、职称评审、绩效考核等方面，加大倾斜支持力度。针对基础研究类人才评价，实行以原创成果和高质量论文为标志的代表作评价，探索建立同行评价的责任机制，加大对重大科学发现和取得原创性突破的基础研究人员的倾斜支持。2022年11月，人力资源社会保障部办公厅印发《关于进一步做好职称评审工作的通知》[②]，提出以下要求：①动态调整职称评审专业；②科学制定职称评审标准；③合理设置论文和科研成果要求；④减少学历、奖项等限制性条件；⑤完善同行评价机制；⑥畅通职称评审绿色通道；⑦开展好职称“定向评价、定向使用”；⑧发挥用人单位主体作用；⑨优化职称评审服务。努力破解职称评审中的“一刀切”、简单化问题，持续深化职称制度改革。

在强化科普能力方面，2022年8月，科技部、中央宣传部、中国科学技术协会（以下简称中国科协）发布《“十四五”国家科学技术普及发展规划》[③]，制定了“十四五”期间的重点任务：①强化新时代科普工作价值引领功能；②加强国家科普能力建设，强化科普理论研究，增强科普创作能力，完善科普设施布局，构建全媒体科学传播矩阵，持续推进科普信息化建设，促进科普领域市场化发展；③推动科普工作全面发展，开展群众性科普活动，推动科普与学校教育深度融合，加强重点领域科普工作，加强应急科普工作，加强针对社会热点的科普，加强国防科普工作；④推动科学普及与科技创新协同发展；⑤抓好公民科学素质提升工作；⑥开展科普交流与合作，拓展国际科普交流机制，深入开展青少年国际科普交流，加强重点领域国际科普合作，促进与港澳台科普合作。2022年9月，中共中央办公厅、国务院办公厅印发了《关于新时代进一步加强科学技术普及

---

① 科技部，教育部，工业和信息化部，等．关于开展科技人才评价改革试点的工作方案［Z］．（2022-09-23）．https://www.gov.cn/zhengce/zhengceku/2022-11/10/content_5725957.htm.

② 人力资源社会保障部办公厅．人力资源社会保障部办公厅关于进一步做好职称评审工作的通知［Z］．（2022-11-30）．https://www.gov.cn/zhengce/zhengceku/2022-12/24/content_5733407.htm.

③ 中国政府网．科技部　中央宣传部　中国科协关于印发《“十四五”国家科学技术普及发展规划》的通知［EB/OL］．（2022-08-04）．https://www.gov.cn/zhengce/zhengceku/2022-08/16/content_5705580.htm.

工作的意见》[①]。为进一步加强科普工作，制定了“到 2035 年，公民具备科学素质比例达到 25%，科普服务高质量发展能效显著，科学文化软实力显著增强，为世界科技强国建设提供有力支撑”的发展目标，并围绕强化全社会科普责任，加强科普能力建设，促进科普与科技创新协同发展，强化科普在终身学习体系中的作用，营造热爱科学、崇尚创新的社会氛围，加强制度保障 6 个方面作了重要部署。2023 年 9 月，为全面加强科学基金的科普工作，国家自然科学基金委员会发布了《国家自然科学基金委员会关于新时代加强科学普及工作的意见》[②]，提出：①多措并举，加强科学基金科普能力建设，加强项目支持、打造“科学基金科普在行动”品牌、构建科普宣传矩阵；②树立大科普理念，提升科学基金资助创新项目资源科普化效能，强化科普协作联动、促进成果应用贯通；③加强组织领导，为科学基金科普工作提供有力支撑，完善组织体制、塑造齐抓共管合力、充分发挥科学传播与成果转化中心的承接主体作用、加强科普队伍建设、构建评价激励机制。

二、学科建设投入稳步增长

近年来，我国学科建设经费投入稳步增长，国家自然科学基金等项目资助力度连续提升，部委及全国高校多点布局，进一步推动各学科的建设和发展。本部分从学科建设投入的经费数据出发进行定量分析，并结合基金委、教育部、各高校的布局举措，力求客观反映我国学科建设投入现状。

（一）基础研究经费突破两千亿元，总体投入稳步增长

国际上通常采用研究与试验发展（Research and Development，以下简称 R&D）活动的规模和强度指标反映一国的科技实力和核心竞争力。

《中华人民共和国国民经济和社会发展第十四个五年规划和 2035 年远景目标纲要》[③]（以下简称《“十四五”规划》）提出目标：“十四五”时期全社会研发经费投入年均增长 7% 以上、力争投入强度高于“十三五”时期实际。《2022 年全国科技经费投入统计公报》显示，2022 全年 R&D 经费支出 30782.9 亿元，比 2021 年增长 10.11%，R&D 经费投入强度（与国内生产总值之比）为 2.54%，比上年提高 0.11 个百分点。2018—2022 年我国 R&D 经费支出及投入强度见图 1-1。

① 中国政府网．中共中央办公厅　国务院办公厅印发《关于新时代进一步加强科学技术普及工作的意见》[EB/OL] .（2022-09-04）. https://www.gov.cn/zhengce/2022-09/04/content_5708260.htm.

② 国家自然科学基金委员会．国家自然科学基金委员会关于新时代加强科学普及工作的意见 [EB/OL] .（2023-09-15）. https://www.nsfc.gov.cn/publish/portal0/tab442/info90268.htm.

③ 中国政府网．中华人民共和国国民经济和社会发展第十四个五年规划和 2035 年远景目标纲要 [EB/OL] .（2021-03-13）. http://www.gov.cn/xinwen/2021-03/13/content_5592681.htm.

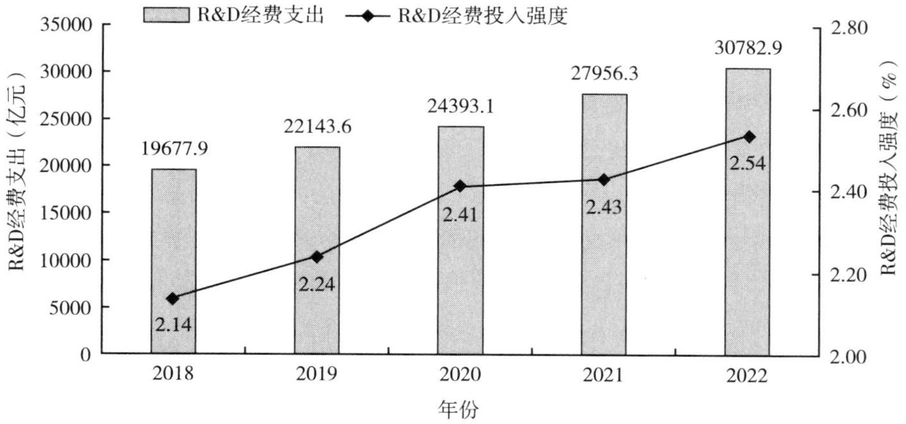

图 1-1　我国 R&D 经费支出和投入强度（2018—2022 年）

数据来源：《2022 年全国科技经费投入统计公报》。

对比世界上主要发达国家和地区，我国在研发经费投入强度方面仍然存在一定差距。根据经济合作与发展组织（Organization for Economic Cooperation and Development，以下简称 OECD）最新的数据显示，2021 年，以色列 R&D 经费投入强度位居首位达到 5.56%，韩国居第二为 4.93%（表 1-1）。我国 R&D 经费强度在 2021 年为 2.43%（图 1-1），超过欧盟的平均水平（2.15%），但不及 OECD 成员国的整体平均水平（2.70%）。

表 1-1　部分国家 R&D 经费投入强度　（单位：%）

序号	国家	2017 年	2018 年	2019 年	2020 年	2021 年
1	以色列	4.62	4.78	5.22	5.71	5.56
2	韩国	4.29	4.52	4.63	4.80	4.93
3	美国	2.90	3.01	3.17	3.47	3.46
4	瑞典	3.36	3.32	3.39	3.49	3.35
5	日本	3.17	3.22	3.22	3.27	3.30
6	比利时	2.67	2.86	3.16	3.35	3.22
7	奥地利	3.06	3.09	3.13	3.20	3.19
8	德国	3.05	3.11	3.17	3.13	3.13
9	芬兰	2.73	2.76	2.80	2.91	2.99
10	丹麦	2.93	2.97	2.90	2.97	2.81
11	冰岛	2.08	2.00	2.33	2.47	2.80
12	中国	2.12	2.14	2.23	2.41	2.43

数据来源：2023 年 9 月 OECD 公布统计数据。

根据《2022 年全国科技经费投入统计公报》，从投入类型看，2022 年全国基础研究经费突破2000亿元，达到2023.5亿元，比2021年增长11.36%；应用研究经费3482.5亿元，比 2021 年增长 10.72%；试验发展经费 25276.9 亿元，比 2021 年增长 9.92%。基础研究经费所占比重为 6.57%，比上年提升 0.07 个百分点；应用研究和试验发展经费所占比重分别为 11.31% 和 82.11%。

2018—2022 年，在 R&D 经费投入大幅增长的背景下，基础研究的投入经费和 R&D 经费占比也在持续增加（图 1–2）。保障基础科学研究领域的经费投入，有利于推动学科均衡协调可持续发展。

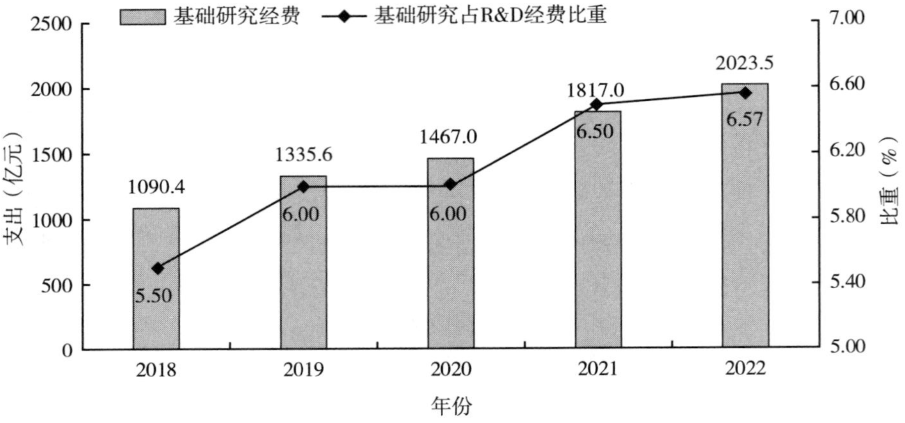

图 1–2　基础研究经费及占比（2018—2022 年）

数据来源:《2022 年全国科技经费投入统计公报》。

从研发主体看，2022 年各类企业 R&D 经费 23878.6 亿元，比上年增长 11.04%；政府属研究机构经费 3814.4 亿元，增长 2.60%；高等学校经费 2412.4 亿元，增长 10.64%；其他主体经费 677.5 亿元，增长 22.34%。企业、政府属研究机构、高等学校经费所占比重分别为 77.57%、12.39% 和 7.84%。由此可见，企业是技术创新的主体，是全社会 R&D 经费增长的主要拉动力量。

从产业部门看，2022 年高技术制造业 R&D 经费 6507.7 亿元，投入强度（与营业收入之比）为 2.91%，比上年提高 0.20 个百分点。规模以上工业企业共涉及 41 个行业大类，R&D 经费投入 19361.8 亿元，R&D 经费投入超过千亿元的行业大类有 7 个，比上年增加 2 个，这 7 个行业的经费占全部规模以上工业企业 R&D 经费的比重为 63.24%（表 1–2）。

表 1–2　2022 年 R&D 经费投入超过千亿元的行业

序号	行业	R&D 经费（亿元）	R&D 经费投入强度（%）
1	计算机、通信和其他电子设备制造业	4099.9	2.63
2	电气机械和器材制造业	2098.5	2.02

续表

序号	行业	R&D 经费（亿元）	R&D 经费投入强度（%）
3	汽车制造业	1651.7	1.83
4	通用设备制造业	1190.6	2.46
5	专用设备制造业	1150.1	2.96
6	医药制造业	1048.9	3.57
7	化学原料和化学制品制造业	1004.9	1.06

数据来源:《2022 年全国科技经费投入统计公报》。

（二）项目资助力度进一步加大

1. 国家自然科学基金

国家自然科学基金是我国基础研究的主要资助渠道之一，重点关注基础前沿问题，注重创新团队和学科交叉，为我国源头创新能力的全面培育作出了重要贡献。

2022 年，自然科学基金财政预算 330.1 亿元（图 1–3），比 2021 年增加 21.0 亿元。共受理来自全国 2405 家单位的 30.69 万份申请，批准资助各类项目 51593 项，项目批准总经费为 389.09 亿元，其中：资助项目直接费用 326.99 亿元，核定 1200 个依托单位间接费用 62.10 亿元。2018—2022 年，国家自然科学基金投入预算由 2018 年的 280.5 亿元增长至 2022 年的 330.1 亿元，资助投入预算年均增长率为 4.2%。

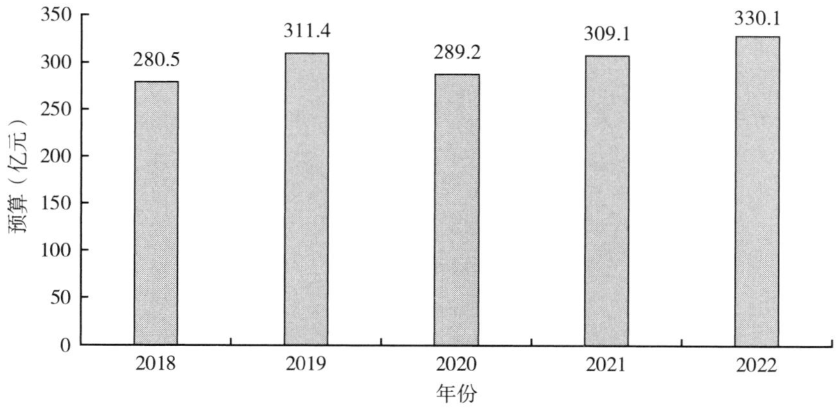

图 1–3　国家自然科学基金历年财政预算情况（2018—2022 年）

数据来源:《国家自然科学基金委员会 2022 年度报告》。

2022 年，国家自然科学基金资助包括面上项目、重点项目等 17 个项目类型，图 1–4 总结了各类型项目资助额度情况。

从资助领域方面来看，2022 年国家自然科学基金资助最多的 3 个学部分别是医学科学部、工程与材料学部和生命科学部（图 1–5）。同时，为引导广大科研人员从国家重大需求和世界科学前沿出发，凝练提出并解决科学问题，“十四五”期间，基金委积极布局115个具有前瞻性、战略性的发展方向，鼓励探索和提出新概念、新理论、新方法，促进科研范式变革和学科交叉融合（表1-3）。

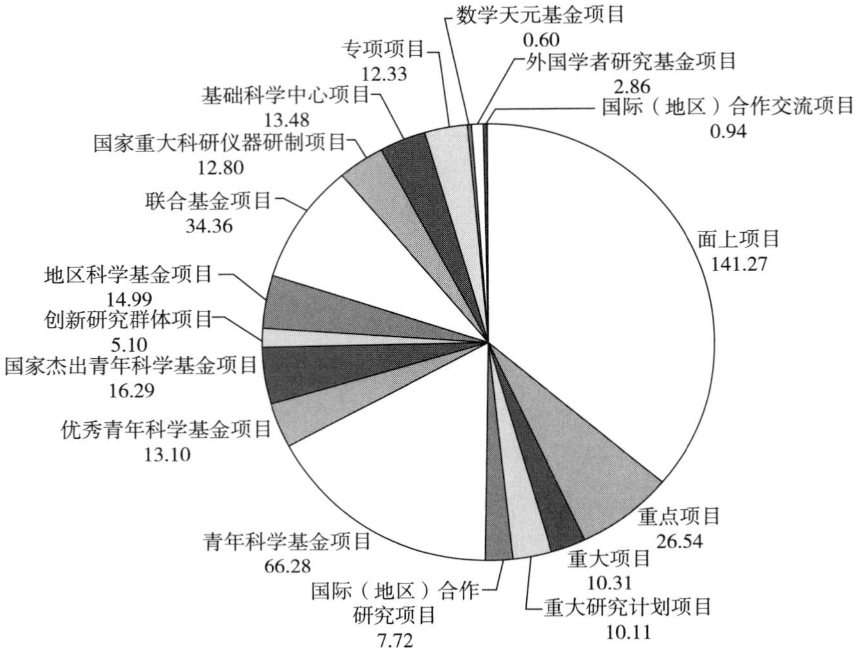

图1-4　2022年国家自然科学基金各类型项目资助额度情况（单位：亿元）

数据来源:《国家自然科学基金委员会2022年度报告》。

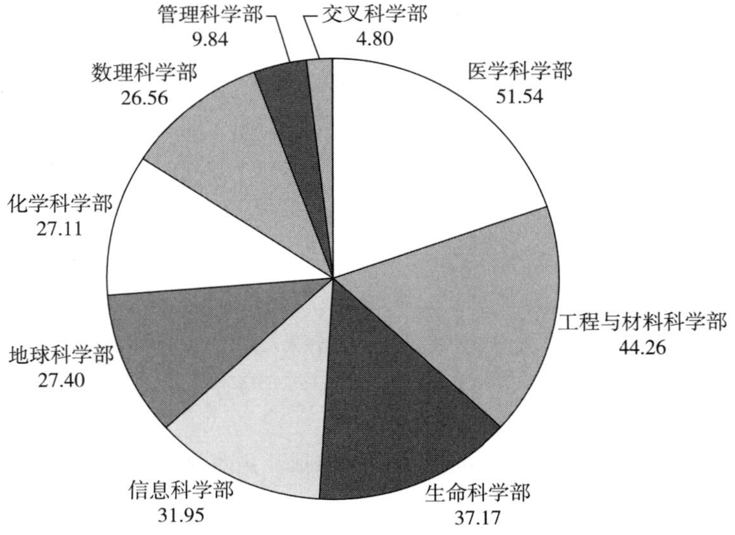

图1-5　2022年国家自然科学基金各个学部批准项目资助资金分布情况（单位：亿元）

数据来源:《国家自然科学基金委员会2022年度报告》。

表 1–3 “十四五”期间优先发展方向

序号	优先发展领域	序号	优先发展领域
1	代数与几何的现代理论	32	地球深部过程与动力学
2	现代分析理论及其应用	33	海洋过程与极地环境
3	问题驱动的应用数学前沿理论与方法	34	地球系统过程与全球变化
4	复杂系统动力学机理认知、设计与调控	35	天气与气候系统与可持续发展
5	新材料与新结构的力学	36	资源能源形成理论及供给潜力
6	高速流动的理论、方法与控制	37	轻质金属材料前沿基础
7	暗物质、暗能量以及星系巡天研究	38	面向 5G/6G 通信的信息功能材料
8	银河系、恒星、太阳及行星系统的多信使探测及研究	39	生物医用高分子材料基础
9	近地小行星动力学特性及监测研究	40	材料多功能集成与器件设计理论基础
10	面向下一代望远镜的关键技术研究	41	战略性关键金属资源开发利用基础理论
11	量子材料与器件	42	低碳能源电力系统与电能高效高质利用理论与技术
12	量子信息和量子精密测量	43	高性能机电装备设计与制造的科学基础
13	复杂结构与介质中的电磁场和声场的机理与调控	44	高效农机装备设计与理论
14	基本费米子及其相互作用	45	土木工程基础设施智能化建造、安全服役与功能提升理论基础
15	强相互作用力的本质	46	巨型水网安全基础理论
16	热核聚变中的关键科学问题	47	城市水循环过程的水质安全保障
17	分子功能体系的精确构筑	48	深海与极地工程装备设计和运维基础理论
18	非常规条件下的传递、反应及测量	49	新型光学技术
19	物质科学的表界面基础	50	光电子器件及集成技术
20	分子选态与动力学	51	宽禁带半导体
21	超越传统体系的电化学能源	52	子器件、射频电路关键技术
22	新范式下的分子化学工程	53	多功能与高效能集成电路
23	多功能耦合的化学传感与成像	54	精准探测与信息融合处理
24	免疫与神经化学生物学	55	新型网络及网络安全
25	绿色合成方法与过程	56	空天地海协同信息网络
26	能源资源高效转化与利用的化学、化工基础	57	工业信息物理系统
27	环境生态体系中关键化学物质的溯源与安全转化	58	安全可信人工智能基础理论
28	大数据与人工智能在化学、化工中的应用	59	类脑模型与类脑信息处理
29	新材料的化学创制	60	智能无人系统技术
30	地球与行星观测的新理论、新技术和新方法	61	生物与医学电子信息获取和处理
31	地球和行星宜居性及演化	62	生物重要性状与环境适应的进化机制

续表

序号	优先发展领域	序号	优先发展领域
63	病原微生物致病及与宿主互作机制及免疫调节	90	人类活动与环境
64	细胞命运可塑性与器官发生、衰老和再生的分子基础	91	面向碳达峰碳中和的能源高效利用与节能减排的科学基础
65	机体功能活动的生物信息流	92	智能运载系统人－机共享驾驶与车－路－云协同技术
66	生态系统对全球变化的响应与适应	93	面向复杂应用场景的计算理论和软硬件基础
67	林草生物质定向培育与高效利用	94	大数据与交互计算技术
68	食品安全与营养、品质的生物学基础与调控机制	95	认知和感知的神经生物学基础
69	农作物重要遗传资源基因发掘及分子设计育种的理论基础	96	跨时空、跨尺度生物分子事件探测与解析
70	园艺作物品质性状形成与调控机理	97	生命体的精准设计、改造与模拟
71	农业动物重要性状形成的生物学基础	98	农作物有害生物成灾与演变机制及其控制基础
72	农业动物重要疫病病原的生物学	99	重大外来入侵物种发生机制与防控技术
73	重大疾病的共性病理机制	100	多学科交叉新型诊疗技术
74	免疫异常与重大疾病	101	复杂系统管理
75	肿瘤发生与演进机制及防治	102	可持续发展中的能源资源与生态环境管理
76	重大慢性病发病机制与防治	103	决策智能与人机融合管理
77	重大传染病发病机制、预测预警与防控	104	政府治理及其规律
78	脑科学与重大脑疾病	105	全球变局下的风险管理
79	衰老与健康增龄	106	巨变中的全球治理
80	生殖健康及遗传与罕见疾病	107	全球性公共危机管理新问题
81	儿童重大疾病的发病机制与防治	108	数字经济的新规律
82	急重症、器官移植、康复和特种医学	109	中国经济发展规律
83	公共卫生与预防医学	110	企业的数字化转型与管理
84	中医理论与中药现代化研究	111	中国企业的管理和新全球化
85	创新药物及生物治疗新技术	112	城市管理的智能化转型
86	智能化医疗的基础理论与关键技术	113	中国乡村振兴与区域协调发展规律
87	大数据与人工智能时代的计算新理论与新方法	114	人口结构与经济社会发展
88	软物质功能体系的设计、调控与理论	115	智慧健康医疗管理
89	生命体系多层次交互通讯的分子基础		

数据来源：国家自然科学基金委网站。

2. 国家重点研发计划

国家重点研发计划由原来的国家重点基础研究发展计划（“973”计划）、国家高技术研究发展计划（“863”计划）、国家科技支撑计划、国际科技合作与交流专项、产业技术研究与开发基金和公益性行业科研专项等整合而成，是针对事关国计民生的重大社会公益性研究，以及事关产业核心竞争力、整体自主创新能力和国家安全的战略性、基础性、前瞻性重大科学问题、重大共性关键技术和产品开展的研究计划，为国民经济和社会发展主要领域提供持续性的支撑和引领。党的二十大以来，国家重点研发计划的资助工作坚持“四个面向”总要求，持续推进“揭榜挂帅”、青年科学家项目等科技管理改革举措，着力提升科研投入绩效，加快实现高水平科技自立自强。截至 2023 年 9 月，科技部本年度共启动 64 项重点专项的申报工作。2022 年，科技部启动了 9 项重点专项的申报工作。各重点专项如表 1–4 所示。

表 1–4　2022—2023 年国家重点研发计划重点专项申报情况

年份	重点专项
2022	交通基础设施，绿色生物制造，智能机器人，网络空间安全治理，储能与智能电网技术，政府间国际科技创新合作，区块链，稀土新材料，战略性科技创新合作
2023	数学和应用研究，催化科学，物态调控，发育编程及其代谢调节，常见多发病防治研究，生物与信息融合（BT 与 IT 融合），政府间国际科技创新合作，合成生物学，氢能技术，煤炭清洁高效利用技术，储能与智能电网技术，可再生能源技术，新能源汽车，交通载运装备与智能交通技术，交通基础设施，长江黄河等重点流域水资源与水环境综合治理，深海和极地关键技术与装备，海洋环境安全保障与岛礁可持续发展，主动健康和人口老龄化科技应对，前沿生物技术，中医药现代化，生物安全关键技术研究，社会治理与智慧社会科技支撑，城镇可持续发展关键技术与装备，高性能制造技术与重大装备，智能传感器，工业软件，增材制造与激光制造，智能机器人，网络空间安全治理，文化科技与现代服务业，地球观测与导航，多模态网络与通信，先进计算与新兴软件，高性能计算，信息光子技术，微纳电子技术，区块链，典型脆弱生态系统保护与修复，农业生物重要性状形成与环境适应性基础研究，农业生物种质资源挖掘与创新利用，主要作物丰产增效科技创新工程，北方干旱半干旱与南方红黄壤等中低产田能力提升科技创新，黑土地保护与利用科技创新，畜禽新品种培育与现代牧场科技创新，林业种质资源培育与质量提升，农业面源、重金属污染防控和绿色投入品研发，重大病虫害防控综合技术研发与示范，动物疫病综合防控关键技术研发与应用，工厂化农业关键技术与智能农机装备，食品制造与农产品物流科技支撑，食品营养与安全关键技术研发，海洋农业与淡水渔业科技创新，乡村产业共性关键技术研发与集成应用，战略性矿产资源开发利用，大气与土壤、地下水污染综合治理，战略性科技创新合作，循环经济关键技术与装备，生育健康及妇女儿童健康保障，先进结构与复合材料，高端功能与智能材料，新型显示与战略性电子材料，稀土新材料，诊疗装备与生物医用材料

数据来源：科技部官方网站整理。

3. 国际大科学计划和国际科技合作

国际大科学计划和大科学工程是聚焦全球共同面临的复杂科学技术问题、由多个国家联合开展的科学研究活动，是人类开拓知识前沿、探索未知世界和解决重大全球性问题的重要手段。

2022 年 11 月 18 日，在北京举行的中关村论坛新闻发布会上，科技部战略规划司司长梁颖达介绍，党的十八大以来，为推进国际科技创新交流合作，我国积极参与全球创新治理，深度参与国际大科学计划和大科学工程，已参与国际热核聚变实验堆、平方公里射电望远镜等近 60 个国际大科学计划和大科学工程。

在加强政府间科技合作方面，中国已与 160 多个国家和地区建立科技合作关系，签订

114 个政府间科技合作协定。通过国家重点研发计划政府间国际科技创新合作重点专项，中国共支持与 60 多个国家、地区、国际组织和多边机制开展联合研究，涉及农业、能源、环境、资源、信息通信、生命健康等领域，共支持立项项目近 2000 项，项目总经费近 100 亿元人民币①。

2023 年，科技部共发布三批国家重点研发计划“政府间国际科技创新合作”重点专项项目的申报指南，拟支持项目数 299 个，国拨经费总概算 8.42 亿元（表 1–5、表 1–6）；发布两批“战略性科技创新合作”重点专项港澳台项目的申报指南，拟支持项目数 40 个，国拨经费总概算 0.8 亿元（表 1–7）。

表 1–5　2023 年“政府间国际科技创新合作”重点专项项目情况

批次	指南方向（个）	合作组织（个）	项目数（个）	国拨经费（亿元）	项目实施周期
第一批	15	14	138	4.96	2~3 年
第二批	12	16	157	3.15	2~3 年
第三批	2	2	4	0.31	2~3 年

数据来源：科技部官方网站整理。

表 1–6　2023 年“政府间国际科技创新合作”重点专项部分项目概览

合作组织	领域方向	项目数（个）	经费（万元）
美国政府	医药卫生、能源、环境、农业技术、基础科学	70	10500
西班牙政府	智慧城市、生产制造技术、生物医学与医疗技术、清洁技术、现代农业、先进材料	20	6000
日本理化学研究所	医药卫生、环境、基础科学等	10	3000
意大利政府	农业与食品科学、应用于文化遗产的人工智能、天体物理学与物理学、绿色能源相关研究、生物医学	10	2000
新西兰政府	食品科学、健康和生物医学、环境科学	6	1080

数据来源：科技部官方网站整理。

表 1–7　2023 年“战略性科技创新合作”重点专项港澳台项目情况

	指南方向（个）	项目数（个）	国拨经费（万元）	项目实施周期	领域方向
第一批	1	15	3000	2~3 年	电子信息、生物医药、节能环保、新材料科学、航空航天、海洋科学
第二批	1	25	5000	2 年	生物技术、人工智能、新材料

数据来源：科技部官方网站整理。

① 中国政府网．我国积极推进全球科技交流合作［EB/OL］.（2022–11–19）. https://www.gov.cn/xinwen/2022–11/19/content_5727817.htm.

（三）基础、交叉学科的布局变化

学科布局是科研的“软基础设施”。2023 年 3 月，中共中央、国务院印发了《党和国家机构改革方案》[①]。其中规定将科学技术部下属的两个项目资助管理专业机构：中国21世纪议程管理中心、科学技术部高技术研究发展中心，划入国家自然科学基金委员会。这一改革举措，是党中央把握基础研究发展规律，结合科学基金新时代职责定位，统筹基础研究和应用基础研究作出的系统性部署，对整合科技资源、打通创新全链条、推动基础研究高质量发展、夯实科技创新根基具有十分重大的意义。除国家改革举措外，近年来，基金委、教育部及各高校也在纷纷布局，进一步推动我国基础、交叉学科建设。

1. 基金委深化改革，推动基础研究高质量发展

2021 年，基金委推动科学基金资助布局改革，依据“源于知识体系逻辑结构、促进知识与应用融通、突出学科交叉融合”的原则，按照基础科学、技术科学、生命与医学、交叉融合 4 个板块构筑资助布局，加强对基础学科的倾斜支持力度，保持优势学科的国际引领地位，扶持传统学科、薄弱学科和濒危学科，关注学科交叉领域中可能产生重大突破的方向（表 1–8）。夯实学科发展基础，打破学科交叉壁垒，构建全面协调可持续发展的高质量学科体系。

表 1–8 基金委资助板块布局情况

板块	构成学科
基础科学	数学、力学、天文、物理、化学、地学等
技术科学	工程、材料科学、信息等
生命与医学	生物学、医学、农业科学等
交叉融合	交叉科学、管理科学等

数据来源:《国家自然科学基金“十四五”发展规划》。

学科交叉融合是未来科学发展的必然趋势，是加速科技创新的重要驱动力。促进交叉学科的发展，对于积极培育新的学科增长极和创新生长点，加快知识生产方式变革和人才培养模式的创新，都具有十分重要的意义。

经中央编办复字〔2020〕46 号文件批准，2020 年 11 月，基金委成立交叉科学部，负责统筹国家自然科学基金交叉科学领域整体资助工作；组织拟定跨科学部领域的发展战略和资助政策；提出交叉科学优先资助方向，组织编写项目指南；负责受理、评审和管理跨学部交叉科学领域项目；负责相关领域重大国际合作研究的组织和管理；负责相关领域专

① 中国政府网．中共中央 国务院印发《党和国家机构改革方案》[EB/OL]．(2023-03-16) . https://www.gov.cn/gongbao/content/2023/content_5748649.htm.

家评审系统的组织与建设；承担交叉科学相关问题的咨询。交叉科学部设有综合与战略规划处以及 4 个科学处，4 个科学处分别聚焦物质科学、智能与智造、生命与健康、融合科学领域。

交叉科学部作为基金委成立的第九个科学部，被赋予了破除学科藩篱，推进原创性、颠覆性创新的特殊使命，在组织架构和资助机制上有多项创新性探索[①]。例如：①交叉科学部打破了传统学部处室以明确学科划分的模式，提出按照交叉领域命名的机构设置架构，采用“矩阵式”扁平化管理模式；项目申请与管理，不设项目申请代码，只设立受理代码 T01~T04；下设的 4 个科学处紧密围绕“四个面向”，组织和部署交叉科学研究类项目；②为充分调动不同学科领域科学家交叉合作积极性，促进深度交叉融合，交叉科学部推动了项目共同申请（CO–PI）立项机制探索；2021 年和 2022 年，交叉科学部在基础科学中心项目开展了“CO–PI”申请模式试点，取消了该类项目原本只能允许一个项目负责人的限制，改为可由两位科学家共同作为项目负责人申请，鼓励组织开展实质性交叉研究，统计数据显示，2021 年和 2022 年“CO–PI”申请项目占比分别达到 16% 与 20%，反映出“CO–PI”申请模式对鼓励和推动部分交叉科研领域项目的重要性和必要性；③交叉科学部结合交叉科学研究本征内涵，设计了适用于交叉科学研究类项目的专用申请书和评审表，重点关注申请人多学科教育背景或跨学科交叉研究经历；④在会议评审环节，发展了“研讨式”而非“陈述式”的答辩模式，提出了兼顾知识维度层次和知识逻辑结构的会议评价机制，评审专家由单一学科领域优秀学者和具有跨学科领域研究经历、学术视野开阔的科学家共同组成。

交叉科学部自成立以来积极布局，2022 年共受理项目申请 2115 项[②]，相较于 2021 年的 754 项[③]有较大幅度增长。得益于交叉科学部的项目资助，我国近两年在芯片设计自动化 EDA、人工智能药物设计、糖合成等众多交叉、前沿研究领域产生了原创性重要性科学成果（表 1–9）。

基金委交叉科学部的成立和发展对各高校交叉科学研究院的成立也起到了很好的示范、引领和带动作用。例如 2023 年 9 月，在国家自然科学基金委交叉科学部、北京大学、湖北省教育厅、湖北省科学技术厅、在汉“双一流”高校等单位专家学者的共同见证下，华中地区首个前沿交叉学科研究院在武汉大学成立。前沿交叉学科研究院的主要任务是：为跨学科研究机构提供精准的管理服务；培育学科交叉文化氛围；交叉学科的人才培养；交叉学科建设机制探索；构建交流与提升机制。

---

① 戴亚飞，张强强，吴飞，等．国家自然科学基金委员会交叉科学部成立、发展与展望［J］．科学通报，2023，68（1）：32–38.

② 赵宋焘，申茜，戴亚飞，等．2022 年度交叉科学部基金项目评审工作综述［J］．中国科学基金，2023，37（1）：54–56.

③ 李江涛，刘雷，王征，等．2021 年度管理科学部基金项目评审工作综述［J］．中国科学基金，2023，37（1）：43–48.

表 1-9　2022—2023 年交叉科学部项目资助成果领域

序号	研究领域	序号	研究领域
1	芯片设计自动化 EDA	20	光控增效型全肿瘤细胞疫苗
2	人工智能药物设计	21	超细内窥镜动态超分辨成像
3	糖合成	22	柔性触觉电子皮肤
4	稠密氢的深度变分自由能方法	23	千公里无中继光纤量子密钥分发
5	基于最优传输的神经网络隐式层建模	24	人造神经
6	基于物理编码学习反应扩散过程	25	细胞有丝分裂期转录调控
7	面向机器学习的材料数据量治理	26	基于分子组装的电子器件
8	人造神经形态视觉和味觉反射系统	27	基于冷冻电子断层扫描的细胞原位成像
9	有机分子 – 稀土纳米晶复合光电功能材料	28	二维半导体欧姆接触
10	伪随机混态的纠缠相变观测	29	光子张量处理芯片
11	新奇拓扑零模式的量子模拟	30	骨关节炎
12	重型车辆氨氢融合零碳动力系统	31	空间转录组整合分析算法
13	基于器件无关量子随机数信标的零知识证明	32	长距离自由空间高精度时间频率传递
14	侗台语起源与扩散	33	破解“全新世温度谜题”
15	外泌体眼内缓释剂型	34	二维材料磁电耦合理论
16	大气科学与环境健康	35	融合远距离量子密钥分发和光纤振动传感
17	DNA 计算	36	廉价高效的低品位废热发电
18	高通量超分辨荧光显微技术	37	低缺陷核酸自组装
19	无液氦低温扫描探针显微镜研制		

数据来源：根据国家自然科学基金委网站整理。

2. 教育部精准施策，优化高等教育学科体系

2022 年 1 月，教育部、财政部、国家发展改革委发布《关于深入推进世界一流大学和一流学科建设的若干意见》①，指出要“夯实基础学科建设。实施‘基础学科深化建设行动’，稳定支持一批立足前沿、自由探索的基础学科，重点布局一批基础学科研究中心。加强数理化生等基础理论研究，扶持一批‘绝学’、冷门学科，改善学科发展生态。根据基础学科特点和创新发展规律，实行建设学科长周期评价，为基础性、前瞻性研究创造宽松包容环境”。同年 2 月，评出第二轮“双一流”建设高校及建设学科名单，其中基础学科相关建设学科共 124 个（不含北大、清华），占全部“双一流”建设学科的 29%。

2020—2021 年，为贯彻落实《教育部等六部门关于实施基础学科拔尖学生培养计划

① 教育部，财政部，国家发展改革委. 关于深入推进世界一流大学和一流学科建设的若干意见［Z］.（2022-01-29）. http://www.moe.gov.cn/srcsite/A22/s7065/202202/t20220211_598706.html.

2.0 的意见》，教育部公布了 3 批基础学科拔尖学生培养计划 2.0 基地名单。截至 2022 年 3 月，在 77 所高水平大学布局建设了 288 个培养基地，吸引 1 万余名优秀学生投身基础学科，96% 的本科毕业生继续在基础学科深造。288 个培养基地涉及生物科学、计算机科学、物理学等 20 个学科，各学科培养基地数量见表 1–10。各高校中入选基地数量最多的 3 所高校是北京大学（19 个）、南京大学（14 个）和浙江大学（12 个）。

表 1–10　基础学科拔尖学生培养计划 2.0 基地学科分布　（单位：个）

序号	学科	数量	序号	学科	数量
1	生物科学	33	11	基础医学	11
2	计算机科学	33	12	地理科学	5
3	物理学	32	13	地质学	5
4	化学	30	14	药学	5
5	数学	29	15	大气科学	4
6	经济学	20	16	地球物理学	4
7	中国语言文学	19	17	心理学	4
8	历史学	17	18	海洋科学	3
9	力学	15	19	天文学	3
10	哲学	14	20	中药学	2

数据来源：教育部网站。

此外，《关于深入推进世界一流大学和一流学科建设的若干意见》还指出要面向集成电路、人工智能、储能技术、数字经济等关键领域加强交叉学科人才培养；要服务新发展格局，优化学科专业布局，布局交叉学科专业，培育学科增长点；以问题为中心，建立交叉学科发展引导机制，搭建交叉学科的国家级平台；以跨学科高水平团队为依托，以国家科技创新基地、重大科技基础设施为支撑，加强资源供给和政策支持，建设交叉学科发展第一方阵；创新交叉融合机制，打破学科专业壁垒，促进自然科学之间、自然科学与人文社会科学之间交叉融合，围绕人工智能、国家安全、国家治理等领域培育新兴交叉学科。

继 2021 年 11 月国务院学位委员会发布关于印发《交叉学科设置与管理办法（试行）》的通知后，国务院学位委员会、教育部 2022 年 9 月公布了《研究生教育学科专业目录》①。从 2023 年起，交叉学科作为一个门类正式“入驻”，增设 7 个学术学位一级学科，分别是集成电路科学与工程、国家安全学、设计学、遥感科学与技术、智能科学与技术、纳米科学与工程和区域国别学；2 个专业学位一级学科，分别是文物、密码。本次修订在统筹一级学科和专业学位类别设置方面迈出了重要步伐，强化了对学术型和应用型两类高层次

① 国务院学位委员会，教育部. 研究生教育学科专业目录（2022 年）[Z].（2022-09-13）. http://www.moe.gov.cn/srcsite/A22/moe_833/202209/t20220914_660828.html.

人才培养的基础支撑，也是夯实分类培养、分类发展的重要举措。

2023 年 4 月，教育部公布了《普通高等学校本科专业目录》，该目录在 2012 年版的基础上，增补了近年来批准增设、列入目录的新专业，共计 286 种。2022 年增设的专业 21 种，其中科技类新专业有地球系统科学、生物统计学、未来机器人等 11 种（表 1–11）。

表 1–11　2022 年新增设的 11 种普通高等学校科技类本科专业

序号	门类、专业类	专业代码	专业名称	学位授予门类	修业年限
1	化学类	070307T	资源化学	理学	四年
2	大气科学类	070604T	地球系统科学	理学	四年
3	统计学类	071203T	数据科学	理学	四年
4	统计学类	071204T	生物统计学	理学	四年
5	材料类	080419T	生物材料	工学	四年
6	电气类	080609T	电动载运工程	工学	四年
7	航空航天类	082012T	飞行器运维工程	工学	四年
8	安全科学与工程类	082904T	安全生产监管	工学	四年
9	交叉工程类	083201TK	未来机器人	工学	四年
10	自然保护与环境生态类	090207TK	国家公园建设与管理	农学、管理学	四年
11	医学技术类	101014TK	医工学	工学	四年

数据来源：教育部网站。

3. 各高校因地制宜，积极布局基础、交叉学科建设

各地高校在创新培养模式、夯实基础研究、促进交叉融合方面，纷纷根据本校的实际情况综合施策，不断推动高等教育和学科建设迈上新台阶。

部分高校在加强基础学科建设和人才培养方面的举措如下。

北京大学：①坚持因材施教，畅通基础学科人才培养本土化路径，扎实开展强基计划、数学英才班、物理学科卓越人才培养计划等项目，设计“3+X”本研贯通培养方案，每年为本科生提供近 500 项高质量研究课题，落实学校各类重点实验室和其他科研实验平台向本科生开放；②加大资源投入，强化对基础学科的支撑保障，持续扩大“关键领域急需人才支持计划”，给予数学、基础医学等国家急需、人才培养基础好的基础学科招生计划增量支持；③深化学科交叉，推动基础学科的前沿创新，打造“区域与国别研究”“临床医学+X”“碳中和核心科学与技术”“数智化 +”等学校层面的学科交叉平台，努力孵化创新成果；设置数据科学与工程、整合生命科学、纳米科学与工程等一级交叉学科，积极培育新的学科增长极和创新生长点。

浙江大学：①对基础学科建设给予长期、稳定、可预期的经费支持和政策配套，优先配置办学资源；②针对国家急需领域的基础学科，持续扩大招生计划增量支持；③设立研究生教育强基基金，进一步提高基础学科研究生资助水平。

武汉大学：①加强组织领导，制定学校《"基础学科拔尖学生培养计划 2.0"工作方案》，在建好自然科学基地班和人文社会科学基地班的基础上，整合各类优质资源，成立"弘毅学堂"并作为统筹单位，牵头推进基础学科拔尖学生培养计划；②促进交叉融合，以基础学科建设为牵引，通过建立双学位试验班、复合型专业、跨学科课程等多种形式，不断强化学科交叉融合，成立人文高等研究院等一批交叉研究平台，聘请知名专家学者参与创新培养理念、完善教学内容、优化课程体系等工作。

随着我国高等教育的快速发展，国内的"双一流"建设高校越来越重视交叉学科建设，通过设立学部、书院等方式建立多种交叉学科研究中心。国内 C9 高校联盟部分交叉学科研究机构如表 1–12 所示。

表 1–12　C9 高校部分交叉学科研究机构

序号	学校名称	研究机构名称
1	北京大学	分子医学研究所、北京国际数学研究中心、燕京学堂、前沿交叉学科研究院、核科学与技术研究院、现代农学院、元培学院等
2	复旦大学	代谢与整合生物学研究院、工程与应用技术研究院、类脑智能科学与技术研究院、智能机器人研究院、大数据研究院、微电子学院、上海数学中心等
3	哈尔滨工业大学	网络空间安全学院、空间环境与物质科学研究院、人工智能研究院、生物信息技术研究院、基础与交叉科学研究院等
4	南京大学	现代工程与应用科学学院、人工智能学院、匡亚明学院等
5	清华大学	全球创新学院、新雅学院、高等研究院、交叉信息研究院、生物医学交叉研究院、未来实验室、脑与智能实验室、核能与新能源技术研究院、碳中和研究院等
6	上海交通大学	能源研究院、人工智能研究院、碳中和发展研究院、个性化医学研究院、转化医学研究院、系统生物医学研究院等
7	西安交通大学	高端装备研究院、精准医疗研究院、转化医学研究院、空天与力学研究院、生物证据研究院、前沿科学技术研究院等
8	浙江大学	生物医学工程与仪器科学学院、求是高等研究院、国际创新研究院、转化医学研究院、数据科学研究中心、金融研究院等
9	中国科学技术大学	大数据学院、网络空间安全学院、核科学技术学院、国家同步辐射实验室、先进技术研究院、合肥微尺度物质科学国家研究中心、环境科学与光电技术学院等

数据来源：各高校官网。

三、学科研究成果产出丰硕

（一）国际科研论文从量的积累迈向质的提高，科研能力稳步提升

为更全面了解我国近年来的国际论文产出情况，从以下多个世界著名科技文献检索数据库统计相关信息。科学引文索引（Science Citation Index，以下简称 SCI）是进行科学统

计与科学评价的主要检索工具，收录世界权威的、高影响力的科技期刊，主要反映基础研究状况；工程索引（Engineering Index，以下简称 EI）主要覆盖工程、应用科学相关研究领域的期刊；科技会议录引文索引（Conference Proceedings Citation Index–Science，以下简称 CPCI–S）则收录了自然科学、医学、农业科学和工程技术等领域的大部分会议文献，在一定程度上反映了科学前沿和最新研究动向。Scopus 是一个全球性文献摘要与科研信息引用数据库，可以满足不同学科领域的用户需求，是进行学术研究和文献检索的重要工具之一。

1. 总体产出情况

根据 SCI 数据库收录的论文进行统计，2022 年世界科技论文总数为 255.44 万篇，比 2021 年增加了 2.2%。2022 年收录中国科技论文为 73.56 万篇，位居世界第一位，占世界份额的 28.9%，所占份额比 2021 年提升了 4.4 个百分点。排在第二位的美国，论文数量为 54.78 万篇，占世界份额的 21.4%。中国学者作为第一作者共计发表论文 68.19 万篇，比 2021 年增加 22.4%，占世界总数的 26.7%。按第一作者论文数排序，中国也排在世界第一位。

EI 数据库 2022 年收录期刊论文总数为 116.26 万篇，比 2021 年增长 11.8%，其中中国论文为 46.10 万篇，占世界论文总数的 39.7%，数量比 2021 年增长 25.3%，所占份额增长 4.3%，排在世界第一位。中国学者作为第一作者共计发表 43.55 万篇 EI 论文，比 2021 年增长了 20.7%，占世界总数的 37.5%，较上一年度增长了 2.8 个百分点。

CPCI–S 数据库 2022 年收录世界重要会议论文为 26.34 万篇，比 2021 年增加了 44.9%，其中收录了中国作者论文 3.46 万篇，比 2021 年增加 13.4%，占世界总数的 13.1%，排在第二位。CPCI–S 收录第一作者单位为中国机构的科技会议论文共计 3.04 万篇，占世界总数的 11.5%。2022 年，有 68 个国家（地区）主办的 1389 场国际会议收录了中国科技人员发表的论文。

Scopus 数据库 2022 年收录的科技论文总数为 313.39 万篇，其中中国科技论文数量为 87.38 万篇，占世界总数的 27.9%，排在第一位，保持 2019 年、2020 年、2021 年的排名（表 1–13）。

表 1–13　2022 年中国国际论文产出及占比

统计源数据库	全球论文总数（万篇）	中国论文数量（万篇）	中国论文量占世界份额（%）	中国论文量在世界排名	中国第一作者论文数量（万篇）	中国第一作者论文世界份额（%）
SCI	255.44	73.56	28.9	1	68.19	26.7
EI	116.26	46.10	39.7	1	43.55	37.5
CPCI–S	26.34	3.46	13.1	2	3.04	11.5
Scopus	313.39	87.38	27.9	1	—	—

数据来源:《中国科技论文统计报告 2023》[①]。

① 中国科学技术信息研究所. 中国科技论文统计结果 2023［R］. 北京：中国科学技术信息研究所，2023.

以 SCI 数据库收录的原创论文（Article）和综述论文（Review）为统计口径，表 1–14 为我国与 G7 国家[①]2021—2022 年论文产出量及增长趋势。我国 2022 年 SCI 论文数量为 73 万余篇，较 2021 年的 63 万余篇，增长了 15.06%。G7 国家 2022 年及 2021 年 SCI 论文数量均低于我国，且都呈负增长趋势。

表 1–14　中国与 G7 国家 SCI 论文产出情况

国家	2022 年论文数（篇）	2021 年论文数（篇）	增长率（%）
中国	731018	635312	15.06
美国	397058	449912	–11.75
德国	120846	135764	–10.99
英国	112702	128549	–12.33
意大利	91268	99487	–8.26
日本	90859	99972	–9.12
法国	77973	87824	–11.22
加拿大	77460	86020	–9.95

数据来源：SCI 数据库，检索时间为 2023 年 11 月。

2013—2023 年（截至 2023 年 8 月），中国科技人员共发表 SCI 国际论文 444.40 万篇，共被引用 6748.23 万次，比 2022 年统计时分别增加了 11.7% 和 18.3%，排在世界第二位。中国平均每篇论文被引用 15.19 次，比上年度统计时的 14.34 次提高了 5.9%，不及美国论文的篇均被引频次（21.43 次），与世界篇均被引频次（15.85 次）也还存在一定差距（表 1–15）。

在 2013—2023 年发表科技论文累计超过 20 万篇以上的国家共有 23 个，按篇均被引频次排序，中国排在第 16 位。篇均被引频次大于世界平均水平的有瑞士、荷兰、丹麦、比利时、英国等 14 个国家（表 1–15）。说明我国发表论文的学术影响力虽在不断提升，但与欧美等发达国家相比仍存在一定差距。

表 1–15　2013—2023 年发表 SCI 国际科技论文数 20 万篇以上的国家论文情况

国家	论文数		被引频次		篇均被引频次	
	篇数	位次	次数	位次	次 / 篇	位次
瑞士	358048	17	9388895	14	26.22	1
荷兰	469138	14	11874097	11	25.31	2

① 西方七国集团（Group of Seven），是主要工业国家会晤和讨论政策的论坛，成员国包括美国、英国、法国、德国、日本、意大利和加拿大 7 个发达国家。

续表

国家	论文数		被引频次		篇均被引频次	
	篇数	位次	次数	位次	次/篇	位次
丹麦	221068	23	5485822	19	24.82	3
比利时	259174	21	6176302	17	23.83	4
英国	1212355	4	27898625	3	23.01	5
瑞典	323039	20	7428464	15	23	6
澳大利亚	757139	10	16617387	7	21.95	7
美国	4502454	1	96474093	1	21.43	8
加拿大	796359	9	16751977	6	21.04	9
德国	1240813	3	25753665	4	20.76	10
法国	822426	8	16959808	5	20.62	11
意大利	827114	7	16082475	8	19.44	12
西班牙	696459	11	13073597	10	18.77	13
沙特阿拉伯	228060	22	3845589	22	16.86	14
韩国	677474	12	10381968	13	15.32	15
中国	4444003	2	67482257	2	15.19	16
日本	892415	5	13469355	9	15.09	17
伊朗	410422	15	5519680	18	13.45	18
波兰	336272	19	4406281	20	13.1	19
印度	854415	6	11005953	12	12.88	20
巴西	539338	13	6849830	16	12.7	21
土耳其	348317	18	3790271	23	10.88	22
俄罗斯	405151	16	4014603	21	9.91	23

数据来源:《中国科技论文统计报告 2023》。

2. 各学科产出情况

图 1-6 为我国 2021 年和 2022 年 SCI 论文数量最多的 10 个学科。可以看出，近两年来我国发表SCI论文的TOP10学科基本保持不变，论文数量最多的3个学科分别为工程学、临床医学和化学，农业科学在 2022 年的发文量超过了分子生物与遗传学，进入了 TOP10 学科。2022 年发文 TOP10 学科论文数量较 2021 年均呈增长趋势，增长最快的学科为环境科学与生态学，增长率为 30.45%。

从 SCI 发文量的世界份额占有率来看，我国份额超过 40% 的学科有 3 个，分别为材料科学（47.26%）、计算机科学（43.08%）和工程学（42.24%）。除此之外，农业科学、分子生物与遗传学、药理学与毒物学、数学等学科，虽然发文量相对不高，但是其占世界份额却较高，相反临床医学虽然发文量很高，但仅占世界份额的 14.21%（表 1-16）。

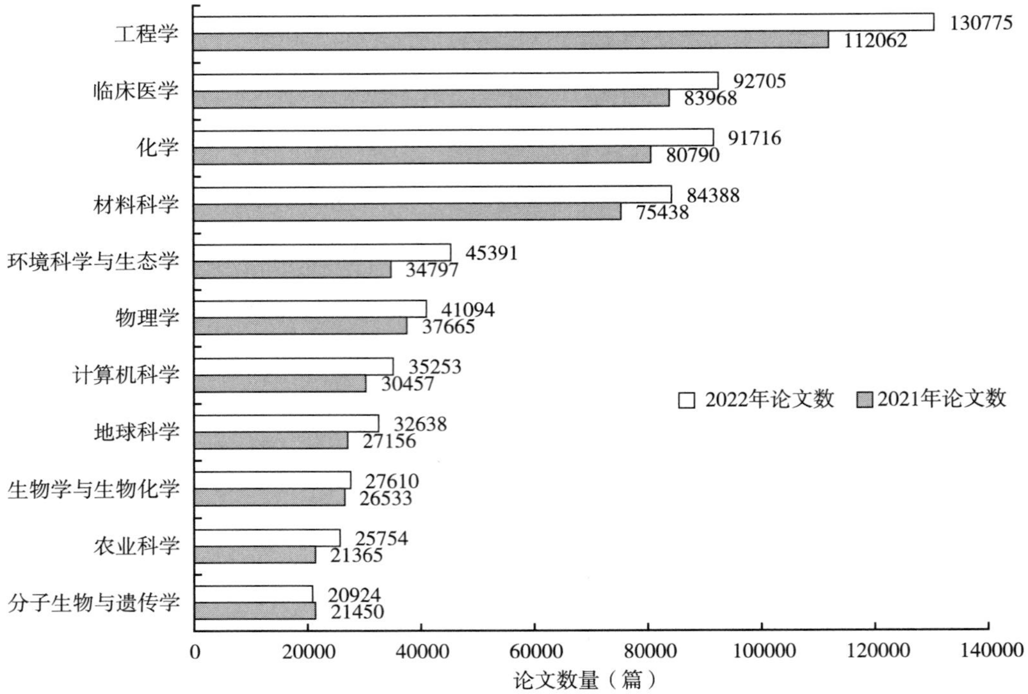

图 1-6　2021—2022 年我国 SCI 论文数量最多的 10 个学科

数据来源：InCites 数据库，检索时间为 2023 年 11 月。

表 1-16　2021—2022 年我国各学科产出 SCI 论文与世界平均水平比较

学科	论文数量（篇）	占世界份额（%）	被引频次（次）	占世界份额（%）	被引频次世界排位	篇均被引（次）	相对影响
工程学	242837	42.24	1733967	49.66	1	7.14	1.18
临床医学	176673	14.21	664978	16.29	2	3.76	1.15
化学	172506	35.98	1363922	44.85	1	7.91	1.25
材料科学	159826	47.26	1537638	58.05	1	9.62	1.23
环境科学与生态学	80188	30.86	624140	39.68	1	7.78	1.29
物理学	78759	33.3	420228	38.15	1	5.34	1.15
计算机科学	65710	43.08	420114	50.33	1	6.39	1.17
地球科学	59794	38.22	330623	45.2	1	5.53	1.18
生物学与生物化学	54143	25.18	323031	26.86	2	5.97	1.07
农业科学	47119	30.05	325421	41.03	1	6.91	1.37
植物学与动物学	45884	21.33	201889	29.22	1	4.4	1.37
分子生物与遗传学	42374	33.18	284804	33.87	2	6.72	1.02
药理学与毒理学	38539	26.36	198028	30.2	1	5.14	1.15
数学	34947	28.08	89262	36.25	1	2.55	1.29
神经科学及行为学	26486	15.55	108118	16.26	2	4.08	1.05

续表

学科	论文数量（篇）	占世界份额（%）	被引频次（次）	占世界份额（%）	被引频次世界排位	篇均被引（次）	相对影响
神经病学与心理学	20800	12.29	65417	12.08	4	3.15	0.98
免疫学	18328	18.21	112790	18.93	2	6.15	1.04
微生物学	16847	24.07	84178	21.87	2	5	0.91
空间科学	6094	16.82	37684	16.99	8	6.18	1.01

数据来源：InCites 数据库，检索时间为 2023 年 11 月。

备注：相对影响为我国篇均被引用次数与该学科世界平均值的比值。

影响力方面，从被引频次来看，我国工程学、化学、材料科学等 11 个学科世界排名第一，临床医学、生物学与生物化学、分子生物与遗传学等 6 个学科排名第二，材料科学与计算机科学被引频次占世界份额超过了一半。从篇均被引及其相对影响来看，我国农业科学、植物学与动物学、数学等 12 个学科的篇均被引频次高于世界平均水平 10% 以上，生物学与生物化学、神经科学及行为学等 5 个学科略高于世界平均水平，而神经病学与心理学、微生物学则低于世界平均水平。

根据 ESI 数据库 2023 年 11 月更新数据，我国在 ESI 各科技学科发表高水平论文数量（包含 ESI 数据库统计的高被引论文和热点论文）及全球排名如表 1–17 所示。从高水平论文量排名来看，在 19 个科技学科领域中，除临床医学、神经科学及行为学、神经病学与心理学、免疫学、空间科学 5 个领域我国位居世界第四至十三位，其余 14 个学科均处于世界前两名的位置。其中，工程学、化学、材料科学、环境科学与生态学、计算机科学、数学、农业科学高水平论文量位居世界首位。

表 1–17 中国在 ESI 学科领域中高水平论文数量和世界排名

学科	论文量	排名	学科	论文量	排名
工程学	10586	1	农业科学	1956	1
化学	9940	1	生物学与生物化学	1625	2
材料科学	7874	1	分子生物与遗传学	1076	2
物理学	4054	2	药理学与毒理学	1030	2
临床医学	4026	8	神经科学及行为学	614	5
环境科学与生态学	3421	1	微生物学	428	2
计算机科学	2783	1	神经病学与心理学	402	7
地球科学	2639	2	免疫学	386	4
植物学与动物学	2591	2	空间科学	240	13
数学	2210	1			

数据来源：ESI 数据库，检索时间为 2023 年 11 月。

注：本表中的排名仅代表我国在该学科 ESI 高水平论文量的世界排名，并不能等同于学科发展水平。

3. 前沿主题分析

（1）研究前沿

中国科学院科技战略咨询研究院、文献情报中心和科睿唯安每年联合发布研究前沿数据和报告，对世界主要国家在主要科学领域的研究活跃程度进行评估。通过持续跟踪全球最重要的科研论文，研究分析论文被引用的模式和聚类，特别是成簇的高被引论文频繁的共同被引用情况来发现研究前沿。而通过对该研究前沿的施引论文的分析，则可以发现该领域的最新进展和发展方向。研究前沿热度指数是衡量研究前沿活跃程度的综合评估指标，分别从核心论文和施引论文的数量和被引频次的份额角度，设计贡献度和影响度两个指标，二者加和构成研究前沿热度指数。

根据《2023 研究前沿热度指数》，美国、中国、英国、德国、法国分别是研究前沿热度指数排名前五的国家。在十一大学科领域的 110 个热点前沿和 18 个新兴前沿中，美国研究前沿热度指数排名第一的前沿数为 69 个，占全部 128 个前沿的 53.91%；中国排名第一的前沿数为 31 个，占 24.22%；英国 8 个前沿排名第一；德国 7 个前沿排名第一；法国没有排名第一的研究前沿。分领域来看，中国在“化学与材料科学”“农业科学、植物学和动物学”“经济学、心理学以及其他社会科学”领域排名第一前沿数量高于其他各国，在“生态与环境科学”领域排名第一前沿数量与美国并列榜首，这些是我国的相对优势领域，其中在“化学与材料科学”领域我国有 8 个前沿排名第一，占比达到 66.67%，表现最为活跃。但是在“地球科学”和“临床医学”领域，中国没有前沿排名第一（表 1–18）。表 1–19 给出了中国研究前沿热度指数排名第一的 31 个研究前沿。

表 1–18　十一大学科分领域层面的 TOP5 国家研究前沿排名第一的数量和比例

领域	研究前沿数量	排名第一前沿数					比例（%）				
		美国	中国	英国	德国	法国	美国	中国	英国	德国	法国
农业科学、植物学和动物学	11	4	5	0	0	0	36.36	45.45	0.00	0.00	0.00
生态和环境科学	11	4	4	1	1	0	36.36	36.36	9.09	9.09	0.00
地球科学	11	10	0	0	0	0	90.91	0.00	0.00	0.00	0.00
临床医学	15	11	0	1	2	0	73.33	0.00	6.67	13.33	0.00
生物科学	14	9	1	1	0	0	64.29	7.14	7.14	0.00	0.00
化学与材料科学	12	3	8	0	1	0	25.00	66.67	0.00	8.33	0.00
物理学	11	9	2	0	0	0	81.82	18.18	0.00	0.00	0.00
天文学与天体物理学	12	6	2	2	2	0	50.00	16.67	16.67	16.67	0.00
数学	10	7	3	0	0	0	70.00	30.00	0.00	0.00	0.00
信息科学	10	4	3	1	0	0	40.00	30.00	10.00	0.00	0.00

续表

领域	研究前沿数量	排名第一前沿数					比例（%）				
		美国	中国	英国	德国	法国	美国	中国	英国	德国	法国
经济学、心理学以及其他社会科学	11	2	3	2	1	0	18.18	27.27	18.18	9.09	0.00

数据来源：《2023 研究前沿热度指数》①。

表 1-19　中国研究前沿热度指数排名第一的前沿

序号	领域	前沿名	前沿类型
1	农业科学、植物学和动物学	NLR 免疫受体介导的植物免疫机制	热点前沿
2		食品中益生菌的微胶囊化研究	热点前沿
3		食物蛋白生物活性肽的结构与功能	热点前沿
4		作物泛基因组研究	热点前沿
5		水果采摘机器人的识别与定位方法	新兴前沿
6	生态与环境科学	利用单原子催化剂活化过氧单硫酸盐	热点前沿
7		环境微塑料颗粒对污染物的吸附	热点前沿
8		土壤微塑料的环境归趋和生态毒理	热点前沿
9		中国臭氧污染状况及健康风险	热点前沿
10	生物科学	铜死亡：铜诱导肿瘤细胞死亡机制	新兴前沿
11	化学与材料科学	海水电解催化剂	热点前沿
12		电催化硝酸根还原合成氨	热点前沿
13		量子点发光二极管	热点前沿
14		电催化合成过氧化氢	热点前沿
15		人工分子机器	热点前沿
16		超分子黏合剂	热点前沿
17		高性能 HER 和 ORR 光催化剂的开发及其在太阳能燃料合成中的应用	新兴前沿
18		聚合物介质电容器的制备	新兴前沿
19	物理学	笼目超导材料 AV_3Sb_5 的特性研究	热点前沿
20		双场量子密钥分发	热点前沿
21	天文学与天体物理学	黑洞阴影和四维 Einstein-Gauss-Bonnet 引力理论	热点前沿
22		事件视界望远镜对人马座 A 超大质量黑洞的观测	新兴前沿
23	数学	非线性时间分数阶反应扩散方程	热点前沿
24		样本均数最优估计方法研究	热点前沿
25		二阶能量稳定 BDF 数值格式	热点前沿

① 中国科学院科技战略咨询研究院. 2023 研究前沿热度指数［R］. 北京：中国科学院科技战略咨询研究院，2023.

续表

序号	领域	前沿名	前沿类型
26	信息科学	用于边缘计算的联邦学习	热点前沿
27		宽度学习系统	热点前沿
28		深度学习在物理层通信中的应用	热点前沿
29	经济学、心理学及其他社会科学	绿色能源消费和经济政策的不确定性研究	热点前沿
30		土地利用效率及可持续发展问题	热点前沿
31		以人为本、可持续性和富有弹性的工业 5.0 发展	新兴前沿

数据来源:《2023 研究前沿热度指数》。

（2）工程前沿

中国工程院作为国家工程科技界最高荣誉性、咨询性学术机构，自 2017 年起联合科睿唯安开展全球工程前沿研究项目。2023 年度全球工程前沿研究项目以数据分析为基础，以专家研判为核心，遵从定量分析与定性研究相结合、数据挖掘与专家论证相佐证、工程研究前沿与工程开发前沿并重的原则，凝练获得 93 个工程研究前沿和 94 个工程开发前沿，重点解读 28 个工程研究前沿和 28 个工程开发前沿，并研判重点工程前沿未来 5~10 年的发展方向和趋势。

表 1–20 为我国在 9 大领域 TOP3 工程研究前沿情况。总的来看，我国在 9 大领域 TOP3 工程研究前沿中，有 25 个工程研究前沿的核心论文量处于世界前三位，其中 16 个前沿的核心论文量居于世界首位。我国在医药卫生领域复杂疾病的多组学特征研究前沿核心论文量最多，为 2488 篇；在能源与矿业工程领域能源资源遥感成像变化检测方法前沿核心论文占全球核心论文比例最高，达 97.22%；在化学、冶金与材料工程领域面向二氧化碳转化利用的高效电催化剂与反应体系前沿篇均被引频次最高，为 251.39 次。然而，我国在一些前沿中核心论文占比较低，例如工程管理领域物流无人机调度与路径优化研究前沿和环境与轻纺工程领域精准营养与健康工程前沿，核心论文占比仅为 11.11% 和 11.79%。

表 1–20　中国在各领域 TOP3 工程研究前沿情况

序号	领域	工程研究前沿	全球核心论文量	中国核心论文				
				论文量	排名	论文比例（%）	被引频次	篇均被引
1	机械与运载工程	高超声速飞行器技术	50	43	1	86.00	4183	97.28
		低碳及零碳燃料发动机技术	47	16	1	34.04	1755	109.69
		动态可重构移动微型机器人集群	11	10	1	90.91	803	80.30
2	信息与电子工程	大模型及其计算系统理论与技术	34	9	2	26.47	419	46.56
		卫星互联网组网理论与关键技术	31	21	1	67.74	1097	52.24
		超大规模硅基量子芯片	56	11	5	19.64	1603	145.73

续表

序号	领域	工程研究前沿	全球核心论文量	中国核心论文				
				论文量	排名	论文比例（%）	被引频次	篇均被引
3	化工、冶金与材料工程	可再生能源驱动生物催化转化二氧化碳合成化学品、能源及材料	92	38	1	41.30	5087	133.87
		冶金流场混沌非线性强化技术研究	120	40	1	33.33	3192	79.80
		面向二氧化碳转化利用的高效电催化剂与反应体系	107	71	1	66.36	17849	251.39
4	能源与矿业工程	海水直接制氢技术研究	455	249	1	54.73	7803	31.34
		紧凑型聚变堆高温超导磁体	468	130	1	27.78	959	7.38
		能源资源遥感成像变化检测方法	36	35	1	97.22	2330	66.57
5	土木、水利与建筑工程	基于人工智能的结构损伤识别及性能预测	54	14	3	25.93	929	66.36
		城市更新中的减碳方法与技术	45	25	1	55.56	1560	62.40
		巨型地质灾害链时空分布与智能化评估	109	84	1	77.06	4731	56.32
6	环境与轻纺工程	土壤中新污染物的环境风险	86	29	1	33.72	2276	78.48
		基于神经网络的集合预报方法	50	22	1	44.00	1978	89.91
		精准营养与健康工程	246	29	6	11.79	2556	88.14
7	农业	园艺作物品质性状形成的遗传学基础与调控网络	58	37	1	63.79	2288	61.84
		动物多组学功能基因挖掘	50	19	2	38.00	924	48.63
		基于深度学习的林木病虫害诊断	18	5	2	27.78	263	52.60
8	医药卫生	复杂疾病的多组学特征研究	9428	2488	2	26.39	196360	78.92
		持续性病毒感染和再激活机制及干预靶点解析	504	173	2	34.33	12280	70.98
		人体核心微生物组及其与宿主互作机制	82	23	2	28.05	2065	89.78
9	工程管理	工业 5.0 环境下人机共融智能制造研究	29	7	2	24.14	569	81.29
		物流无人机调度与路径优化研究	45	5	2	11.11	444	88.80
		重大工程创新生态系统共生逻辑及治理研究	45	21	1	46.67	397	18.90

数据来源：《全球工程前沿 2023》①。

① 中国工程院全球工程前沿项目组．全球工程前沿 2023［M］．北京：高等教育出版社，2023.

（二）国内科研论文总量回升，部分学科表现亮眼

1. 总体产出情况

根据中国知网中国知识资源总库数据显示，我国期刊 2013—2022 年共发表科技论文 1551.09 万篇，各年度论文数量如图 1-7 所示。2013—2018 年，我国国内论文产出数量整体呈上升趋势，2018—2021 年连续下降，2021 年为 10 年来最低值，仅有发文 127.1 万篇，2022 年有所回升，发文量为 136.6 万篇。

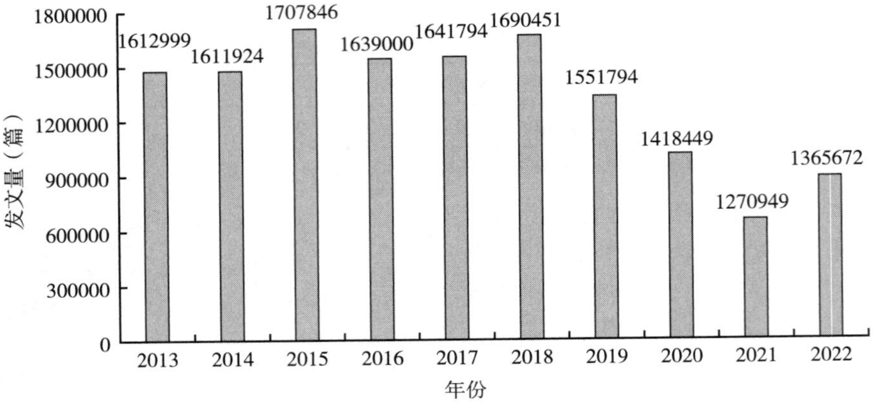

图 1-7　2013—2022 年中国国内论文产出情况

数据来源：中国知识资源总库。

2. 各学科产出情况

根据中国知网 168 专题分类体系，我国 2022 年国内论文发文量最多的 20 个学科如表 1-21 所示。发文量前三名的学科是建筑科学与工程、计算机软件及计算机应用和环境科学与资源利用，它们在 2022 年的发文量分别为 95233 篇、85686 篇、70305 篇。相比 2021 年这些学科中发文增长最多的是环境科学与资源利用（+22.9%）、轻工业手工业（+16.0%）和有机化工（+14.8%）；同比负增长最多的是计算机软件及计算机应用（-14.6%）、临床医学（-12.5%）和外科学（-10.8%）。

表 1-21　2022 年中国国内论文发文量 TOP20 学科

序号	CNKI 专题	2022 年					2021 年		发文量增长率（%）
		发文量	总被引	总下载	篇均被引	篇均下载	发文量	排名	
1	建筑科学与工程	95233	59456	20311540	0.62	213.28	86147	2	10.5
2	计算机软件及计算机应用	85686	82859	27324728	0.97	318.89	100281	1	-14.6
3	环境科学与资源利用	70305	80896	28558501	1.15	406.21	57214	4	22.9

续表

序号	CNKI 专题	2022 年					2021 年		发文量增长率（%）
		发文量	总被引	总下载	篇均被引	篇均下载	发文量	排名	
4	临床医学	65329	43403	12613964	0.66	193.08	74693	3	-12.5
5	电力工业	63944	55448	14897292	0.87	232.97	56667	5	12.8
6	肿瘤学	47620	26462	7100810	0.56	149.11	49830	6	-4.4
7	自动化技术	47024	51501	16232274	1.10	345.19	46132	7	1.9
8	轻工业手工业	46697	36447	13060011	0.78	279.68	40240	9	16.0
9	公路与水路运输	44055	26648	6999497	0.60	158.88	42646	8	3.3
10	中医学	39380	36791	10673494	0.93	271.04	37500	11	5.0
11	外科学	35144	20742	3757553	0.59	106.92	39418	10	-10.8
12	有机化工	33085	19059	7456902	0.58	225.39	28817	14	14.8
13	矿业工程	33059	24472	4560295	0.74	137.94	33746	12	-2.0
14	畜牧与动物医学	30988	22175	5391275	0.72	173.98	31100	13	-0.4
15	石油天然气工业	28548	16659	4685989	0.58	164.14	26562	15	7.5
16	金属学及金属工艺	25473	13973	4321747	0.55	169.66	24858	16	2.5
17	地质学	24224	19213	4676613	0.79	193.06	23503	19	3.1
18	电信技术	24166	15507	4805910	0.64	198.87	24608	18	-1.8
19	医药卫生方针政策与法律	23832	21378	5913511	0.90	248.13	22714	21	4.9
20	中药学	23028	31576	10405230	1.37	451.85	22708	22	1.4

数据来源：中国知识资源总库，评价指标统计时间截至 2023 年 7 月 31 日。

（三）专利数量再创新高，科技创新生态进一步优化

1. 专利申请

专利是技术创新的产物，专利的数量和质量可以反映一个国家或机构的创新活跃度、技术产出水平及经济发展潜力。根据世界知识产权组织（WIPO）发布的《世界知识产权指标2023》（*World Intellectual Property Indicators 2023*）①，从世界五大知识产权局各年度受理的专利申请数量变化趋势（图 1–8）可以看出，早在 2013 年，中国受理的专利申请数量就已经超过了美国等世界四大知识产权局，随后一直是世界上受理专利申请最多的国家，并保持较快增长。

① 世界知识产权组织．世界知识产权指标 2023［R］．瑞士：世界知识产权组织，2023.

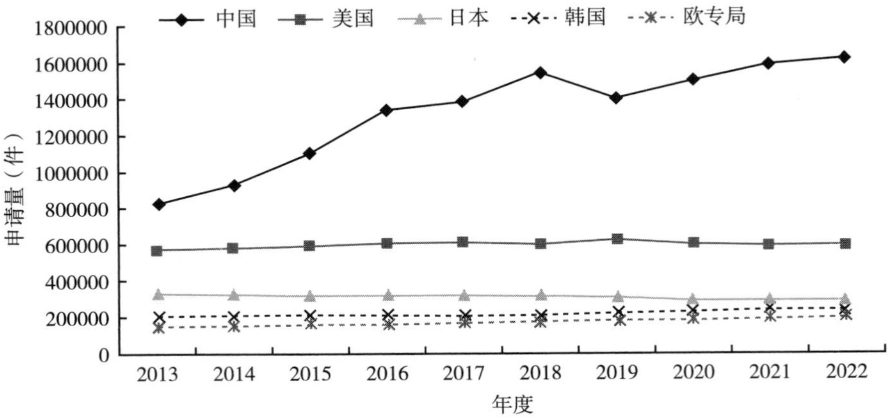

图 1–8　五大专利申请国专利申请年度变化趋势

数据来源:《世界知识产权指标 2023》。

根据《世界知识产权指标 2023》，2022 年全球共有 345 万余件专利提交申请，同比增长了 1.7%。其中，向世界五大知识产权局申请专利的数量占世界总量的 84.87%。2022 年，中国国家知识产权局（CNIPA）共收到 161.9 万件专利申请，较 2021 年增加了 2.12%。其余知识产权局数据依次为美国专利商标局（USPTO）（594340 件）、日本特许厅（JPO）（289530 件）、韩国特许厅（KIPO）（237633 件）和欧洲专利局（EPO）（193610 件）（表 1–22）。

表 1–22　2021—2022 年全球知识产权申请情况

		2021 年	2022 年	增长率（%）
世界范围知识产权申请量		3400500	3457400	1.7
TOP5	中国国家知识产权局（CNIPA）	1585663	1619268	2.12
	美国专利商标局（USPTO）	591473	594340	0.48
	日本特许厅（JPO）	289200	289530	0.11
	韩国特许厅（KIPO）	237998	237633	–0.15
	欧洲专利局（EPO）	188778	193610	2.56

数据来源:《世界知识产权指标 2023》。

从申请人的国别属性来看，中国的申请人在 2022 年提交了约 158 万件专利申请，覆盖国内和国外司法管辖区。紧随其后的是美国（505539 件）、日本（405361 件）、韩国（272315 件）和德国（155896 件）。从增长率来看，中国（+3.1%）、韩国（+1.9%）和美国（+1.1%）在 2022 年的申请量高于 2021 年；与之相反，德国（–4.8%）和日本（–1.6%）在 2022 年的申请量较 2021 年有所下降。

2021 年，计算机技术成为全球公布的专利申请中出现频率最高的技术，占全球总量的 11.1%，位列其后的是电气机械（6.4%）、测量（5.8%）、医疗技术（5.2%）和数字通信（4.9%）。在排名前 15 位的技术领域中，化学工程（+11.4%）、计算机技术（+11.0%）和信息技术管理方法（+13.7%）是 2011—2021 年实现两位数增长的 3 个技术领域。

根据 2019—2021 年已公布申请专利的技术领域分布情况，中国在计算机技术（12.1%）领域申请的专利数量最多，其次是测量（7.1%）和电机、仪器、能源（6.2%）领域；美国专利主要集中在计算机技术（12.9%）、医疗技术（9.4%）、数字通信（7.8%）领域；日本专利则主要是电机、仪器、能源（9.9%）领域；韩国和英国专利主要集中在计算机技术（8.9%、8.6%）领域；德国和法国专利则主要是运输（11.7%、11.3%）领域（表 1–23）。

表 1–23　2019—2021 年已公布申请专利的技术领域分布情况

技术领域	占比（%）						
	中国	美国	日本	韩国	德国	法国	英国
电机、仪器、能源	6.2	4.1	9.9	8.2	9.4	6.5	5.3
视听技术	2.6	2.8	4.2	4.9	1.5	1.7	1.9
电信	1.6	2.2	2.0	2.4	0.9	1.4	1.4
数字通信	5.0	7.8	3.0	5.7	2.1	3.1	2.7
基本沟通流程	0.4	0.8	0.7	0.5	0.6	0.6	0.7
计算机技术	12.1	12.9	6.1	8.9	4.0	5.0	8.6
IT 管理方法	2.7	2.8	2.1	3.7	0.7	0.8	1.3
半导体	1.8	2.6	5.2	6.7	1.8	2.4	1.4
光学	1.5	2.0	5.4	2.7	1.8	2.1	1.9
测量	7.1	3.9	4.9	3.6	6.5	5.0	4.5
生物材料分析	0.5	0.9	0.4	0.5	0.6	0.8	1.1
控制	2.6	2.2	2.7	2.0	2.5	1.5	1.6
医疗技术	3.4	9.4	3.7	4.3	4.5	5.4	7.6
有机精细化学	1.6	2.6	1.4	2.0	2.7	4.5	3.7
生物技术	1.6	4.5	1.2	1.7	2.0	3.1	5.3
制药	1.8	6.4	1.3	2.1	2.4	4.4	7.6
高分子化学、聚合物	1.5	1.2	2.5	1.5	1.9	2.0	0.7
食品化学	1.8	1.0	0.9	2.0	0.5	1.1	0.7
基础材料化学	2.4	2.1	2.2	1.7	2.7	2.4	2.3
材料、冶金	2.8	1.1	2.4	1.8	1.9	2.5	1.4
表面技术、涂层	1.4	1.2	2.5	1.4	1.6	1.8	1.0
微结构与纳米技术	0.2	0.1	0.1	0.1	0.2	0.2	0.2

续表

技术领域	占比（%）						
	中国	美国	日本	韩国	德国	法国	英国
化学工程	4.2	2.2	1.5	2.3	2.6	2.9	3.0
环境技术	2.8	1.1	1.1	1.6	1.4	1.4	1.6
操纵	3.6	2.1	3.3	2.2	3.4	2.5	2.5
机床	4.8	1.5	2.5	1.8	3.6	1.3	1.2
发动机、泵、涡轮机	1.3	1.7	2.4	1.5	4.5	4.5	3.6
纺织和造纸机	1.4	0.9	2.4	0.8	1.4	0.7	0.9
其他专用机器	4.3	3.2	3.0	3.0	4.1	4.5	2.9
热工艺和设备	2.0	0.9	1.8	1.9	1.5	1.6	1.0
机械元件	2.0	1.7	2.8	2.2	6.4	4.2	3.0
运输	3.2	3.6	6.0	4.6	11.7	11.3	4.9
家具、游戏	1.8	2.1	4.6	2.6	1.8	1.4	2.5
其他消费品	1.6	1.8	1.6	3.1	1.7	2.6	6.0
土木工程	4.6	2.8	2.3	3.7	3.2	3.0	4.4

数据来源：《世界知识产权指标 2023》。

2. 成果转化

根据《2022 年中国专利调查报告》内容，《“十四五”国家知识产权保护和运用规划》明确提出要“提高知识产权转移转化成效，支撑实体经济创新发展”。2022 年专利调查显示，我国专利产业化率稳中有升，专利许可、转让比例较上年有所提高，产业化收益水平保持平稳，转移转化整体水平稳步提升。

从我国专利产业化情况分析，2022 年我国发明专利产业化率为 36.7%，较 2021 年提高 1.3 个百分点，自 2018 年以来逐年稳步上升。从不同专利权人来看，企业专利产业化率最高，为 49.3%；科研单位其次，为 14.3%；高校最低，仅为 3.5%。

从我国专利的许可和转让情况分析，2022 年我国有效专利许可率为 9.5%，较 2021 年提高 4.2 个百分点。从不同专利权人来看，企业有效专利许可率为 9.9%，高校 6.5%，科研单位为 5.3%。2022 年我国发明专利转让率为 11.5%，较上年提高 4.1 个百分点，呈快速上升态势。

从发明专利自行产业化收益金额的分布情况看，2022 年发明专利产业化收益水平略有提升。收益金额在 100 万元 / 件以上和 500 万元 / 件以上的发明专利比例分别为 56.4% 和 34.7%，分别较 2021 年高出 11.5 和 9.3 个百分点。

3. 工程开发前沿核心专利公开比例

在《2023 全球工程前沿》公布的 27 个各领域 TOP3 工程开发前沿中，我国在 19 个前沿核心专利公开量排名第一位，其中在水下无人救援机器人等 5 个前沿，我国核心专利公开量占比大于 90%。我国在环境与轻纺工程领域低碳源污水脱氮工艺工程开发前沿的核心专利公开量最多，为 927 件，占比为 98.51%，且该前沿专利被引数也最高，达 2543 次（表 1-24）。

表 1-24　中国在各领域 TOP3 工程开发前沿中核心专利的公开情况

领域	工程研究前沿	公开量		公开量比例（%）	被引数	被引数比例（%）
		数量	排名			
机械与运载工程	多机器人协同作业优化技术	416	1	89.46	1518	78.13
	低成本可回收复用航天器	72	1	50.70	327	53.43
	水下无人救援机器人	178	1	96.22	454	88.67
信息与电子工程	光控相控阵天线技术	184	1	70.77	529	49.76
	基于脑机接口的无人系统控制技术	370	1	79.74	1047	77.90
	面向多样性计算的算力网络构建技术	113	2	17.71	328	10.81
化工、冶金与材料工程	基于可再生能源的冶金低碳化利用	513	1	98.65	218	97.76
	基于人工智能大规模语言模型的化工新材料设计与制备	177	2	37.34	877	64.77
	面向高温环境的金属基复合材料设计与制备	529	1	88.76	892	86.35
能源与矿业工程	动力电池快速充电及管理技术	217	1	46.77	5840	36.67
	快堆金属燃料和氮化物、碳化物核燃料及循环应用	162	1	51.92	237	46.75
	地面高精度重力测量找矿技术	87	2	25.44	952	17.40
土木、水利与建筑工程	排水管道漏损智能探测与修复技术	22	1	100.00	43	100.00
	毫米级全球和区域坐标框架建立技术	46	1	83.64	85	89.47
	城市历史文化资源保护与利用的数字化技术体系	15	1	78.95	114	93.44
环境与轻纺工程	低碳源污水脱氮工艺	927	1	98.51	2543	86.70
	海洋上层水体生物光学剖面激光探测技术	193	1	44.78	364	26.13
	纤维素基抗菌纺织材料	722	1	72.20	1002	25.43
农业	基于结构生物学的绿色农药分子设计	7	5	6.31	21	0.96
	饲料的预消化发酵生物加工制备	64	1	87.67	62	81.58
	水产动物生态化繁育技术	52	1	100.00	25	100.00
医药卫生	T 细胞受体工程化 T 细胞疗法	278	2	33.37	691	17.45
	抗体偶联药物免疫联合治疗恶性肿瘤	61	2	18.26	240	13.22
	单细胞空间转录组技术	61	2	37.65	190	20.17
工程管理	线性规划和整数规划求解器	61	1	59.22	182	55.49
	基于工业互联网和大数据的智能工厂运维系统	35	1	53.03	88	8.07
	基于深度学习的建筑方案自动生成方法与系统	1	3	3.85	0	0.00

数据来源：《全球工程前沿 2023》。

四、基础条件支撑效应显现

国家重大科技基础设施、高端科研仪器设备、科学数据资源是国家创新体系的重要组成部分，也是支持学科创新发展的重要基础保障。《“十四五”规划》提出要“适度超前布局国家重大科技基础设施，提高共享水平和使用效率。集约化建设自然科技资源库、国家野外科学观测研究站（网）和科学大数据中心，加强高端科研仪器设备研发制造，构建国家科研论文和科技信息高端交流平台”，对提升国家创新整体效能、服务国家重大科技创新意义重大。

（一）国家重大科技基础设施服务科学研究效应显现

重大科技基础设施是为探索未知世界、发现自然规律、引领技术变革提供极限研究手段的大型复杂科学技术研究装置或系统。重大科技基础设施是解决重点产业“卡脖子”问题、支撑关键核心技术攻关、保障经济社会发展和国家安全的物质技术基础，是抢占全球科技制高点、构筑竞争新优势的战略必争之地。截至 2023 年 8 月，我国共建成国家重大科技基础设施 84 台，其中专用研究设施 41 台，公共实施设施 31 台，公益服务设施 12 台。目前，我国重大科技基础设施已覆盖了 39 个学科领域，对我国各领域的科技发展起着重要的支撑作用。我国重大科技基础设施学科分布情况如表 1–25 所示，拥有重大科技基础设施较多的学科为地球科学、物理学、交通运输工程、环境科学技术及资源科学技术、生物学等。

表 1–25　我国重大科技基础设施学科分布

序号	学科领域	数量（个）	序号	学科领域	数量（个）
1	地球科学	22	14	其他	6
2	物理学	16	15	能源科学技术	6
3	交通运输工程	10	16	信息科学与系统科学	5
4	环境科学技术及资源科学技术	10	17	土木建设工程	4
5	生物学	10	18	航空航天领域	4
6	力学	9	19	临床医学	3
7	材料科学	9	20	地球观测	3
8	化学	8	21	测绘科学技术	3
9	自然科学相关工程与技术	8	22	空间科学领域	3
10	航空、航天科学技术	8	23	药学	3
11	天文学	7	24	信息与系统科学相关工程与技术	2
12	工程与技术科学基础学科	7	25	基础医学	2
13	水产学	7	26	核科学技术	2

续表

序号	学科领域	数量（个）	序号	学科领域	数量（个）
27	计算机科学技术	2	34	机械工程	1
28	中医学与中药学	1	35	林学	1
29	军事学	1	36	水利工程	1
30	农学	1	37	畜牧、兽医科学	1
31	动力与电气工程	1	38	空间地球系统科学	1
32	化学工程	1	39	食品科学技术	1
33	数学	1			

数据来源：重大科研基础设施和大型科研仪器国家网络管理平台，跨学科基础设施将被分别统计。

“十四五”期间布局建设的重大科技基础设施分为四类，分别为：战略导向型、应用支撑型、前瞻引领型、民生改善型，各项设施如表 1–26 所示。

表 1–26 《“十四五”规划》国家重大科技基础设施

类型	设施名称
战略导向型	空间环境地基监测网、高精度地基授时系统、大型低速风洞、海底科学观测网、空间环境地面模拟装置、聚变堆主机关键系统综合研究设施等
应用支撑型	高能同步辐射光源、高效低碳燃气轮机试验装置、超重力离心模拟与试验装置、加速器驱动嬗变研究装置、未来网络试验设施等
前瞻引领型	硬 X 射线自由电子激光装置、高海拔宇宙线观测站、综合极端条件实验装置、极深地下极低辐射本底前沿物理实验设施、精密重力测量研究设施、强流重离子加速器装置等
民生改善型	转化医学研究设施、多模态跨尺度生物医学成像设施、模式动物表型与遗传研究设施、地震科学实验场、地球系统数值模拟器等

数据来源：《中华人民共和国国民经济和社会发展第十四个五年规划和 2035 年远景目标纲要》。

“十四五”以来，建成启用的重大科技基础设施共 4 台，分别为：航空遥感系统、高海拔宇宙线观测站、鹏城云脑、蓝海 101 科学调查船，它们在各领域科学研究和发展中发挥了重要的支撑作用。

1. 航空遥感系统

航空遥感系统，就是把多种对地观测载荷集成在一架飞机上，通过航空飞行来实现对地观测①。经过十多年的建设，2021 年 7 月 22 日，由中国科学院空天信息创新研究院承担的大科学装置航空遥感系统顺利通过国家验收，投入正式运行，其综合性能与美国国家航空航天局戈达德飞行中心运行的航空遥感系统相当，整体技术居国内领先、国际先进水

① 光明日报. 航空遥感：重装上阵 跨越山海［N/OL］.（2023-08-17）. https://epaper.gmw.cn/gmrb/html/2023-08/17/nw.D110000gmrb_20230817_1-16.htm.

平，并对各领域用户开放。

2023 年 5 月，中国科学院青藏高原研究所、中国科学院空天信息创新研究院、中国科学院西北生态环境研究院、中国科学院精密测量科学与技术创新研究院以及武汉大学在青海省海北藏族自治州八一冰川地区利用航空遥感系统组织实施冰川透视航空与地面联合科学实验。这是国际上首次开展基于航空平台的 P/L/VHF 三波段（P 波段、L 波段、甚高频段）雷达联合冰川探测实验。

2. 高海拔宇宙线观测站

高海拔宇宙线观测站（LHAASO，拉索）位于中国四川省稻城县海子山，占地面积约 1.36 平方千米，是世界上海拔最高、规模最大、灵敏度最强的宇宙射线探测装置。其核心科学目标是探索高能宇宙线起源以及相关的宇宙演化和高能天体活动，并寻找暗物质；广泛搜索宇宙中尤其是银河系内部的伽马射线源等。2023 年 5 月 10 日，拉索顺利通过国家验收。

基于超高的探测灵敏度，拉索在初步运行期间已取得多项突破性科学成果，包括：在银河系内发现大量超高能宇宙加速器候选天体，记录到人类观测到的最高能量光子，精确测定了标准烛光蟹状星云的超高能段亮度，发现 1000 万亿电子伏伽马辐射等。拉索面向国内外全面开放共享，2023 年 6 月，中国科学院高能物理研究所科研人员通过拉索对宇宙中一次伽马射线暴进行了完整监测，这是人类首次完整记录到这一高能爆发现象的全过程，相关研究成果发表在《科学》（*Science*）杂志。

3. 鹏城云脑

鹏城云脑网络智能重大科技基础设施是面向国家重大战略需求，支撑新一代网络通信智能关键技术创新突破，推动数字经济快速健康发展的关键大科学装置。鹏城云脑Ⅱ已完成主体建设，基于自主可控的国产 AI 芯片，采用高效能计算体系结构，是国内首个全面自主可控的 E 级（即百亿亿次）智能算力平台。

“鹏城云脑”的性能世界领先。2022 年 11 月，鹏城云脑Ⅱ连续在 Graph500SSSP 性能榜单、世界人工智能算力性能 500 排行榜（AIPerf500）中位居榜首；同月，又以绝对技术优势获得全球 IO500 排行历史性的五连冠。在 MLPerf Training v1.0 基准测试中，鹏城云脑Ⅱ获得自然语言处理赛道（256 卡同等规模）第一名。此外，作为国之重器，鹏城云脑Ⅱ建立了支持千亿参数超大规模 AI 模型的并行训练平台，支持了“鹏程 · 盘古”“鹏程 · 神农”“鹏程 · 大圣”“鹏城 – 百度 · 文心”“悟道 2.0”等一系列 AI 大模型训练，支撑了疫情防控、智能交通等场景的实际应用，形成了具有影响力的 AI 大模型应用技术体系，推动了产业升级和技术体系并行发展。这是我国在国际 AI 技术竞争主赛道上一个新的里程碑，而且是用自主软硬件技术构筑。

正在研制的鹏城云脑下一代设施，算力将是鹏城云脑Ⅱ的 16 倍，建成后将继续助力我国人工智能国产自主产业生态发展。

4. 蓝海 101 科学调查船

蓝海 101 号调查船是我国海洋渔业科学研究的国之重器和农业现代化标志性工程。调查船拥有渔业资源调查、环境生态调查、海洋理化分析、声像评估与遥感四大科学调查系统，配备国际先进的科考设备和甲板机械操控系统，能高效完成海洋渔业资源与环境综合调查，海基数据系统信息采集、传输、处理与集成，调查能力和技术水平达到国内领先，并跻身国际先进科考船行列。

蓝海 101 号调查船自入列以来，先后完成了国家自然科学基金共享航次计划渤黄海共享航次、黄海冬季越冬场调查、黄海渔业资源调查、中韩暂定措施水域渔业资源调查任务等任务，为各高校和科研院所涉海科研工作，提供了必要硬性支撑。

（二）高端科研仪器设备国产化情况有所改善

根据重大科研基础设施和大型科研仪器国家网络管理平台上登记的数据，截至 2023 年 8 月，在高等学校和科研院所，我国共有大型科研仪器 12.8 万余台，其中进口仪器为 9.4 万台，占比 72.92%。高校有 53.41%（68882 台）的共享仪器。启用于 2019—2023 年的仪器为 2.9 万余台，其中国产仪器 8717 台，国产率为 29.20%，这一比例相比于 2014—2018 年的 18.13%[①]，提高了 11.07%，表明我国高端仪器在国产化方面虽然有所改善，但还有很长的路要走。各类共享仪器情况如表 1-27 所示，可以看出，近五年我国启用的各类高端仪器中，数量最多的是分析仪器，超过 15000 台，占比超过 50%。国产率最高的是计算机及其配套设备（国产率 79.52%），国产率最低的是分析仪器（国产率 14.40%）。分析仪器作为科研仪器的关键大类，于科研、工业发展都有着十分关键的作用，但是由于其技术门槛高，研发涉及诸多领域，开发周期漫长，当前全球垄断格局稳固。我国近五年来在这一领域仪器的国产率虽然相较于 2014—2018 年周期的 7.3% 提高了接近一倍，但还是主要依赖进口，仍然是需要努力破局的关键技术问题。

表 1-27　启用于 2019—2023 年各类共享仪器数量统计

仪器类别	国产仪器（台）	仪器全部总量（台）	国产率（%）
分析仪器	2224	15445	14.40
工艺试验仪器	1335	2322	57.49
物理性能测试仪器	851	2280	37.32
电子测量仪器	476	1386	34.34
计量仪器	570	1357	42.00
医学科研仪器	204	894	22.82
计算机及其配套设备	330	415	79.52

① 中国科学技术协会. 2018—2019 学科发展报告综合卷［M］. 北京：中国科学技术出版社，2021.

续表

仪器类别	国产仪器（台）	仪器全部总量（台）	国产率（%）
地球探测仪器	138	351	39.32
特种检测仪器	145	322	45.03
海洋仪器	149	334	44.61
激光器	100	362	27.62
大气探测仪器	160	336	47.62
核仪器	24	79	30.38
天文仪器	50	69	72.46
其他仪器	1961	3897	50.32
合计	8717	29849	29.20

数据来源：重大科研基础设施和大型科研仪器国家网络管理平台。

2022 年 7—8 月，科技部、财政部会同有关部门，委托国家科技基础条件平台中心，组织开展了 2022 年中央级高校和科研院所等单位科研设施与仪器开放共享评价考核工作。共有 24 个部门 345 家单位参加评价考核，涉及原值 50 万元以上科研仪器共计 4.7 万台（套），重大科研基础设施 85 个[①]。总体看来，与 2021 年相比，参评单位对开放共享更加重视，开放共享意识显著增强，对外服务成效明显，管理和共享应用水平进一步提升。参评的科研仪器年平均有效工作机时为 1351 小时，相比 2021 年的 1278 小时提升了 5.71%，纳入国家网络管理平台统一管理的仪器入网比例为 100%，较 2021 年度（98%）提高了 2 个百分点。参评的 85 个重大科研基础设施运行和开放共享情况较好，在支撑国家重大科研任务、推动产业技术创新、服务国家重大战略需求和国民经济持续发展等方面取得了显著成效。

（三）科学数据资源体系发展迅速，数据要素市场建设加快

数据要素具有独特性，是数字经济的关键组成部分。在当今大数据时代，科技创新越来越依赖于对科学数据的分析挖掘和综合利用。科学数据是开展科学研究和创新发现的重要基础性战略资源，开展科技计划项目科学数据汇交，规范科学数据的汇交管理、长期保存和共享应用，将有效解决科学数据分散重复的问题，促进科学数据的流转、利用和增值，推动科学研究和科技成果产出，提升数据生产者和持有者的影响力，提高国家财政投入产出效益和国家安全支撑保障能力。

近年来，国家和省市出台一系列政策、文件，提出探索推进政府数据开发利用、挖掘

① 科技部. 科技部办公厅　财政部办公厅关于发布 2022 年中央级高校和科研院所等单位重大科研基础设施和大型科研仪器开放共享评价考核结果的通知［EB/OL］.（2022-09-23）. https://www.most.gov.cn/xxgk/xinxifenlei/fdzdgknr/qtwj/qtwj2022/202209/t20220927_182645.html.

数据价值、促进数据要素市场化配置改革等工作要求。2022 年 12 月，《中共中央　国务院关于构建数据基础制度更好发挥数据要素作用的意见》发布①，其中明确要建立公共数据的分类分级确权授权制度，建立数据资源持有权、数据加工使用权、数据产品经营权等分置的产权运行机制，其为市场化方式探索“共同使用、共享收益”的新模式指明方向和奠定基础，初步搭建我国数据基础制度体系，更是对数据交易所的发展起到了积极作用。2023 年 2 月，中共中央、国务院印发了《数字中国建设整体布局规划》，高度重视公共数据的作用，提出推动公共数据汇聚利用，建设公共卫生、科技、教育等重要领域国家数据资源库，并强调“增强数据安全保障能力”。各地区也纷纷研究制定专门政策，不断推进政府数据共享开放，积极探索数据资源开发利用。2022 年 1 月，山东省发布《山东省公共数据开放办法》，促进和规范公共数据开放，提高社会治理能力和公共服务水平。2023 年 3 月，上海市发布《上海市公共数据共享实施办法（试行）》，进一步完善数据管理制度体系、深化公共数据规范治理和共享应用。2023 年 3 月，深圳市发改委印发《深圳市数据交易管理暂行办法》，拟从数据交易主体、交易场所运营机构、交易标的、交易行为、交易安全等方面明确管理要求。

近年来，我国数据资源规模快速增长。根据国家互联网信息办公室发布的《数字中国发展报告（2022 年）》显示，2022 年我国数据产量达 8.1ZB，同比增长 22.7%，全球占比达 10.5%，位居世界第二，截至 2022 年年底，我国数据存储量达 724.5EB，同比增长 21.1%，全球占比达 14.4%。

数据要素市场建设加快。截至 2023 年 6 月底，全国各地由政府发起、主导或批复的数据交易所达到 44 家，头部数据交易所交易规模已达到亿元至十亿元级别，且呈现爆发式增长趋势。2023 中国国际大数据产业博览会上公布数据显示，2022 年全年我国数据交易所交易规模达 40 亿元左右。广州数据交易所于 2022 年 9 月揭牌，截至 2023 年 8 月，累计交易金额突破 16 亿元。深圳数据交易所于 2022 年 11 月正式揭牌成立，截至 2023 年 3 月，累计交易量破 16 亿元。

科学数据共享是国家科技创新体系建设的重要内容，提升科学数据资源共享服务能力，建立科学数据共享的可持续发展机制，可以为大数据时代的科学研究和经济社会发展提供基础支撑。科学数据中心是促进科学数据开放共享的重要载体，我国的科学数据共享平台可以分为国家级数据中心、省级数据平台、高校科研院所数据中心以及企业建设的科学数据平台四大类。在这其中，国家级数据中心扮演着重要的角色，2019 年，科技部、财政部首批成立了 20 个国家科学数据中心，近两年来各数据中心在推动不同学科领域科学数据汇交采集、存储管理、处理加工、分析挖掘与开放共享工作取得了重要的进展，特别是国家对地观测科学数据中心、国家青藏高原科学数据中心、国家人口健康科学数据中

① 新华社．中共中央　国务院关于构建数据基础制度更好发挥数据要素作用的意见［N/OL］．（2022-12-19）. https://www.gov.cn/zhengce/2022-12/19/content_5732695.htm.

心、国家冰川冻土沙漠科学数据中心等。

1. 国家对地观测科学数据中心

国家对地观测科学数据中心依托中国科学院空天信息创新研究院组建，由十几个国家级和行业性遥感数据中心共同参与建设，其数据服务门户为国家综合地球观测数据共享平台（China GEOSS–DSNET，简称共享平台），共享平台总体数据资源量达到 2.5PB。

2022 年 11 月，国家对地观测科学数据中心商业（天仪卫星）数据资源分中心正式成立，该中心将依托天仪研究院的在轨 SAR 卫星海丝一号、巢湖一号和未来即将发射并组网运行的商业 SAR 卫星，开展 SAR 遥感影像数据获取、处理、存储和分发工作，并将在国土、交通、农林、水利及应急等领域发挥 SAR 遥感卫星实时性高、主动成像的特点，为各行业提供支撑。2023 年 9 月，由国家对地观测科学数据中心自主研制的我国首个汛情动态预警监测综合服务平台上线，该平台基于人工智能和大数据技术实时采集和动态更新全国县级以上水利数据中心发布的水情预警信息，先后为重庆、北京、西安、宜昌、福州等城市的重大洪涝灾害应急工作提供遥感影像、社交媒体、基础地理等数据支撑。

2. 国家青藏高原科学数据中心

国家青藏高原科学数据中心是我国唯一针对青藏高原及周边地区的数据中心，负责青藏高原及周边地区各类科学数据的收集 / 汇交、存储、管理、集成、挖掘、分析、共享和应用推广，是该区域科学数据门类最全、最权威的数据中心。国家青藏高原科学数据中心是青藏科考等国家重大需求的数据平台、青藏高原研究项目的数据银行、青藏高原国际合作的数据枢纽和青藏高原公众科学的数据窗口。数据中心已整合的数据资源涵盖大气、冰冻圈、水文、生态、地质、地球物理、自然资源、基础地理、社会经济等学科和领域，并开发了在线大数据分析、模型应用等功能，推动青藏高原科学数据、方法、模型与服务的广泛集成。两次青藏高原综合科学考察，使大量的创新性科考数据成果被汇聚到国家青藏高原科学数据中心，截至 2023 年 9 月，国家青藏高原科学数据中心已经集成发布了青藏高原及周边地区的科学数据集 6207 个，开放数据集 4008 个，数据文件量 340.83TB，用户达 75565 个，数据浏览量为 2551.6 万次。

3. 国家人口健康科学数据中心

国家人口健康科学数据中心面向全社会开放，提供人口健康领域数据资源支撑和开放共享服务。截至 2023 年 9 月，中心汇交项目 1197 个，数据集总数 17667 个，数据总量高达 1.20PB，数据记录数达到 1823.95 千万条。数据资源涉及生物医学、基础医学、临床医学、药学、公共卫生、中医药学、人口与生殖健康等多方面。

国家人口健康科学数据中心于 2022 年 7 月起与北京航空航天大学联合开发科学数据算力共享的安全计算平台，平台内配置各类数据加工处理软件，支持高并发用户在线浏览、加工处理、分析挖掘数据，为科研工作者提供安全运算环境。该平台有利于解决目前生物医学大数据分析挖掘所需的数据可用不可得、动态大算力配置、多源数据联合等应用需求。2022 年 7—10 月，数据中心为首都师范大学管理学院学生提供实习培养，支持了 56 名本科生、硕士研究生完成实习，积极支持了高校的人才培养。2022 年 9 月，由中国医学科学院基础医学研究所赵秀丽研究室和国家人口健康科学数据中心联合建设的中国人成骨不全症变异数据库正式上线，该数据库提供致病变异、病例查询及知识科普等服务，可作为致病性验证的重要参考，为成骨不全症的遗传学诊断提供数据支撑。

4. 国家冰川冻土沙漠科学数据中心

国家冰川冻土沙漠科学数据中心从寒区旱区研究战略需求出发，面向冰川、冻土、沙漠、黄土、灾害及其作用区长期开展数据收集整理，对科研项目产出数据进行永久备份和存档，开展科学数据全生命周期相关理论方法、关键技术、产品研发研制和数据工程体系建设，服务国家科技创新、战略需求和区域经济社会发展。数据中心自成立以来为“渝昆高速铁路选线”等多个重要项目提供数据支撑。

为深入贯彻习近平总书记关于黄河流域生态保护和高质量发展重要指示，国家冰川冻土沙漠科学数据中心于 2022 年启动了黄河流域科学数据开放共享联盟建设，该联盟由中国科学院西北生态环境资源研究院等国内 60 余家兄弟单位共建。联盟以黄河数据中心建设为抓手，全面开展黄河流域生态保护与高质量发展基础科学数据和专题数据产品建设，推进黄河流域科学数据实现从被动共享到主动共享转变，通过建立数据、计算与服务一体化的数据共享平台，为黄河流域科学数据共享与应用提供服务新模式。2022 年 11 月，黄河数据中心正式上线开展服务。目前，黄河数据中心已构建了 20 大类、174 个小类的 1000 多个专题数据集，总体包含了黄河流域基础地理数据、气候数据、水文水资源数据、冰川冻土积雪数据、流域湿地等数据。

五、学科平台建设协同引领

国际科技创新中心、综合性国家科学中心、全国重点实验室、科技产业创新平台、新型研发机构等各类科研平台是国家科技领域竞争的重要战略科技力量，是国家创新体系建设的基础平台，它们在推动科技创新、促进技术进步和支持产业升级方面发挥着关键作用。近年来，各类平台的规模和数量持续增长，覆盖领域不断广泛，正日益成为各领域的科技创新要素汇聚地，为我国的科技发展和科技领域国际竞争作出了重要贡献。

（一）国际科技创新中心：加快建设，抢占全球创新战略高地

国际科技创新中心是以打造世界科学前沿和新兴产业技术创新策源地、全球创新要素汇聚地为目标。平台的定位为科学新发现、技术新发明、产业新方向、发展新理念的重要策源地（大型城市或城市集群）。国际科技创新中心对创新资源流动具有显著的引导、组织和控制能力，是国家科技领域竞争的重要阵地，对实现高水平科技自立自强、加快形成发展格局具有重要意义，将有力支撑科技强国和中国式现代化建设。《“十四五”规划》明

确提出支持北京、上海、粤港澳大湾区形成国际科技创新中心。

1. 北京国际科技创新中心

北京国际科技创新中心是科技部与北京市共建项目。北京作为我国科技基础最为雄厚、创新资源最为集聚、创新主体最为活跃的区域之一，在教育、科技、人才方面具有独特优势。完全有基础、有底气、有信心、有能力建成国际科技创新中心。

近年来，国家对于北京建设国际科技创新中心的定位和举措进一步明确。2023 年 5 月 11—12 日，习近平总书记在河北考察，主持召开深入推进京津冀协同发展座谈会并发表重要讲话指出，“要加快建设北京国际科技创新中心和高水平人才高地，着力打造我国自主创新的重要源头和原始创新的主要策源地”。2023 年 5 月 17 日，科技部等 12 部门共同制定了《深入贯彻落实习近平总书记重要批示精神　加快推动北京国际科技创新中心建设的工作方案》，进一步明确了北京国际科技中心的建设目标和主要任务。

“三城一区”作为北京国际科技创新中心建设的主平台，包括中关村科学城、怀柔科学城、未来科学城、创新型产业集群示范区。中关村科学城系统布局基础前沿技术，集聚全球高端创新要素。着力发展新一代信息技术、节能环保、航空航天、生物、新材料、新能源、新能源汽车、高端装备制造等八大战略性新兴产业的高端环节，形成一批具有全球影响力的原创成果、国际标准、技术创新中心和创新型领军企业集群。怀柔科学城体系化布局一批重大科技设施平台，建设综合性国家科学中心，成为世界级原始创新的承载区。围绕物质、空间、地球系统、生命、智能等五大科学方向的成果孵化，着力培育科技服务业、新材料、生命健康、智能信息与精密仪器、太空与地球探测、节能环保等高精尖产业，形成战略性创新突破。未来科学城构建多元主体协同创新格局，建设重大共性技术研发创新平台。增强创新要素活力，聚焦先进能源，构筑全球创新高地；发展先进制造，构建融合创新体系；强化医药健康，打造科学创新中心。创新型产业集群示范区（北京经济技术开发区和顺义区）承接三大科学城科技成果转化，以首都创新驱动发展前沿阵地、科技成果转化与产业化承载地、智能制造创新发展示范区为定位，着力打造高精尖产业主阵地和成果转化示范区。

北京国际科技创新中心建设取得扎实进展。根据施普林格 · 自然集团、清华大学产业发展与环境治理研究中心面向全球发布的《国际科技创新中心指数 2022》《国际科技创新中心指数 2023》显示，北京在 2022 年首次超越伦敦，在全球国际科技创新中心中位列第三，2023 年继续位居第三。

北京国际科技创新中心已经汇聚了大量原创性、引领性的科技成果。2023 年 5 月 30 日，在 2023 中关村论坛重大成果发布会上，北京国际科技创新中心发布面向世界科技前沿、面向经济主战场、面向国家重大需求、面向人民生命健康的 20 项重大科技成果（表 1–28），其中不乏一些前沿领域的技术突破。比如，2022 年 10 月 9 日，中国首颗综合性太阳探测卫星“夸父一号”成功发射。

表 1-28　20 项重大科技成果

面向世界科技前沿	硅基光电子集成芯片与多功能系统
	夸父卫星在轨获得世界一流天基太阳硬 X 射线图像等系列成果
	通用视觉大模型 SegGPT
	高能同步辐射光源直线加速器满能量出束
	下一代云化开放无线网络新型空口试验验证平台
面向经济主战场	30 微米厚度柔性可折叠玻璃
	百兆瓦级先进压缩空气储能技术
	己内酰胺绿色生产成套新技术
	180 千瓦高效率氢燃料电池发动机系统
	高能量密度钠离子电池
面向国家重大需求	随钻成像测井仪器及井地数据传输系统
	集成电路用 12 英寸高纯钴靶材及阳极
	低温法烟气污染物近零排放控制（COAP）技术
	基因编辑新型核酸酶
	新一代“人造太阳”
面向人民生命健康	颅内病灶磁共振引导激光消融治疗系统
	深脑成像微型化三光子显微镜
	北斗卫星通信融入大众智能手机及实现产业化项目
	基于国际首创技术的基因测序仪
	国产体外膜肺氧合治疗（ECMO）产品

数据来源：北京国际科技创新中心网站。

2. 上海国际科技创新中心

上海国际科技创新中心是党中央、国务院决策的重大战略部署，由国家发展改革委牵头推进建设。2021 年 9 月 10 日，国家发展改革委会同上海市印发《上海市建设具有全球影响力的科技创新中心“十四五”规划》① 明确指出，到 2025 年，上海将努力成为科学新发现、技术新发明、产业新方向、发展新理念的重要策源地。为 2030 年形成具有全球影响力的科技创新中心城市的核心功能奠定坚实基础，为提升上海“五个中心”能级和城市核心竞争力提供重要支撑。2023 年 2 月 27 日，为深入贯彻习近平总书记关于“提升上海科技创新中心科技创新策源能力”的重要指示精神，全面落实上海科创中心建设国家战略，上海发布《推进“大零号湾”科技创新策源功能区建设方案》②，定位世界级“科创湾

① 上海市人民政府.《上海市建设具有全球影响力的科技创新中心“十四五”规划》[EB/OL].（2021-09-10）. https://www.shanghai.gov.cn/hqkjcx1/20230627/cc3e2e0adb0548f6b535138e2bc8a353.html.

② 上海市人民政府. 关于印发《推进“大零号湾”科技创新策源功能区建设方案》的通知 [EB/OL].（2023-02-27）. https://www.shanghai.gov.cn/gwk/search/content/9bf60e9c9b1c48f19cf92163fabc506e.

区”，打造上海科创中心重要策源地和区域经济社会发展增长极。

2022 年以来，上海国际科技创新中心建设顺利推进，一批高水平创新平台建成投用。李政道研究所投入使用，上海交大张江科学园开园启用。张江复旦国际创新中心、同济自主智能无人科学中心建设加快推进。清华国际创新中心、浙大高等研究院积极承接战略任务。未来车脑芯片、AI 辅助创新药物研发、人类表型组二期等新一批市级科技重大专项依托创新平台加快推动。

3. 粤港澳大湾区国际科技创新中心

2019 年 2 月，《粤港澳大湾区发展规划纲要》明确提出建设国际科技创新中心。近年来，广东锚定粤港澳大湾区国际科技创新中心建设的任务，携手港澳打造具有全球影响力的科技和产业创新高地[①]。2023 年 5 月 29 日，中共广东省委、广东省人民政府发布《关于新时代广东高质量发展的若干意见》[②]。意见指出，纵深推进粤港澳大湾区建设，打造高质量发展的重要动力源。加快建设粤港澳大湾区国际科技创新中心，推进大湾区综合性国家科学中心建设，构建以广深港、广珠澳科技创新走廊为主轴，其他城市协同支撑的创新格局；加快建设国家实验室、重大科技基础设施，争创更多全国重点实验室，推动省实验室提质增效；协同港澳推进建设一批重大创新平台，做大做强粤港澳大湾区国家技术创新中心，打造若干产业创新高地。

随着粤港澳大湾区国际科技创新中心加快建设，大湾区创新驱动力不断增强。当前，广东省正携手港澳高水平推进新阶段大湾区建设。广东省人民政府新闻办公室举办的 2023 年大湾区科学论坛新闻发布会介绍，自粤港澳大湾区国际科技创新中心建设启动以来，从要素、平台、项目、人才等方面加快推动粤港澳科技交流合作，各项工作取得新进展[③]。

携手共建粤港澳科技创新平台。依托横琴、前海、南沙等重大平台促进三地合作，在横琴布局建设各类创新平台 31 家、前海集聚创新载体 125 家、南沙建成高端创新平台 132 家。推动广州实验室与香港中文大学签署战略合作协议，南方海洋科学与工程省实验室在香港设立分部。携手港澳建设 20 家粤港澳联合实验室，在河套建设粤港澳大湾区（广东）量子科学中心。

支持三地创新主体开展联合科研攻关。实施科技创新联合资助计划，累计支持 300 多个项目，约 3 亿元。省重点领域研发计划、基础研究重大项目、省自然科学基金面上项目、青年提升项目以及深圳、珠海等专项向港澳开放。

---

① 人民日报．大湾区国际科技创新中心建设扎实推进［EB/OL］．（2023-05-19）. http://www.scio.gov.cn/live/2023/32570/xgbd/202309/t20230904_767255.html.

② 广东省人民政府．中共广东省委 广东省人民政府关于新时代广东高质量发展的若干意见［EB/OL］．（2023-05-29）. https://www.gd.gov.cn/gdywdt/zwzt/kdyxtz/zcsd/content/post_4188148.html.

③ 广东省科技厅．广东省科技厅厅长：大湾区国际科技创新中心建设取得新进展［EB/OL］．（2023-05-16）. https://gdstc.gd.gov.cn/kjzx_n/gdkj_n/content/post_4181921.html.

加快大湾区高水平人才高地建设。建成面向港澳的科技孵化载体超 130 家，在孵港澳创业团队和企业近 1100 个。广东省承办的中国创新创业大赛（港澳台赛）带动 3400 多家港澳台企业来粤同台竞技，300 多家港澳台企业落地广东。推进外国人来华工作许可、外籍和港澳台高层次人才认定，推动各地市设立“国际人才一站式服务专区”等。

（二）综合性国家科学中心：集聚资源，固牢强化原始创新根基

综合性国家科学中心是国际科技创新中心的核心支撑，是国家创新体系建设的基础平台。平台定位为具备一定科研原始创新实力，并具备汇聚世界一流科学家，突破重大科学难题和前沿科技瓶颈的潜力的地区（一般城市或大型城市规划的创新集聚区）。2016 年以来，获批建设的综合性国家科学中心有 5 个，分别是：上海张江综合性国家科学中心、合肥综合性国家科学中心、北京怀柔综合性国家科学中心、大湾区综合性国家科学中心和西安综合性国家科学中心。五大综合性国家科学中心的基本情况与建设进展见表 1–29。

表 1–29　五大综合性国家科学中心基本情况与建设进展

	上海张江综合性国家科学中心	合肥综合性国家科学中心	北京怀柔综合性国家科学中心	大湾区综合性国家科学中心	西安综合性国家科学中心
获批时间	2016 年 2 月	2017 年 1 月	2017 年 5 月	2020 年 1 月	2023 年 1 月
目标使命	全球规模最大、种类最全、综合能力最强的光子大科学设施集聚地，成为中国乃至全球新知识、新技术的创造之地，新产业的培育之地	建设国家实验室、重大科技基础设施集群、交叉前沿研究平台、产业创新平台、“双一流”大学和学科“2+8+N+3”多类型 / 多层次的创新体系	聚焦战略性、前瞻性基础研究，聚焦关键共性技术、前沿引领技术、颠覆性技术创新，强化基础研究、应用研究与产业化的衔接，实现引领性原创成果重大突破，打造国际科技创新中心的核心支撑	以深圳为主阵地建设综合性国家科学中心，加快深港科技创新合作区、光明科学城、西丽湖国际科教城等平台建设，力争在关键核心技术攻关、战略性新兴产业发展等方面实现新突破	打造具有全球影响力的硬科技创新策源地、具有前沿引领性的新兴产业衍生地和“一带一路”顶尖人才首选地
领域方向	光子科学、计算科学、纳米科技、能源科技、生命科学、类脑智能	基础物理、量子科学、新能源、人工智能	物质、空间、地球系统、生命、智能	生命科学、新材料、医学	大数据、人工智能、增材制造、机器人、卫星应用
空间载体	一心一核、多圈多点、森林绕城，总面积约 94 平方千米	三区：国家实验室核心区和成果转化区、大科学装置集中区、教育科研区，规划面积约 102 平方千米	“一核四区”，即核心区，科学教育区、科研转化区、综合服务配套区、生态保障区，规划面积约 100.9 平方千米	光明科学城：一心两区，规划面积约 99 平方千米	“一核两翼”构成，“一核”是丝路科学城（高新 CID），“两翼”分别是沣西交大创新港和长安大学城，规划面积约 220 平方千米

续表

	上海张江综合性国家科学中心	合肥综合性国家科学中心	北京怀柔综合性国家科学中心	大湾区综合性国家科学中心	西安综合性国家科学中心
科技基础设施	已建成、在建和规划的国家重大科技基础设施共有 12 个，包括目前我国用户最多、开放度最高、综合成果最显著的重大科技基础设施——上海光源	已有、在建、预研的大科学装置 10 余个，其中已有的大科学装置 3 个：全超导托克马克核聚变实验装置、稳态强磁场实验装置、同步辐射光源	包含在用、在建的 31 个科学设施平台：高能同步辐射光源等 6 个大科学装置、11 个科教基础设施、13 个交叉研究平台和 1 个科技创新基地（机械研究总院怀柔科技创新基地）	未来网络基础设施、深圳国家基因库、国家超级计算深圳中心、材料基因组、精准医学影像、空间引力波探测、空间环境与物质作用研究、脑解析与脑模拟设施、合成生物研究设施等	先进阿秒激光设施、高精度地基授时系统、分子医学转化中心、国家超算（西安）中心和人工智能计算中心二期等

数据来源：根据五大科学中心规划建设相关通知信息整理（截至 2023 年 9 月）。

1. 上海张江综合性国家科学中心

建设上海张江综合性国家科学中心，是上海加快建设具有全球影响力的科技创新中心的关键举措和核心任务，是为了构建代表世界先进水平的重大科技基础设施群，提升我国在交叉前沿领域的源头创新能力和科技综合实力，代表国家在更高层次上参与全球科技竞争与合作。2022 年 12 月，浦东新区人民政府印发《张江科学城扩区提质三年行动方案（2022—2024 年）》[①]，推动张江科学城扩区提质，促进更大范围产城融合，提升科学城创新策源功能，加快把张江科学城建设成为“科学特征明显、科技要素集聚、环境人文生态、充满创新活力”的国际一流科学城。

2023 年 1 月，上海市科委发布了《2022 上海科技进步报告》[②]（以下简称《报告》）。《报告》显示，2022 年上海推进张江综合性国家科学中心建设，全力构建国家实验室体系，加快建成重大科技基础设施集群，持续打造高能级科技创新平台，加速高水平人才高地建设步伐，布局实施一批面向未来的重大战略性、前沿性科学研究项目，取得一系列基础研究和应用研究的原创性成果。集成电路、生物医药、人工智能三大主导产业优化提升，其中：集成电路重大关键装备取得阶段性进展；生物医药百余种创新药、创新器械产品获批；人工智能创新成果不断涌现，大规模视觉模型与算法开源开放平台达到世界先进水平。

在高能级科技创新平台和机构加快集聚发展方面，上海光源二期基本建成，硬 X 射线自由电子激光装置等一批国家重大科技基础设施加快建设；启动建设上海前瞻物质研究

① 浦东新区人民政府. 关于印发《张江科学城扩区提质三年行动方案（2022—2024 年）》的通知［EB/OL］.（2022-12-29）. https://www.pudong.gov.cn/zwgk/ghjh-qzf/2023/53/307257.html.

② 上海市科学技术委员会. 2022 上海科技进步报告［EB/OL］.（2023-02-17）. https://stcsm.sh.gov.cn/newspecial/2022jb/list.html.

院，持续推进李政道研究所、上海期智研究院和上海脑科学与类脑研究中心等高水平研究机构建设；上海光源、国家蛋白质科学研究（上海）设施、上海超级计算中心等一批已建成大科学设施服务效能不断提升；新一批“十四五”国家重大科技基础设施规划正式项目和储备项目稳步推进。

2. 安徽合肥综合性国家科学中心

2017 年 1 月，合肥综合性国家科学中心正式获批，成为全国第二个综合性国家科学中心[①]。合肥综合性国家科学中心聚焦信息、能源、健康、环境四大领域，建设由国家实验室、重大科技基础设施集群、交叉前沿研究平台和产业创新转化平台、“双一流”大学和学科组成的多类型、多层次的创新体系，使之成为代表国家水平、体现国家意志、承载国家使命的综合性国家科学中心。

近两年来，一批重大创新平台建设取得新进展。聚变堆主机园区工程建成交付使用，合肥先进光源等 2 个设施获批启动建设。加快强光磁集成实验装置、超级陶粲装置关键技术预研。形成大科学装置建成、在建、预研梯次推进的格局，获批“十四五”国家大科学装置数量、建成在建装置数量均位居全国第三。合肥综合性国家科学中心建成运行能源、人工智能、大健康、数据空间 4 个研究院，登记成立环境研究院。

原始创新力不断加强，重大科研成果不断涌现。取得国内首创纯氨燃烧器点火成功、参与世界首颗量子微纳卫星发射、相关技术助力神舟系列飞船顺利返回等重大科技成果，科学中心项目单位 14 项成果在国际三大顶级期刊发表，稳态强磁场刷新磁场强度世界纪录等 2 项成果入选年度中国科技十大新闻。2023 年 4 月，正在运行的世界首个全超导托卡马克 EAST 装置获重大成果，成功实现了 403 秒稳态长脉冲高约束模等离子体运行，创造了托卡马克装置高约束模式运行新的世界纪录。

促进科研成果实现转移转化。2022 年 5 月，合肥出台《合肥市进一步加强科技成果转化若干措施（试行）》[②]，进一步健全科技成果发现、对接、转化等制度。目前，合肥已组建 5 个科技成果转化专班，常态化登“门”（校门）入“室”（实验室）对接高校院所，跟踪科研项目进展，协调推进成果转化落地，支持太赫兹激光主动成像等一批成果转化项目，共催生 30 余家高新技术企业。制定《合肥市可转化科技成果分类评价办法》，设立总规模 5 亿元的种子基金，组建全国首个城市“场景创新促进中心”，制定“三新”产品认定及推广实施方案，打造“三新”示范应用场景 272 个。

3. 北京怀柔综合性国家科学中心

北京怀柔综合性国家科学中心是北京建设具有全球影响力的科技创新中心的重要支撑，是世界级原始创新承载区。主要聚焦原始创新、基础研究，聚力攻关“卡脖子”关键

---

① 安徽省人民政府网. 国家点名大力推进合肥综合性国家科学中心建设［EB/OL］.（2021-03-10）. https://www.ah.gov.cn/zwyw/jryw/553964181.html.

② 合肥市人民政府. 关于印发合肥市进一步加强科技成果转化若干措施（试行）的通知［EB/OL］.（2022-07-04）. https://www.hefei.gov.cn/xxgk/szfgb/2022/dlh/szfwj/107847360.html.

核心技术难题，实现更多“从 0 到 1”的突破。

怀柔科学城已从建设为主全面进入建设与运行并重的新阶段。截至 2023 年 7 月，在央地协同、部市共建、院市合作大力推动下，“十三五”布局的 29 个科学设施平台土建工程已全部完工，已投入运行的综合极端条件实验装置、地球系统数值模拟装置及 5 个第一批交叉研究平台，共吸引剑桥大学、北京大学、国家海洋环境预报中心、百度等国内外高校院所和企业用户 380 余个，涉及课题 870 余项，累计获批机时 58 万小时、3 亿核小时（算力）。同时，正在加快推进“十四五”新布局的科学设施建设，加快人类器官生理病理模拟装置和 4 个设施平台布局实施。

北京怀柔综合性国家科学中心的国家战略科技力量逐步形成。全国重点实验室方面，央企、科研院所全国重点实验室重组后，怀柔全国重点实验室将达到 6 家。国家科研机构方面，中国科学院共有 19 家科研院所在怀柔落地。新型研发机构方面，全市 8 家新型研发机构有 3 家落地怀柔。2023 年 5 月，雁栖湖应用数学研究院已整体入驻怀柔。高水平研究型大学方面，中国科学院大学已将注册地迁址怀柔，在怀柔师生达到 1.5 万人，成为科学城人才的活水源头。科技领军企业方面，机械科学研究总院的轻量化院、基础院将整建制入驻，已落地先进成形技术与装备全国重点实验室等一批国家级创新平台；有研科技集团与怀柔区共建有色金属新材料科创园，已入驻 36 家仪器和传感器相关企业，拥有 9 个国家级工程中心和实验室。

北京怀柔综合性国家科学中心的科技创新生态体系已初步形成。目前在怀柔的科研人员达到 1.8 万人，其中两院院士 71 人，全球高被引科学家 20 人，集聚各领域的外籍人才 600 余人。国际科技交流合作方面，怀柔科学城依托国家科学中心国际合作联盟等国际组织，成功举办四届科学中心国际研讨会。一批国际高端学术活动永久落户，其中中国干细胞与再生医学协同创新平台大会连续在怀柔举办三届，制定并发布了首个干细胞国际标准；北京大学“怀柔论坛”连续举办两届，推动形成生物医学成像技术创新联盟；纳米能源与纳米系统会议连续举办六届，集聚来自全球 20 个国家和地区的近 2000 位纳米能源研究领域专家学者及各界人士。引领性原创成果方面，累计产出重要科研成果 102 项，其中“新型复合折射透镜”等 35 项“卡脖子”技术取得突破；“新一代数据确权与交易技术”“微型化三光子显微镜”等 26 项原始创新成果引领世界技术潮流；“高品质因数超导腔”等 14 项科研成果打破国外技术垄断，跨入世界前列。

4. 粤港澳大湾区综合性国家科学中心

2020 年 7 月，国家发展改革委、科技部批复同意建设粤港澳大湾区综合性国家科学中心。这是继上海张江、安徽合肥、北京怀柔之后的我国第四个综合性国家科学中心。粤港澳大湾区综合性国家科学中心建设立足大湾区城市群发展优势和特点，充分依托大湾区的设施、学科和产业基础，突出强调科学性、经济性和开放性，以光明科学城、松山湖科学城、南沙科学城为核心载体，构建“原始创新→技术创新→产业应用”全链条创新体系[①]。

作为粤港澳大湾区综合性国家科学中心先行启动区，光明科学城立足全球视野，以打造世界一流科学城为奋斗目标，探索“有为政府 + 有效市场”的科学城发展“深圳路径”，为加快实现高水平科技自立自强贡献力量[②]。近年来，光明科学城围绕信息、生命、新材料三大学科领域，集中布局 9 个重大科技基础设施、11 个前沿交叉研究平台、2 所省级重点实验室、2 所研究型高校，共 24 个重大科技创新载体，并全力以赴加快建设。其中，合成生物研究、脑解析与脑模拟、材料基因组等大科学装置于 2023 年 4 月入驻；深圳建市以来投资最大的科学设施——深圳中能高重复频率 X 射线自由电子激光装置，2023 年 6 月底启动建设；深圳医学科学院及深圳湾实验室院区永久场地一体化项目 2023 年 8 月正式开工。合成生物产业是光明区重点布局的未来产业之一，经过多年深耕，目前已形成较为完善的产业链，近 80 家合成生物企业集聚光明区，估值超 280 亿元。

松山湖科学城以高新区部分区域为主体，整合大朗、大岭山和黄江三镇相关地段，是广深港澳科技创新走廊的重要节点，是新时期东莞参与粤港澳大湾区综合性国家科学中心建设的重要战略平台。广东省推进大湾区综合性科学中心先行启动区实施方案中涉及东莞任务共 41 项，目前已完成或启动 31 项，取得了令人瞩目的扎实进展。大装置、大平台正加速聚势成群。中国散裂中子源、南方光源研究测试平台投用。总建筑面积 12.68 万平方米的松山湖材料实验室新园区一期正式启用，二期正在筹建。阿秒科学中心揭牌成立，推动阿秒激光设施建设，从而提升我国在超快科学领域的综合竞争力。松山湖科学城的进展与成功离不开得天独厚的研发土壤与日臻成熟的研发体系。除了香港城市大学（东莞）、大湾区大学（松山湖校区）等高水平研究型大学加快建设外，还培育引进新型研发机构近 30 家，市级以上工程中心和重点实验室超 300 家。搭建起“莞仪在线”科研仪器共享平台，目前已上线高精尖仪器 6522 台，不仅让企业能在家门口轻松完成测试，而且还能降低创新成本。

广州南沙科学城是广州市和中国科学院共同谋划、共同建设的科创资源集聚高地，将建设成为粤港澳大湾区综合性国家科学中心主要承载区。2022 年 6 月，国务院正式印发《广州南沙深化面向世界的粤港澳全面合作总体方案》，提出要高水平建设南沙科学城。2023 年 6 月，广州市人民政府正式印发《广州南沙科学城总体发展规划（2022—2035 年）》，标志着广州加快推动南沙科学城建成大湾区综合性国家科学中心主要承载区、打造一流科学城的步伐更进一步。2022 年以来，南沙区高新技术企业数量已突破 900 家，新入库科技型中小企业数量突破 2000 家，已联合中国科学院和相关高校携手建立重大科技创新平台 22 个，建立以科技金融为主线的产业创新支持体系，为南沙区 301 家科技型

---

① 刘洋，盘思桃，张寒旭，等. 加快建设粤港澳大湾区综合性国家科学中心［J］. 宏观经济管理，2023（2）：50-58.

② 人民日报. 光明科学城：探索“有为政府 + 有效市场”的科学城发展“深圳路径”［N/OL］.（2023-09-19）. http://gdstc.gd.gov.cn/kjzx_n/gdkj_n/content/post_4255691.html.

企业提供授信金额 21.26 亿元。冷泉生态系统、高超声速风洞、极端海洋科考设施、大洋钻探船等重大科技基础设施，正在加快推进。“年轻”的南沙科学城，凭借自身先天的区位优势，在湾区范围内，将进一步强化广深“双城联动”，与光明科学城、松山湖科学城联动协同发展，打造粤港澳创新合作生态，实现价值共创、利益共享。

5. 陕西西安综合性国家科学中心

2023 年 1 月，陕西省两会召开，其中 2023 年陕西省政府工作报告显示，西安市获批建设综合性科学中心和科技创新中心。此次获批，意味着西安成为继北京、上海以及粤港澳大湾区后第四个获批建设“双中心”的城市，丝路科学城成为“双中心”核心承载区。2023 年 4 月，西安高新区举行“双中心”核心区建设启动仪式，20 个高能级创新平台项目签约落地西安科学园。

西安科学园布局的各项重大科技基础设施和各类高能级创新平台均建设卓有成效。高精度地基授时系统、先进阿秒激光设施两个大科学装置在顺利推进；2022 年国家超算西安中心一期建成投用，未来人工智能计算中心算力达到 300P，位列全国第二，算力填充率达到 98.3%，助力 260 家创新主体实现智能化升级①。2023 年上半年建成投用了西安电子科技大学宽禁带半导体国家工程中心、先进光子器件工程创新平台、奕斯伟集成电路创新中心等多个科创项目，进一步激发了科创动能。另外，还有陕西空天动力创新中心、北斗星基增强系统等项目正在加速建设，建成后将为光子、空天、新能源新材料等产业发展提供更强创新驱动力。

（三）全国重点实验室：优化重组，焕活科技战略力量新机

全国重点实验室是原国家重点实验室体系重组调整后形成的创新平台。全国重点实验室体系依托国家科研机构、高水平研究型大学、科技领军企业设立，是中国特色国家实验室体系的组成部分，按照我国“四个面向”的要求，紧跟世界科技发展大势，适应国家发展对科技发展提出的任务要求，多出战略性、关键性重大科技成果，形成体系化、结构合理、运行高效的实验室体系。

2022 年 2 月 11 日，科技部办公厅印发《关于贯彻落实〈重组国家重点实验室体系方案〉的通知》，提出了《全国重点实验室建设标准（试行）》和《全国重点实验室建设“五问”》，正式开启全国重点实验室优化重组。同年 5 月，科技部遴选出首批 20 家标杆全国重点实验室，涉及生物农业、能源、人工智能、高端仪器等领域（表 1–30）。随后，重组、新建工作加速。相关省份也积极响应跟进，2023 年 8 月，四川省出台《科技创新支撑服务“四化同步、城乡融合、五区共兴”实施方案（2023—2027 年）》②，明确到 2027

① 三秦网. 2022 年西安研发投入强度达 5.18%，副省级城市排名第二！［EB/OL］.（2023–01–13）. https://www.sanqin.com/2023-01/13/content_9981317.html.

② 四川省科学技术厅. 关于印发《科技创新支撑服务“四化同步、城乡融合、五区共兴”实施方案（2023—2027 年）》的通知［EB/OL］.（2023–08–28）. https://kjt.sc.gov.cn/kjt/gstz/2023/8/28/ecb3dac32fc94b0d87b5867ea86ff0cd.shtml.

年，四川将建成15家全国重点实验室。8月17日，福建省政府官宣，推动现有4家国家重点实验室重组纳入全国重点实验室序列。《山东省建设绿色低碳高质量发展先行区三年行动计划（2023—2025年）》[①]中则明确写道，在新一代信息技术、新材料、高端装备、农业、绿色矿山等领域按程序稳步重组一批全国重点实验室，力争2025年全国重点实验室达到30家。2023年浙江全省科技工作会议也提出，今后五年，力争建设国家实验室（基地）4家、全国重点实验室16家。

表1-30 首批20家标杆全国重点实验室建设情况

序号	依托单位	实验室名称
1	北京大学	微纳电子器件与集成技术全国重点实验室
2	清华大学	高端装备界面科学与技术全国重点实验室
3	清华大学	新型电力系统运行与控制全国重点实验室
4	北京航空航天大学	虚拟现实技术与系统全国重点实验室
5	北京理工大学、同济大学	自主智能无人系统全国重点实验室
6	中国农业大学	畜禽生物育种全国重点实验室
7	复旦大学	集成芯片与系统全国重点实验室
8	浙江大学	脑机智能全国重点实验室
9	科大讯飞股份有限公司、中国科学技术大学	认知智能全国重点实验室
10	湖南省农业科学院、武汉大学	杂交水稻全国重点实验室
11	华中农业大学	作物遗传改良全国重点实验室
12	兰州大学	草种创新与草地农业生态系统全国重点实验室
13	中国科学院计算技术研究所	处理器芯片全国重点实验室
14	中国科学院上海微系统与信息技术研究所	集成电路材料全国重点实验室
15	人民网	传播内容认知全国重点实验室
16	中国科学院自动化研究所	多模态人工智能系统全国重点实验室
17	中国华能集团有限公司	高效灵活煤电及碳捕集利用封存全国重点实验室
18	中国科学院大连化学物理研究所	能源催化转化全国重点实验室
19	中国电力科学研究院有限公司	电网安全全国重点实验室
20	中国石油天然气股份有限公司勘探开发研究院	提高油气采收率全国重点实验室

数据来源：各学校官网、媒体报道。

据不完全统计，截至2023年9月，已有超过200家国家重点实验室完成重组，被纳入全国重点实验室的新序列之中。国家重点实验室体系的重组，是国家战略资源投入的一

① 山东省人民政府. 印发《山东省建设绿色低碳高质量发展先行区三年行动计划（2023—2025年）》[EB/OL].（2023-01-03）. http://www.shandong.gov.cn/art/2023/1/3/art_107860_123166.html?eqid=8dbebf2400019c5100000006648faa95.

次重新分配。优化重组后的全国重点实验室新序列能够接替国家重点实验室，成为新的国家战略科技力量，为国家科技创新和发展注入蓬勃生机和活力。

（四）科技产业创新平台：协同创新，打造产学研用全链通路

1. 国家级创新载体

2020 年，科技部《关于推进国家技术创新中心建设的总体方案（暂行）》提出“围绕国家创新体系建设总体布局，形成国家技术创新中心、国家产业创新中心、国家制造业创新中心等分工明确，与国家实验室、国家重点实验室有机衔接、相互支撑的总体布局”。这也是为了弥补技术创新与产业发展之间的断层，促进实验室技术向实际产品转移转化。各创新载体建设情况统计分析见表 1–31。

表 1–31　部分创新载体建设情况统计分析

创新载体名称	部门	主要方向	至今建设总体数量（个）
国家技术创新中心	科技部	科学到技术的转化	18
国家产业创新中心	国家发展改革委	产业升级和聚集	5
国家制造业创新中心	工信部	制造业升级	26
国家工程研究中心	国家发展改革委	技术创新与成果转化	191

数据来源：根据国家发展改革委、科技部、工信部网站资料整理。

国家技术创新中心分为综合类和领域类两个类别进行布局建设，主要职责是促进重大基础研究成果产业化，为区域和产业发展提供源头技术供给，为科技型中小企业孵化、培育和发展提供创新服务。综合类创新中心围绕落实国家重大区域发展战略和推动重点区域创新发展，开展跨区域、跨领域、跨学科协同创新与开放合作，成为国家技术创新体系的战略节点、高质量发展重大动力源，形成支撑创新型国家建设、提升国家创新能力和核心竞争力的重要增长极。领域类创新中心围绕落实国家科技创新重大战略任务部署，开展关键技术攻关，为行业内企业特别是科技型中小企业提供技术创新与成果转化服务，提升我国重点产业领域创新能力与核心竞争力。2022—2023 年国家技术创新中心组建情况见表 1–32。

表 1–32　2022—2023 年国家技术创新中心组建情况

获批或揭牌时间	中心名称	地点	类型
2022 年 6 月	国家智能设计与数控技术创新中心	湖北武汉	领域类
2022 年 6 月	国家数字建造技术创新中心	湖北武汉	领域类
2022 年 8 月	国家燃料电池技术创新中心	山东潍坊	领域类
2022 年 12 月	国家集成电路设计自动化技术创新中心	江苏南京	领域类

续表

获批或揭牌时间	中心名称	地点	类型
2023 年 5 月	国家区块链技术（能源领域）创新中心	北京	领域类
2023 年 5 月	国家建筑绿色低碳技术创新中心	北京	领域类

数据来源：根据科技部网站相关通知信息整理。

国家产业创新中心是整合联合行业内的创新资源、构建高效协作创新网络的重要载体，是特定战略性领域颠覆性技术创新、先进适用产业技术开发与推广应用、系统性技术解决方案研发供给、高成长型科技企业投资孵化的重要平台，是推动新兴产业集聚发展、培育壮大经济发展新动能的重要力量。2018 年以来，国家发展改革委先后批复在智能铸造、生物育种、先进计算等领域设立国家产业创新中心，但生物医药领域一直是空白。2022 年 3 月，国家精准医学产业创新中心在成都天府国际生物城揭牌。该中心由四川大学华西医院牵头建设，是国家在生物医药领域布局建设的第一个产业创新中心，也是精准医学方向唯一的产业创新中心。

国家制造业创新中心建设是构建国家制造业创新体系的重要举措。制造业是立国之本、强国之基。作为制造强国战略五大工程之首的国家制造业创新中心面向制造业转型升级和培育发展新动能的重大需求，通过聚焦产业薄弱环节，整合各类创新资源，开展关键共性技术攻关，打通技术开发、转移扩散到商业化应用的创新链条，以此来全面提升制造业竞争力。自 2016 年启动制造业创新中心建设工程以来，工信部围绕新一代信息技术、机器人等 36 个重点建设领域（表 1–33），在全国认定批复了 26 家国家级制造业创新中心。

2022 年 11 月，工信部批复组建了 3 家国家制造业创新中心，分别是国家石墨烯创新中心、国家虚拟现实创新中心和国家超高清视频创新中心。其中，国家虚拟现实创新中心是虚拟现实领域唯一的国家级创新中心，由青岛和南昌共建。2022—2023 年获批建设的国家制造业创新中心如表 1–34 所示。

表 1–33　国家制造业创新中心 36 个重点建设领域

序号	领域名称	序号	领域名称
1	新一代信息光电子	9	集成电路先进工艺
2	印刷及柔性显示	10	工业信息安全
3	机器人	11	先进复合材料
4	轻量化材料及成型技术与装备	12	智能语音
5	燃气轮机	13	石墨烯
6	高档数控机床	14	深远海海洋工程装备
7	稀土功能材料	15	数字化设计与制造
8	传感器	16	智能网联汽车

续表

序号	领域名称	序号	领域名称
17	工业云制造	27	先进印染技术
18	工业大数据	28	农业机械装备
19	高性能医疗器械	29	先进轨道交通设备
20	资源循环利用	30	存储器
21	医药高端制剂与绿色制药	31	集成电路特色工艺及封装测试
22	先进功能材料	32	半导体关键装备和材料
23	先进陶瓷材料	33	5G 中高频器件
24	玻璃新材料	34	超高清视频制作技术
25	高性能膜材料	35	虚拟现实
26	高端智能化家用电器	36	先进操作系统

数据来源：根据工信部网站资料整理。

表 1-34　2022—2023 年国家制造业创新中心建设情况

获批时间	中心名称	地点	建设单位
2022 年	国家石墨烯创新中心	浙江宁波	宁波石墨烯创新中心有限公司
2022 年	国家虚拟现实创新中心	江西南昌 山东青岛	南昌虚拟现实研究院有限公司 青岛虚拟现实研究院有限公司
2022 年	国家超高清视频创新中心	四川成都	四川新视创伟超高清科技有限公司

数据来源：根据工信部网站相关通知信息整理。

国家工程研究中心是国家发展改革委根据建设创新型国家和产业结构优化升级的重大战略需求，以提高自主创新能力、增强产业核心竞争能力和发展后劲为目标，组织具有较强研究开发和综合实力的企业、科研单位、高等院校等建设的研究开发实体。2021 年 12 月 20 日，国家发展改革委发布《国家发展改革委办公厅关于印发纳入新序列管理的国家工程研究中心名单的通知》（发改办高技〔2021〕1022 号），按照《国家工程研究中心管理办法》《国家工程研究中心评价工作指南》和有关工作要求，先后分两批对现有 349 家国家工程研究中心和国家工程实验室进行优化整合。经过严格评审，最终有 191 家获准纳入新序列（统称“国家工程研究中心”），全国由高校牵头建设的国家工程研究中心获准纳入新序列的有 53 家，占总数的 28%。

2. 国家产教融合创新平台

国家产教融合创新平台是国家发展改革委“十四五”时期教育强国推进工程中支持高校“双一流”建设、服务国家重大战略的重大项目，是深化产教融合，优化高等教育结构的重要举措。教育部高等教育司在 2023 年工作要点中也提到，加快集成电路、储能、生物育种、医学攻关国家产教融合平台建设。据不完全统计，截至 2023 年 7 月，已有至少

28 所高校获批国家产教融合创新平台。2022—2023 年国家产教融合创新平台建设情况如表 1–35 所示。

表 1–35　2022—2023 年部分国家产教融合创新平台建设情况

平台名称	细分领域	获批高校
国家储能技术产教融合创新平台（第二批）		重庆大学
		哈尔滨工业大学
		上海交通大学
		中国石油大学（北京）
国家生物育种产教融合创新平台	生猪育种	华中农业大学
	玉米育种	中国农业大学
	大豆育种	南京农业大学
国家医学攻关产教融合创新平台	医工结合	东南大学
		西安交通大学
		山东大学
		浙江大学
		北京航空航天大学
		四川大学
	中医药和中西医结合	广州中医药大学
		天津中医药大学
		北京中医药大学
		上海中医药大学
	疫苗研发	南京医科大学
		厦门大学
		复旦大学
	肿瘤及重大疾病治疗	北京大学

数据来源：根据国家发展改革委、教育部、地方政府、高校网站相关通知信息整理。

2022 年年底，中共中央办公厅、国务院办公厅印发《关于深化现代职业教育体系建设改革的意见》[①]，提出从省域、市域、行业 3 个维度“一体两翼”建设现代职业教育体系。其中第三个维度行业层面，由链长企业牵头，整合上下游资源，联合学校、科研机构，共同建设一批跨区域的行业产教融合共同体。

中国中车牵头的国家轨道交通装备行业产教融合共同体，是第一个国家级行业产教融

① 中共中央办公厅，国务院办公厅．中共中央办公厅　国务院办公厅印发《关于深化现代职业教育体系建设改革的意见》[EB/OL]．（2022–12–21）．https://www.gov.cn/gongbao/content/2023/content_5736711.htm.

合共同体。新组建的共同体与过去的职教集团、产业学院等产教融合形式相比，更加注重政府搭台和机制保障。5个产业集聚地区教育行政部门作为支持单位参与了共同体的组建，指导区域产教匹配，政企校共同研究解决建设发展中的重大事项。与职教集团等由职业院校牵头的产教融合形式相比，新组建的共同体更加注重由行业龙头企业把握产教融合的主导权，构建以教促产、以产助教生态圈①。

3. 新型研发机构

新型研发机构是聚焦科技创新需求，主要从事科学研究、技术创新和研发服务，投资主体多元化、管理制度现代化、运行机制市场化、用人机制灵活的独立法人机构，可依法注册为科技类民办非企业单位、事业单位和企业。通过发展新型研发机构能够进一步优化科研力量布局，强化产业技术供给，促进科技成果转移转化，推动科技创新和经济社会发展深度融合。

近年，我国新型研发机构数量逐年增长，群体规模持续扩大。根据2023年5月30日，科技部火炬中心发布的《2022年新型研发机构发展报告》显示，截至2021年年底，全国新型研发机构数量共计2412家，同比增长12.71%。按照地区划分，东部地区有1445家（占59.9%），中部地区553家（占22.9%），西部地区331家（占13.7%），东北地区83家（占3.4%）。此外，新型研发机构广泛分布于多个产业领域。数据显示，新一代信息技术产业领域新型研发机构数量占比最高，达33.15%；其次为高端装备制造产业领域和新材料产业领域，机构数量占比分别为25.62%和25.37%。

2023年以来，新型研发机构建设受到各地高度重视。以高质量发展为目标，上海、广州、青岛、合肥等多地均出台了新型研发机构发展专项新政，大力推进新型研发机构建设。其中上海出台15条高质量发展意见为新型研发机构高质量发展“提速”，广州在组建新型研发机构联盟的同时，提出打造世界一流新型研发机构目标。如今，广州已有各类新型研发机构近80家。其中，广东粤港澳大湾区国家纳米科技创新研究院等9家新型研发机构成立投资基金，基金规模超过8亿元，投资入股18家企业。成都则在7月出台的推动科技成果转化28条举措中，用2条专门部署“培育以成果转化为导向的新型研发机构”，目前，成都新型研发机构数量已超100家。作为继高校、科研院所、企业之后的第四支科技创新队伍，新型研发机构在打通创新链、产业链中发挥异军突起的重要作用。

（五）科技学术期刊：传播交流，建设世界一流期刊成效显著

科技期刊作为各学科学术交流和成果展示的平台，是与学科建设相生相长、相辅相成的重要载体。科技期刊在科技发展过程中，发挥着组织和引导创新、增进学术交流、挖掘科技潜能、促进成果转化等多方面的功能。2019年8月16日，中国科协、中宣部、教育

---

① 中国教育报．首个国家级产教融合共同体的大不同［N/OL］．（2023-07-26）．http://www.moe.gov.cn/jyb_xwfb/s5147/202307/t20230726_1070828.html.

部、科技部联合印发《关于深化改革　培育世界一流科技期刊的意见》[①]，明确了我国科技期刊的发展目标，提出了实现一流期刊建设的措施和途径。在其指导下，近年来我国世界一流科技期刊建设取得了较为显著的成效。

1. 创办英文期刊势头迅猛

英语是国际学术交流的主流语言，对于我国这样的非英语母语国家而言，创办英文科技期刊可减少由语言带来的期刊国际学术交流障碍，对于促进科研成果的传播、提升我国科技期刊的国际话语权，乃至我国在国际舞台的科技竞争力和文化软实力都至关重要。

据中国科学技术信息研究所截至 2022 年 6 月的统计，目前我国 435 种英文版科技期刊的创办时间可分为 5 个阶段：1980 年以前 19 种（占总数的 4.4%），1980—1990 年 79 种（18.2%），1991—2000 年 57 种（13.1%），2001—2010 年 82 种（18.8%），2011—2022 年 198 种（45.5%），显示近 10 年来期刊主管部门对英文科技期刊的高度重视[②]。

根据国家新闻出版总署对于创办期刊的批复时间统计，2020—2022 年国家新闻出版总署共批准了 69 种英文科技期刊的 CN 号，其中 2020 年 15 种、2021 年 42 种、2022 年 12 种。按照《中国图书馆分类法》的一级学科类别分析，2020—2022 年新创英文科技期刊最多的 3 个学科是医药、卫生，综合性医药卫生（27 种）、工业技术总论（16 种）和数理科学和化学（8 种）（表 1–36），这些新创办英文期刊将在国内外各学科的学术成果交流中发挥重要作用。

表 1–36　2020—2022 年新创英文科技期刊学科分布

学科	期刊数（种）	占比（%）
R 医药、卫生，综合性医药卫生	27	39.13
T 工业技术总论	16	23.19
O 数理科学和化学	8	11.59
Q 生物科学	6	8.70
S 农业、林业、综合性农业科学	4	5.80
X 环境科学、安全科学	3	4.35
V 航空、航天	2	2.90
U 交通运输	1	1.45
N 自然科学总论	1	1.45
P 天文学、地球科学	1	1.45
合计	69	100.00

数据来源：国家新闻出版署网站。

① 中国科协．四部门联合印发《关于深化改革　培育世界一流科技期刊的意见》[EB/OL]．https://www.cast.org.cn/xw/TTXW/art/2019/art_b5da1323b57c4d16b779172ad533cd88.html.

② 任胜利，杨洁，宁笔，等．2022 年我国英文科技期刊发展回顾[J]．科技与出版，2023（3）：50–57.

2. 期刊整体发展稳步向前

自 2020 年起，在中国科协的资助下，中国知网与中国科学技术信息研究所等多家单位联合开展了“科技期刊世界影响力指数”（WJCI）研究。该研究旨在建立世界范围内科技期刊公平评价体系，推动中外科技期刊同质等效使用。项目研究成果《科技期刊世界影响力指数（WJCI）报告》（以下简称《WJCI 报告》）已经连续四年通过网络免费向全球发布（https://wjci.cnki.net/）。WJCI 评价体系打破了先有数据库、再有期刊评价的窠臼，从源头建立全新的统计源期刊集，并将网络传播指标纳入学术期刊评价，以便能更加全面地反映新媒体环境下期刊的综合影响力。通过《WJCI 报告》（2021—2023 版）（数据统计年份为 2020—2022 年）可以在世界学科领域内探究我国科技期刊近年来的发展和影响力变化情况。

各项评价指标稳步增长。2022 年 WJCI 收录中国期刊 1772 种，占 WJCI 收录世界期刊总数（15555 种）的 11.39%，比 2021 年增加了 138 种，比 2020 年增加了 188 种。2022 年，WJCI 收录中国期刊的总被引频次为 405.51 万次，相较于 2020 年增加了 42.12%。WJCI 收录中国期刊刊均影响因子从 2020 年的 1.364 增加到 2022 年的 2.139，增加了 56.82%。中国期刊刊均 WJCI 指数从 2020 年的 1.331 增加到了 2022 年的 1.487，增加了 11.72%。组成 WJCI 的两个维度指标：基于引证的 WAJCI，期刊平均值从 2020 年的 1.155 增加到了 2022 年的 1.306，增加了 13.07%；基于网络传播使用情况的 WI，期刊平均值由 2020 年的 0.176 增加到 2022 年的 0.181，增长了 2.84%。各项评价指标三年来都呈稳步增长趋势（表 1–37）。

表 1–37　2020—2022 年 WJCI 中国期刊指标变化

统计年	刊数（种）	WJCI 均值	WAJCI 均值	WI 均值	总被引频次（万次）	国际引用频次（万次）	国际引用占比（%）	总被引频次均值	影响因子均值
2020	1584	1.331	1.155	0.176	285.32	76.42	26.78	1802	1.364
2021	1634	1.378	1.200	0.178	330.28	109.32	33.1	2021	1.805
2022	1772	1.487	1.306	0.181	405.51	152.41	37.58	2288	2.139

数据来源：《WJCI 报告》（2021—2023 版）。

中国期刊在世界学科领域的排名显著提升，分区结构不断优化。2022 年，中国期刊进入 WJCI–Q1 区的有 321 种，占 WJCI 收录中国期刊的 18.12%，占 WJCI–Q1 区世界期刊总数（4151 种）的 7.73%。WJCI–Q1 区中国期刊比 2020 年的 229 种增加了 92 种，占 WJCI 收录中国期刊的比例增加了 3.67%（表 1–38）。

中国期刊学科分析：根据《WJCI 报告》（2023 版），中国期刊入选各学科 Q1Q2 区期刊在 10 种以上的学科有 21 个，表明在这些学科领域，我国科技期刊发展相对较好（表 1–39）。但在 99 个三级学科 Q1Q2 区没有中国期刊入选，其中在 30 个三级学科甚至没有中国期刊入选，说明我国应加强在这些学科领域内办好刊、创新刊的力度（表 1–40）。

表 1–38　2020—2022 年 WJCI 各分区中国期刊数量统计表　（单位：种）

统计年	Q1	Q2	Q3	Q4	合计
2020	229	395	603	555	1584
2021	244	398	588	601	1634
2022	321	440	600	640	1772

数据来源：《WJCI 报告》（2021—2023 版）。

表 1–39　我国科技期刊发展较好的学科期刊分区统计表　（单位：种）

序号	学科领域	世界期刊数	中国期刊数	中国期刊数			
				Q1	Q2	Q3	Q4
1	科学技术综合	276	82	14	28	29	11
2	地球科学综合	129	16	3	8	3	2
3	自然地理学	182	19	8	8	3	
4	地质学	144	41	16	11	11	3
5	植物学	210	22	7	6	6	3
6	医学综合	328	41	8	7	22	4
7	中医学与中药学、结合与补充医学	60	32	6	7	8	11
8	农业科学综合	150	33	17	16		
9	农艺学	141	32	6	8	13	5
10	林学综合	91	20	3	9	7	1
11	工程综合	162	37	13	12	9	3
12	金属学	72	28	7	6	9	6
13	矿山工程技术	64	25	6	11	7	1
14	石油天然气工业	63	41	9	11	9	12
15	冶金工程技术	82	25	9	5	6	5
16	电气工程	231	39	10	7	10	12
17	能源系统工程	91	22	6	8	5	3
18	电子技术	124	45	4	9	17	15
19	计算机科学技术综合	172	16	6	5	3	2
20	航空、航天科学技术	93	43	7	10	18	8
21	环境科学技术综合	236	36	8	5	10	13

数据来源：《WJCI 报告》（2023 版）。

表 1–40 《WJCI 报告》（2023 版）无中国期刊的学科列表

序号	学科领域	序号	学科领域
1	自然科学史	16	实验心理学
2	数理逻辑与数学基础	17	发展心理学
3	数学分析	18	数理心理学、心理统计法
4	函数论	19	生理心理学
5	离散数学	20	应用心理学
6	电磁学	21	教育心理学
7	结构生物学	22	医学信息学
8	呼吸生理学	23	疼痛研究
9	感官生理学	24	医学技术
10	生殖生物学	25	家庭医学、社区医学
11	进化论、生物系统发育	26	材料检测与分析技术
12	细胞与分子神经科学	27	计算机理论与方法
13	动物生态学和动物地理学	28	信息系统与管理
14	认知心理学	29	环境生态学
15	社会心理学、法制心理学	30	环境管理学、环境法学

数据来源：《WJCI 报告》（2023 版）。

从影响力突出的中国期刊来看，根据《WJCI 报告》（2023 版），中国期刊的 WJCI 指数进入全球各学科前 5% 的有 22 种，位居学科前三名的有 24 种，更有 2 种中国期刊的 WJCI 指数位居全球第一（表 1–41）。

表 1–41 2022 年学科排名位居全球前三的中国期刊

序号	期刊名称	学科领域	WJCI	排名
1	*Bioactive materials*	生物材料学	5.384	3/44
		生物医学工程	6.742	3/107
2	*Cell Research*	细胞工程	14.568	1/44
3	*Communications in Nonlinear Science and Numerical Simulation*	非线性科学	2.806	2/20
4	*Engineering*	工程综合	11.044	3/162
5	*Fungal Diversity*	真菌学	4.482	2/40
6	*Geoscience Frontiers*	地质学	6.692	3/144
7	*InfoMat*	工程力学	8.218	2/36
8	*Journal of Magnesium and Alloys*	冶金工程技术	10.970	2/82
9	*Journal of Ocean Engineering and Science*	海洋工程与技术	4.412	3/37

续表

序号	期刊名称	学科	WJCI	排名
10	*Light：Science&Applications*	光学	7.463	3/94
11	*Military Medical Research*	特种医学	9.906	1/19
12	*npj Computational Materials*	材料力学	7.075	2/102
		仿真科学技术	8.460	2/107
13	*Science China Information Sciences*	信息科学	6.499	3/71
14	*Science China Physics，Mechanics&Astronomy*	力学综合	4.281	3/66
		天文学综合	7.361	3/29
15	*Signal Transduction and Targeted Therapy*	医学影像学、医学成像技术	11.736	2/149
16	*Transactions of Nonferrous Metals Society of China*	金属学	3.851	3/72
17	电子测量与仪器学报	计量与标准化	2.580	3/17
18	农业工程学报	农业工程	4.712	3/27
19	农业机械学报	农业生物学	5.378	2/25
20	食品与发酵工业	制糖、食品发酵与酿造技术	2.428	2/14
21	水利学报	水利工程	2.556	3/55
22	系统工程理论与实践	系统科学	2.696	2/25
23	中国人口·资源与环境	人口统计学	2.104	2/10
24	中国中药杂志	中医学与中药学、结合与补充医学	3.533	3/60

数据来源:《WJCI 报告》(2023 版)。

按国家统计 WJCI 均值，可揭示该国期刊总体水平。2022 年，中国期刊的 WJCI 指数均值为 1.487，G7 国家中 WJCI 指数均值最高的是英国（2.854），其次是美国（2.784）。中国与 G7 国家 WJCI 指数均值对比来看，与英国、美国、德国还有较大差距，略低于加拿大，高于法国、意大利和日本（图 1-9）。

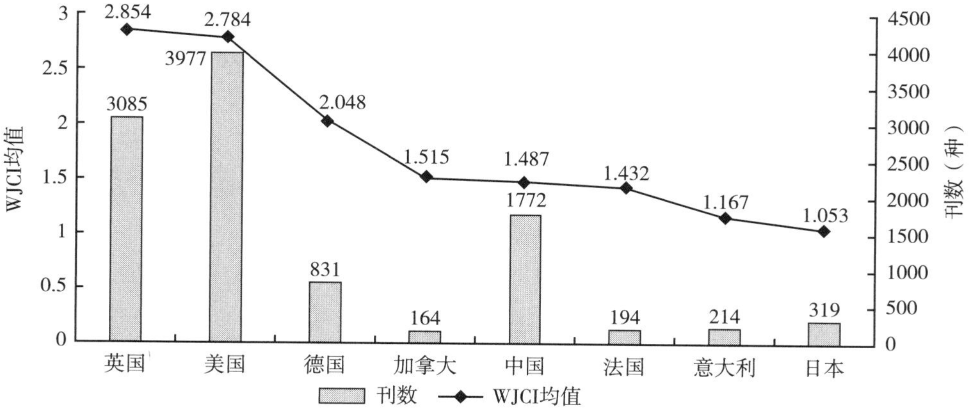

图 1-9　2022 年中国与 G7 国家期刊 WJCI 指数均值

数据来源:《WJCI 报告》(2023 版)。

六、科技人才队伍持续壮大

人才队伍是衡量一个国家综合国力的重要指标。随着新一轮科技革命和产业变革深入发展，全方位谋划基础学科人才培养，重点推进优势学科与特色学科领军人才队伍建设，打造世界重要人才中心和创新高地尤为重要。本部分从 R&D 人员、技能人才、高校专任教师、研究生规模、学科领军人才等方面分析近年来我国各学科人才队伍的发展情况。

（一）研发人员队伍持续扩大

1. R&D 人员投入大幅增长

R&D 人员的规模和素质是支撑我国科技创新能力，实现我国科技发展规划目标的重要基础条件。R&D 人员全时当量指全时人员数加非全时人员按工作量折算为全时人员数的总和，反映投入从事拥有自主知识产权的研究开发活动的人力规模。2018—2022 年，我国 R&D 人员全时当量持续增长，2022 年为 635.4 万人年，比 2021 年增加 63.8 万人年，增速为 11.16%（图 1-10）。

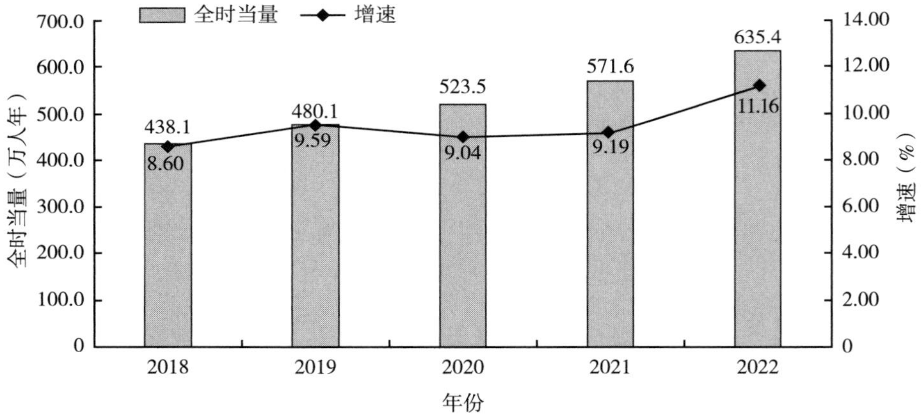

图 1-10　2018—2022 年中国 R&D 人员全时当量

数据来源：《中国统计年鉴 2023》。

按研究性质分，R&D 活动可以分为基础研究、应用研究和试验发展三大类。2022 年 R&D 人员全时当量中，占比最大的是从事试验发展的 R&D 人员，总量为 510.3 万人年，占比 80.32%；从事应用研究的人员为 74.1 万人年，占比 11.66%；从事基础研究的人员为 50.9 万人年，占比 8.01%（图 1-11）。

2018—2022 年，三类研究人员数量保持稳定增长的态势。2021—2022 年，基础研究人员增加 3.7 万人年，增速为 7.84%；应用研究人员增加 5.0 万人年，增速为 7.24%；试验发展人员增加 55 万人年，增速为 12.08%。

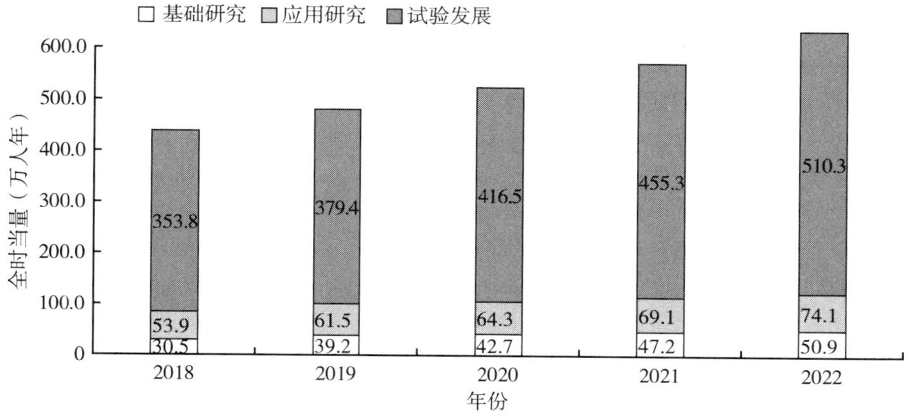

图 1-11 2018—2022 年中国 R&D 人员从事研究类型分布

数据来源:《中国统计年鉴 2023》。

2. 技能人才不断增多

技能人才在各行业领域中具有丰富的业务知识和熟练的操作经验，是强化企业生产管理、提高企业生产效率的中坚力量。根据《人力资源和社会保障事业发展统计公报》①，2018—2022 年，我国取得职业资格证书或职业技能等级证书的人数整体呈增长趋势，2022 年为 1234.3 万人次，较 2021 年的 898.8 万人次增长了 37.33%。2022 年取得技师以上职业资格证书或职业技能等级证书的人数为 35.6 万人次，较 2021 年的 30.2 万人次增长了 17.88%（表 1-42）。

表 1-42 2018—2022 年我国取得职业资格证书或职业技能等级证书人数

（单位：万人次）

年份	取得证书人数	取得技师以上证书人数
2018	903	35
2019	861.9	28.4
2020	962.6	25.8
2021	898.8	30.2
2022	1234.3	35.6

数据来源：2018—2022 年《人力资源和社会保障事业发展统计公报》。

3. 高校专任教师数量稳步提升

高校专任教师是新时代推动我国教育事业发展、人才培养和各学科科研创新的重要力

① 人力资源和社会保障部. 人力资源和社会保障事业发展统计公报［EB/OL］.（2023-06-20）. http://www.mohrss.gov.cn/SYrlzyhshbzb/zwgk/szrs/tjgb/.

量，其规模和水平对提高我国科技创新能力，实现科技高水平自立自强有重要意义。根据《全国教育事业发展统计公报》①，2018—2022 年我国高等学校专任教师数量分别为 167.28 万人、174.01 万人、183.30 万人、188.52 万人、197.78 万人，呈逐年稳步增长趋势，同比增速分别为 4.02%、5.34%、2.85% 和 4.91%（图 1–12）。

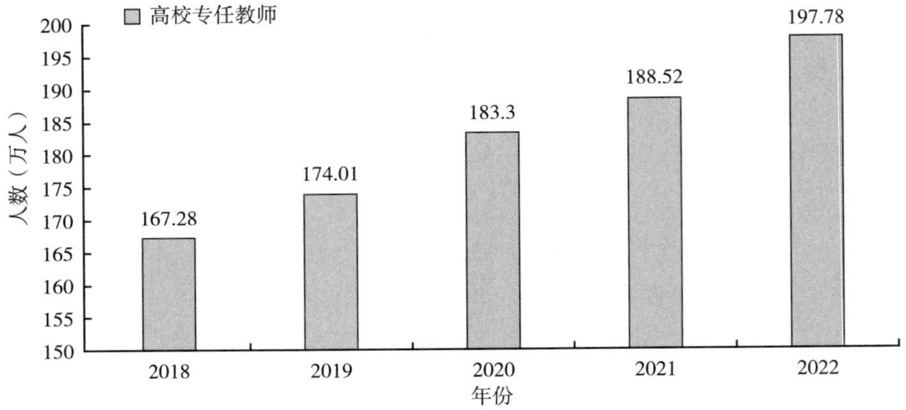

图 1–12　2018—2022 年我国高等学校专任教师数量变化

数据来源：《全国教育事业发展统计公报》。

4. 研究生培养规模逐年扩大

研究生培养规模保持稳定增长，可以不断为我国科研队伍注入新鲜血液。根据 2018—2022 年教育部《全国教育事业发展统计公报》，我国研究生培养规模持续扩大。2022 年招收研究生 124.25 万人，同比增长 5.61%，其中博士生 13.9 万人，硕士生 110.35 万人；毕业研究生 86.22 万人，同比增长 11.57%，其中博士生 8.23 万人，硕士生 77.98 万人，连续 5 年呈现递增的趋势，见表 1–43。

表 1–43　2018—2022 年我国研究生招收与毕业人数变化　（单位：万人）

年份	招生人数			毕业人数		
	研究生	博士研究生	硕士研究生	研究生	博士研究生	硕士研究生
2018	85.80	9.55	76.25	60.44	6.07	54.36
2019	91.65	10.52	81.13	63.97	6.26	57.71
2020	110.66	11.60	99.05	72.86	6.62	66.25
2021	117.65	12.58	105.07	77.28	7.20	70.07
2022	124.25	13.90	110.35	86.22	8.23	77.98

数据来源：2018—2022 年《全国教育事业发展统计公报》。

① 教育部. 全国教育事业发展统计公报［EB/OL］. http://www.moe.gov.cn/jyb_sjzl/sjzl_fztjgb/.

（二）学科领军人才持续涌现

1. 国家 / 地区分析

根据科睿唯安发布的 2022 年度和 2023 年度“高被引科学家”名单，来自全球 60 个国家和地区的 6938 名和 6849 名科学家入选高被引科学家名单，较 2021 年的 6602 人次分别增长 5.09% 和 3.74%。入选科学家在过去十年产生了重要的学术影响力。表 1-44 是 2022 年度和 2023 年度高被引科学家上榜人次排名前十的国家和地区名单，可以看出，中国（不含港澳台地区）科学家上榜人数增长较大，2023 年入选科学家比 2022 年的 1169 人次（占比 16.2%）增长 9.07%，上升到 1275 人次（占比 17.9%），紧随美国之后，居世界第二位。美国科学家在榜单中依然占据主导地位，表明其科学研究水平依然居于世界领先，但所占份额有所下降。

表 1-44　2022—2023 年度高被引科学家上榜人次前 10 国家 / 地区名单

排名	2022 年度			2023 年度		
	国家 / 地区	高被引科学家总人次	入榜比例（%）	国家 / 地区	高被引科学家总人次	入榜比例（%）
1	美国	2764	38.3	美国	2669	37.5
2	中国	1169	16.2	中国	1275	17.9
3	英国	579	8.0	英国	574	8.1
4	德国	369	5.1	德国	336	4.7
5	澳大利亚	337	4.7	澳大利亚	321	4.5
6	加拿大	226	3.1	加拿大	218	3.1
7	荷兰	210	2.9	荷兰	195	2.7
8	法国	134	1.9	法国	139	2
9	瑞士	112	1.6	中国香港	120	1.7
10	新加坡	106	1.5	意大利	115	1.6

数据来源：Highly Cited Researchers 2022①，Highly Cited Researchers 2023②。

2. 学科分析

2022—2023 年度，中国高被引科学家主要科技学科分布情况如图 1-13 所示。2022 年入选的 1169 名学者中，达到 10 人次的学科有 12 个，比 2021 年增加 1 个。2023 年入选的 1275 名学者中，达到 10 人次的学科有 12 个，与 2022 年持平。与 2022 年相比，2023 年入选人数增长超过 10 人次的有地球科学（增长 17 人次）、材料科学（增长 14 人次）、

① 科睿唯安. Highly Cited Researchers 2022［EB/OL］.（2022-11-15）. https://clarivate.com.cn/2022/11/15/.

② 科睿唯安. Highly Cited Researchers 2023［EB/OL］.（2023-11-15）. https://clarivate.com.cn/2023/11/15/hcr2023/.

植物与动物科学（增长 10 人次）、微生物学（增长 10 人次）。高被引科学家的学科分布情况与我国论文的发表情况基本吻合，也表明了我国顶尖水平科学家在相关学科领域的研究实力。

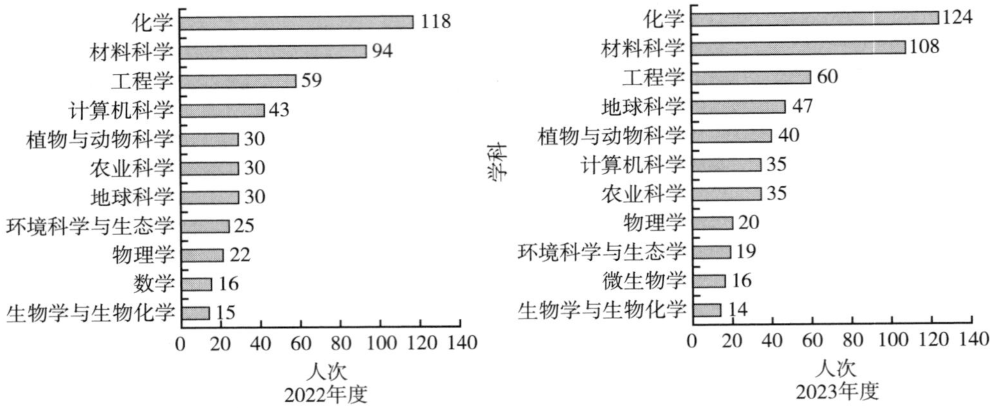

图 1-13　2022—2023 年度中国高被引科学家主要学科分布

数据来源：Highly Cited Researchers 2022，Highly Cited Researchers 2023。

七、国内外交流合作深入推进

（一）国际交流合作

随着全球化的发展，各学科领域的国际交流与合作不断加强，已成为推动科技、文化和经济发展的重要驱动力。通过高水平、高质量的知识交流与国际合作，解决人类发展面临的共同挑战，是增进团结、包容，实现共赢的重要举措。党的二十大以来，习近平总书记对各学科领域国际合作、国际交流学习多次作出重要指示批示，我国各学科国际合作交流的规模和影响力正在逐渐扩大，我国积极扩大科技领域开放合作，主动融入全球科技创新网络。

1. 国际合作论文数量与合作范围持续扩大

随着科学技术全球化进程的推进，国际合作日益广泛多元，国际合作论文的增长表明我国科技发展正在与世界科技快速融合。国际科技合作实施情况通过国际合作论文这一重要载体得以体现。据 SCI 数据库统计，2022 年中国发表的国际论文中，国际合作论文 15.92 万篇，占中国发表论文总数的 21.6%。国际合作论文比 2021 年增加了 1.00 万篇，增长了 6.7%。我国科学家参与国际合作论文的数量正在持续提高。2022 年中国的国际合作论文中，与美国合写的为 4.89 万篇，占国际合作论文总数的 30.75%，比 2021 年减少近 0.59 万篇。

2022 年，中国作者作为第一作者的国际合作论文共 11.67 万篇，占中国全部国际合作论文的 73.3%，合作论文量比 2021 年增加 14.51%，合作伙伴涉及 173 个国家（地区），其中排在前六位的分别是美国、英国、澳大利亚、加拿大、德国和日本（图 1–14）。

其他国家作者为第一作者，中国作者参与工作的国际合作论文为 4.25 万篇，与 2021 年相差不大，合作伙伴涉及 184 个国家（地区），作者来自 3 个及以上国家的论文比例（47.62%）较 2021 年（46.46%）有所增加。

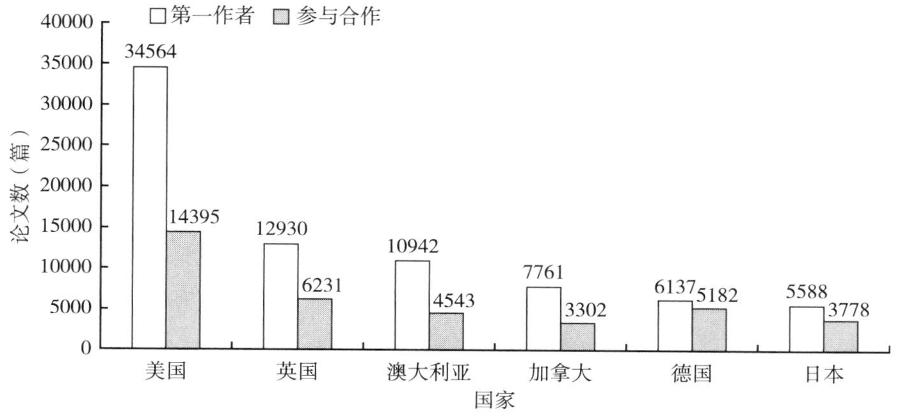

图 1–14　2022 年中国作者作为第一作者和作为参与方产出合作论文较多的 6 个国家

数据来源:《中国科技论文统计报告 2023》[①]。

2. 国际合作学科分布较为集中

根据 SCI 对国际科技论文学科分布的统计数据，2022 年中国学者发表的国际合作论文中，数量较多的学科为生物学，化学，电子、通信与自动控制，材料科学，临床医学，计算技术，物理学，环境科学。这些学科是当前我国科学工作者与世界各国学者合作开展科学研究的主要领域（表 1–45）。

表 1–45　2022 年中国作者国际合作论文较多的学科

排序	中国作者为第一作者			中国作者为参与方		
	学科	论文数（篇）	占本学科论文比例（%）	学科	论文数（篇）	占本学科论文比例（%）
1	生物学	11535	15.53	生物学	5014	6.75
2	化学	10657	13.78	临床医学	4960	5.58
3	电子、通信与自动控制	8890	19.48	化学	4098	5.30
4	材料科学	8494	14.94	物理学	2809	6.44
5	临床医学	8338	9.38	材料科学	2730	4.80
6	计算技术	8335	24.62	环境科学	2632	6.62

数据来源:《中国科技论文统计报告 2023》[②]。

①② 中国科学技术信息研究所. 中国科技论文统计结果 2023［R］. 北京：中国科学技术信息研究所，2023.

3. 国际交流学习触底回升

教育部数据显示，自改革开放到 2021 年年底，我国各类出国留学人员数量在 800 万左右，学成回国留学人员数量 550 万左右，留学人员遍及世界 160 多个国家和地区的 1 万多所高校。2022 年高等教育阶段中国留学生人数较多的国家有美国（290086 人）、澳大利亚（152715 人）、英国（150720 人）、加拿大（105265 人）、日本（94063 人）、德国（40122 人）、俄罗斯（39939 人）、法国（27950 人）、意大利（16754 人）和新西兰（11880 人）①。

根据《中国留学发展报告（2022）》②，受中美关系和疫情影响，中国学生赴美留学热情持续降温，2020—2021 学年，中国在美留学生相比 2019—2020 学年减少 14.6%，出现十年来首次负增长。《2022 年美国门户开放报告》显示，2021—2022 学年在美国接受高等教育的中国学生已经连续两年快速下降，占比也从上一年度最高的 34.7% 下降至 30.6%，但人数仍位居第一位。由于赴美留学的不确定因素，不少计划出国留学的中国留学生更加倾向选择留学环境及签证政策更为友好的国家和地区。根据启德教育发布的《2023 中国留学白皮书》，留学申请量在 2020 年触底后近两年呈持续反弹趋势，2022 年整体申请量同比 2021 年增长幅度达到了 23.4%，其中澳大利亚（涨幅 84.9%）和英国（涨幅 32.8%）同比 2021 年有明显上升。从留学生结构来看，硕士申请仍为中国学生留学选择的主流，占比 81.2%，本科占比 16.6%，博士占比 0.5%。博士留学的申请量平均每年增长 10%~15%，2022 年相比 2021 年增长了 21.8%。

在留学学科专业选择方面，最近几个学年，中国留学生选择专业方向受到目的国签证政策的影响，个别国家限制了部分 STEM 专业中国留学生的签证，但总体来看，中国赴美国、加拿大、法国、德国、澳大利亚的留学生以攻读 STEM 专业和商科、社会科学为主，人文、艺术等领域的留学生也越来越多（表 1–46）。目前，虽然以美国为首的五眼联盟国家从经济和贸易的角度鼓励国际学生前来留学，并在一定程度上放宽了对 STEM 领域留学的限制，但在一些关键技术领域仍对留学生持敏感态度，并可能会受到更严格的安全审查和留学限制。

中国出国留学在进入 21 世纪后迎来高潮，回国留学生数量也随之上升，从 2000 年的 0.9 万人增加到了 2019 年的 58.03 万人。回流率，即当年回国留学生与出国留学生之比，从 2000 年的 23.3% 增加到了 2019 年的 82.5%。2020 年后，国际关系的变化和新冠疫情等因素又进一步加速推动了海外人才回流，留学人员回国总人数呈显著上升趋势。国家信息中心大数据发展部针对 2021 年海归大数据分析，2021 年回国就业学生首次超过百万，达

① 中国日报．2023 中国留学白皮书在京发布报［EB/OL］．https://cn.chinadaily.com.cn/a/202303/18/WS641564e2a3102ada8b2342fd.html.

② 全球化智库，中国银行，西南财经大学发展研究院．中国留学发展报告（2022）［M］．北京：社会科学文献出版社，2022.

到 104.9 万人[①]。

表 1–46　中国留学生赴主要国家留学的专业学科占比

国家	学年	中国留学生赴该国留学的主要专业占比
美国	2020—2021	数学与计算机科学（22.2%）、工程学（17.5%）、工商管理（16.6%）、社会科学（9.6%）、物理和生命科学（9.1%）
加拿大	2019—2020	工商管理（27.3%）、数学、计算机与信息科学（16.9%）、社会与行为科学（14.0%）、工程与工程技术（12.5%）、科学与科技（10.7%）
法国	2018—2019	人文学与语言学（39.6%）、自然科学与运动研究（32.4%）、经济学与社会科学（24.7%）、法律与政治学（2.2%）、医学及相关专业（1.1%）
德国	2019—2020	工程学（41%）、商科（14%）、自然科学（9%）、数学（8%）、艺术（7%）
澳大利亚	2020—2021	管理和贸易（41.8%）、信息技术（11.4%）、社会与文化（11.0%）、工程技术（9.6%）、创意艺术（6.9%）、自然和物理科学（6.4%）

数据来源:《中国留学发展报告（2022）》。

领英发布的《2022 中国留学生归国求职洞察报告》显示，2020 年中国留学生学成回国人数同比增长 33.9%，且在 2021 年持续保持增长态势。智联招聘发布的《2022 中国海归就业调查报告》同样显示，2022 年回国求职留学生数量再创新高，同比增长 8.6%。各项数据表明，中国留学生回国就业的意向正在不断增强[②]。

（二）国内交流合作

1. 国内论文合作规模与影响力不断增强

根据中国知网中国知识资源总库，2021—2022 年，我国期刊共发表科技论文 263.66 万篇，其中由国内多个机构学者合作发表的论文 66.13 万篇，占比 25.08%。截至 2023 年 7 月 31 日，上述多机构合作论文的篇均被引频次为 2.19 次、篇均下载频次为 318.92 次，相比非多机构合作论文的篇均被引频次为 1.33 次、篇均下载频次为 221.95 次，多机构合作论文获得更大的影响力。2022 年我国多个机构学者合作在中国期刊上发表的科技论文 33.74 万篇，比 2021 年增加 1.36 万篇，增长 4.2%。上述数据说明，我国多机构合作在国内期刊发表的科技论文的规模正在不断扩大。

根据我国期刊发表科技类合著论文学科分布的统计数据，2021—2022 年多机构合著论文数量较多的前 10 个学科如表 1–47 所示。多机构合著论文数量和占比最高的 3 个学科分别是电力工业、环境科学与资源利用、中医学。

① 中国教育在线. 中国学生出国留学趋势调查报告 2022［EB/OL］. https://www.eol.cn/e_html/report/Osr2022/content.shtml.

② 光明网. 透过数据，洞悉 2023 年留学趋势［EB/OL］. https://epaper.gmw.cn/lx/html/2023-01/05/nw.D110000lx_202301052-04.htm.

表 1-47　2021—2022 年合著期刊论文最多的前 10 个学科

学科	合著论文量（篇）	占所有学科合著论文量比例（%）	被引频次（次）	篇均被引频次（次）
电力工业	38854	4.28	118686	3.05
环境科学与资源利用	38722	4.27	134412	3.47
中医学	38679	4.26	100831	2.61
建筑科学与工程	36584	4.03	71789	1.96
计算机软件及计算机应用	31776	3.50	87611	2.76
肿瘤学	30003	3.31	42147	1.40
地质学	26641	2.93	60797	2.28
自动化技术	25912	2.85	78345	3.02
临床医学	25188	2.77	50373	2.00
中药学	22909	2.52	76613	3.34

数据来源：中国知识资源总库，评价指标统计时间截至 2023 年 7 月 31 日。

2. 研究生联合培养互动比较频繁

中国知网中国知识资源总库收录的 2021—2022 学位年度科技领域博硕士学位论文共计 52.39 万篇，由多位导师联合指导的论文为 12.66 万篇，占比 24.17%。

根据国内多导师指导的博士、硕士学位论文学科分布的统计数据，2021—2022 学位年博士研究生联合培养较多的学科为材料科学、化学、电力工业、物理学、有机化工、环境科学与资源利用等，学位论文量均在 600 篇以上，占所有学科学位论文量比例均在 4.5% 以上（表 1-48）。硕士研究生联合培养较多的学科为计算机软件及计算机应用、自动化技术、建筑科学与工程、电力工业、环境科学与资源利用等，学位论文量均在 1 万篇以上，占所有学科学位论文量比例均在 6% 以上（表 1-49）。我国导师合作开展研究生培养的学科领域分布也相对比较均匀。

表 1-48　2021—2022 年多导师指导博士学位论文最多的前 10 个学科

学科	多导师指导学位论文量（篇）	占所有学科学位论文量比例（%）	被引频次（次）	篇均被引频次（次）
材料科学	774	5.62	291	0.38
化学	723	5.25	221	0.31
电力工业	704	5.11	434	0.62
物理学	673	4.89	147	0.22
有机化工	673	4.89	319	0.47
环境科学与资源利用	654	4.75	559	0.85

续表

学科	多导师指导学位论文量（篇）	占所有学科学位论文量比例（%）	被引频次（次）	篇均被引频次（次）
自动化技术	578	4.20	417	0.72
生物学	544	3.95	215	0.40
计算机软件及计算机应用	465	3.38	286	0.62
金属学及金属工艺	432	3.14	224	0.52

数据来源：中国知识资源总库，评价指标统计时间截至 2023 年 7 月 31 日。

表 1-49　2021—2022 年多导师指导硕士学位论文最多的前 10 个学科

学科	多导师指导学位论文量（篇）	占所有学科学位论文量比例（%）	被引频次（次）	篇均被引频次（次）
计算机软件及计算机应用	14617	8.13	6231	0.43
自动化技术	13094	7.29	5966	0.46
建筑科学与工程	11912	6.63	5982	0.50
电力工业	11378	6.33	4688	0.41
环境科学与资源利用	10802	6.01	4731	0.44
有机化工	7640	4.25	2131	0.28
轻工业手工业	5881	3.27	3843	0.65
化学	5467	3.04	986	0.18
材料科学	5391	3.00	1209	0.22
金属学及金属工艺	4668	2.60	1711	0.37

数据来源：中国知识资源总库，评价指标统计时间截至 2023 年 7 月 31 日。

第二节　学科发展动态

近年来，我国各学科研究水平不断提升，国际影响力显著增强。国内外学术论文、专利等产出量不断增加，学术交流越发频繁，研究成果支撑引领经济发展的作用不断增强。2022—2023 年，各学科取得了一系列引领世界的重大成果，基础研究为高水平科技自立自强“固本强基”，技术研究与前沿领域驶入创新发展“快速车道”，应用研究为经济发展和人民健康“保驾护航”。本节汇总梳理了本周期内立项的各学科提供的重要成果，并结合其他科学技术领域的重要研究进展，呈现各领域近两年来的重要研究进展和重点关注研究方向，以展示各学科最新发展动态。

一、物理学

在光学领域，浙江大学邱建荣团队及其合作者们发现了飞秒激光诱导复杂体系微纳结构形成的新机制。以含氯溴碘离子的氧化物玻璃体系为例，实现了玻璃中具有成分和带隙可控发光可调的钙钛矿纳米晶三维（3D）直接光刻，呈现红橙黄绿蓝等不同颜色的发光。形成的纳米晶在紫外线辐照、有机溶液浸泡和 250℃高温环境中表现出显著的稳定性。并进一步演示了这种 3D 微纳结构在超大容量长寿命信息存储、高稳定的最小像素尺寸微米级的 Micro-LED 列阵，实现了 1080p 级别动态立体彩色全息显示。该成果揭示了飞秒激光诱导空间选择性介观尺度分相和离子交换的规律，开拓了飞秒激光三维极端制造新技术原理。

在超导物理学领域，我国近两年来有多项重要突破。上海交通大学贾金锋、郑浩团队与麻省理工学院傅亮团队合作，实验证实超导态“分段费米面”的存在，成功验证了 58 年前德国物理学家彼得·富尔德的理论预言。该研究开辟了调控物态、构筑新型拓扑超导的新方法。中山大学王猛团队与清华大学、华南理工大学等单位合作研究，首次发现在 14 吉帕压力下达到液氮温区的镍氧化物超导体。这是由我国科学家率先独立发现的全新高温超导体系，是人类目前发现的第二种液氮温区非常规超导材料，是基础研究领域的重要突破。这一研究成果将有望推动破解高温超导机理，使设计和预测高温超导材料成为可能，使超导在信息技术、工业加工、电力、生物医学和交通运输等领域实现更广泛的应用。电子科技大学李言荣团队与美国布朗大学詹姆斯·瓦莱斯、北京大学谢心澄院士等协同攻关，成功突破了费米子体系的限制，首次在玻色子体系中诱导出奇异金属态。该研究为理解凝聚态物理中奇异金属的物理规律、揭示奇异金属的普适性、完善量子相变理论奠定了科学基础，对推动未来低能耗超导量子计算以及极高灵敏量子探测技术的发展具有重要的理论和实际意义。南方科技大学俞大鹏、徐源，福州大学郑仕标，清华大学孙麓岩等展示了一种基于超导电路量子电动力学架构的量子纠错方法，实现了量子纠错正增益，改善了目前量子纠错面临“越纠越错”的尴尬局面。该研究展示了量子纠错的优越性，表明了硬件高效的离散变量编码在容错量子计算中的潜力。

稳态强磁场是物质科学研究需要的一种极端实验条件，是推动重大科学发现的“利器”。2022 年，国家重大科技基础设施“稳态强磁场实验装置”实现重大突破。8 月 12 日，“稳态强磁场实验装置”混合磁体在 26.9 兆瓦的电源功率下创造场强 45.22 万高斯的稳态强磁场，超越已保持了 23 年之久的 45 万高斯稳态强磁场世界纪录，成为我国科学实验极端条件建设乃至世界强磁场技术发展的重要里程碑。

二、化学

在碳材料领域，中国科学院化学研究所郑健团队创制了一种新的碳同素异形体——单层聚合 C_{60} 单晶。这是一种全新的簇聚二维超结构，由 C_{60} 簇笼在平面上通过 C—C 键相互共价键合形成规则的拓扑结构。这是碳材料领域从 0 到 1 的重大突破，这种以富勒烯团簇为单元构筑碳材料的方法，甚至开辟了一个碳材料研究新方向。单层聚合 C_{60} 具有良好的热力学稳定性，具有接近单晶硅的禁带宽度及各向异性的声子模式和电导，未来有望应用于半导体、量子计算、非线性光学、化学催化等领域。作为我国原创的碳材料，该研究已经被国外科学家反向跟进并引发研究热潮。《自然》(*Nature*) 同期评述认为“这种通用的方法可以开发新的碳材料”，《化学和工程新闻》(*Chemical & Engineering News*) 专题报道了该项工作，碳材料权威期刊《碳》(*Carbon*) 认为“这是碳材料领域最令人兴奋的进展之一，开启了新的碳材料体系”。该工作被中国科协《科技导报》评为“2022 年中国重大科学、技术和工程进展”。

在催化研究领域，2022 年，电子科技大学夏川课题组、中国科学院深圳先进技术研究院于涛课题组和中国科学技术大学曾杰课题组共同创建了一种二氧化碳转化新路径，通过电催化与生物合成相结合，成功以二氧化碳和水为原料合成了葡萄糖和脂肪酸，为人工和半人工合成“粮食”提供了新路径。该研究开辟了电化学结合活细胞催化制备葡萄糖等粮食产物的新策略，为进一步发展基于电力驱动的新型农业与生物制造业提供了新范例，是二氧化碳利用的重要发展方向。该成果 4 月 28 日以封面文章形式在《自然：催化》(*Nature Catalysis*) 发表。

在化学物理领域，中国科学技术大学潘建伟、赵博团队与中国科学院化学研究所白春礼团队合作，在钠钾基态分子和钾原子混合气中，在分子 – 原子 Feshbach 共振附近利用射频合成技术首次相干地合成了超冷三原子分子。这一研究是在化学物理领域取得的重大突破，相关研究成果发表在《自然》(*Nature*) 杂志上，向基于超冷原子分子的量子模拟和超冷量子化学的研究迈出了重要一步，为未来超冷三原子分子的制备和控制开辟出一条道路。厦门大学廖洪钢、孙世刚和北京化工大学陈建峰等开发高时空分辨电化学原位液相透射电镜技术，耦合真实电解液环境和外加电场，实现对锂硫电池界面反应原子尺度动态实时观测和研究。近百年来，电化学界面反应通常被认为仅存在“内球反应”和“外球反应”单分子途径。该研究揭示了电化学界面反应存在第三种“电荷存储聚集反应”机制，加深了对多硫化物演变及其对电池表界面反应动力学影响的认识，为下一代锂硫电池设计提供指导。

在化学工业领域，2022 年 11 月，中国工程院院士谢和平团队在《自然》(*Nature*) 杂志发表论文，建立了可在海水里原位直接电解制氢的新技术。该技术彻底隔绝了海水离子，实现了无淡化过程、无副反应、无额外能耗的高效海水原位直接电解制氢。这一技术未来有望与海上可再生能源相结合，构建无淡化、无额外催化剂工程、无海水输运、无污染处理的海水原位直接电解制氢工厂。2022 年，厦门大学谢素原团队与袁友珠团队，联合中国科学院福建物质结构研究所和厦门福纳新材料科技有限公司的研究人员研发了以 C_{60} 电子缓冲来稳定亚铜的富勒烯—铜—二氧化硅催化剂，实现了富勒烯缓冲的铜催化草酸二甲酯在温和压力条件下数千克规模的乙二醇合成，有望降低目前乙二醇对石油技术路线的依赖。

三、地质学

在地球深部探测领域，我国首个万米深地科探井开钻。2023 年 5 月 30 日，中国石油塔里木油田公司深地塔科 1 井开钻入地。深地塔科 1 井开钻，旨在探索万米级特深层地质、工程科学理论，标志着我国向地球深部探测技术系列取得新的重大突破，钻探能力开启“万米时代”。深地塔科 1 井位于新疆阿克苏地区沙雅县境内，紧邻埋深达 8000 米的富满 10 亿吨级超深油气区。这口井设计井深 1.11 万米，设计钻完井周期 457 天，将创造全球万米深井钻探用时最快纪录。为保障万米级特深井“打成、打快、打好”，中国石油攻关研发智能控制一体化平台、钻井自主决策工控系统、超高重载井架底座等一批关键核心技术装备，自主研制国际领先的智能钻机，成功产出 1.2 万米特深井自动化钻机，为万米深地工程科学探索研究提供装备和技术保障。

四、天文学

天文学是孕育重大原创发现的前沿科学。2022—2023 年依托“中国天眼”500 米口径球面射电望远镜（FAST），我国在天文学取得系列重要进展。

快速射电暴是宇宙无线电波段最剧烈的爆发现象，起源未知，是天文领域重大热点前沿之一。2022 年 6 月 9 日，中国科学院国家天文台李菂领导的国际合作团队利用 FAST 发现了世界首例持续活跃的快速射电暴 FRB20190520B，并确认近源区域拥有目前已知的最大电子密度，有效推进了快速射电暴多波段研究，为最终揭示快速射电暴起源奠定了观测基础。9 月 21 日，FAST 快速射电暴优先和重大项目科学研究团队利用 FAST 对一例位于银河系外的快速射电暴开展了深度观测，首次探测到距离快速射电暴中心仅 1 个天文单位（即太阳到地球的距离）的周边环境的磁场变化，向着揭示快速射电暴中心引擎机制迈出重要一步。10 月 19 日，中国科学院国家天文台徐聪领导的国际合作团队，利用 FAST 对致密星系群“斯蒂芬五重星系”及周围天区的氢原子气体进行了成像观测，发现了一个尺度大约为 200 万光年的巨大原子气体结构，比银河系大 20 倍，这是迄今为止在宇宙中探测到的最大的原子气体结构。上述重要成果均在《自然》（*Nature*）、《科学》（*Science*）上发表。

2023 年 6 月，由中国科学院国家天文台等单位科研人员组成的中国脉冲星测时阵列研究团队，利用FAST探测到纳赫兹引力波存在的关键性证据。国际3个团队也在同一时间宣布了相似的结果，这不仅使得研究结果可以相互印证，进一步提高了成果的准确性，也表明我国纳赫兹引力波研究与国际同步达到领先水平。相关成果入选《科学》（*Science*）杂志2023年度十大科学突破。

五、自然地理学

青藏高原是世界屋脊、亚洲水塔，是地球第三极。我国在2017年启动了第二次青藏高原综合科学考察研究，涉及十大科考任务。“巅峰使命”珠峰科考是第二次青藏科考的实施活动，2022年4月正式启动，2023年5月30日主体任务完成。

“巅峰使命”珠峰科考创造多项新纪录。此次科考在西风－季风协同作用及影响、巅峰海拔的强烈升温、巅峰海拔的冰雪融化、高新技术平台观测的水汽和温室气体、珠峰地区的强大气氧化性过程、珠峰地区人体生理的特殊反应、珠峰地区变绿的生态过程等方面取得了众多亮点成果，创下多项科考新纪录。其中，“巅峰使命”珠峰科考首次建成了梯度联网的巅峰站并实现了数据实时传输，架设了世界上海拔最高的气象站（8830米），建成了从4276~8830米海拔梯度的观测网络，实现了观测数据实时传输；科考首次成功获取了海拔6500米、7028米和8848米的冰雪样品；科考所使用的“极目一号”Ⅲ型系留浮空艇长55米、高19米，体积达9060立方米，创造了海拔9050米浮空艇原位大气环境科学观测的纪录。此外，“巅峰使命”珠峰科考首次利用高精度雷达测量了珠峰顶部的冰雪厚度；首次采用多种先进技术获得地面至39千米高空大气臭氧浓度数据和三维风场；首次获得高原常驻和短居人群的高山生理适应数据等。

六、高端科学仪器与集成电路先进装备

在半导体技术方面，超越硅基极限的二维晶体管问世。传统晶体管因接近物理极限而制约了芯片的进一步发展。2023年北京大学彭练矛院士、邱晨光团队构筑了10纳米超短沟道弹道二维硒化铟晶体管，创造性地提出“稀土钇元素掺杂诱导二维相变理论”，并发明了“原子级可控精准掺杂技术”，成功克服了二维领域金属和半导体接触的国际难题，首次使得二维晶体管实际性能超过业界硅基10纳米节点鳍式场效应晶体管和国际半导体路线图预测的硅极限，并且将二维晶体管的工作电压降到0.5V，室温弹道率提升至所有晶体管最高纪录的83%。该成果是国际上迄今速度最快、能耗最低的二维晶体管。相关成果发表于《自然》（*Nature*）杂志。

集成电路设备行业兼具知识密集、人才密集、技术密集、研发强度高等特征，其学科发展离不开产业的推动。产业需求既是学科研究的课题来源，也是学科发展的技术立足点。相较其他学科，集成电路装备凸显出强烈的产业引导作用，大企业在学科发展上影响

力巨大。在 2022—2023 年，国内诸多集成电路装备企业的技术研究都有亮眼的成绩。

在刻蚀机领域，2023 年北方华创自主研发发布的离子刻蚀机 Accura BE，实现我国干法刻蚀“零”的突破，技术性能已达业界主流水平。另外，该公司还发布 NMC508RIE 介质刻蚀机，实现 CCP（电容耦合高能等离子体）介质刻蚀。同为刻蚀机领域的领先者，中微公司的 12 英寸 CCP 刻蚀设备已应用于 65 纳米到 5 纳米及更先进生产线上，于 28 纳米及以下的一体化大马士革刻蚀进展良好，设备已在 64 层和 128 层 3D NAND 量产线上应用。

在薄膜沉积设备领域，拓荆科技的 PECVD（化学气相沉积）设备覆盖全系列薄膜材料，完成 28 纳米量产应用。微导纳米的 ALD（原子层沉积）设备也于 2022 年实现了在 28 纳米制造关键工艺中的突破。

在化学机械抛光领域，华海清科于 2022 年实现了用于硅通孔技术（TSV）的化学机械抛光技术。

在半导体涂胶显影领域，芯源微电子于 2022 年推出前道浸没式高产能涂胶显影机，在前道晶圆加工环节 28 纳米及以上工艺节点实现全覆盖，并实现集成电路前道芯片制程领域用单片式清洗机研发及产业化。北方华创正式发布 12 英寸去胶机 ACE i300，开拓 12 英寸刻蚀领域全新版图。ACE i300 可实现刻蚀后去胶、离子注入后去胶、去残胶、表面处理等去胶工艺全覆盖。

在清洗装备领域，至纯科技的清洗设备目前已能满足 28 纳米全部湿法工艺需求，其用于 14 纳米及以下的清洗设备也已研发成功并交付多个客户。盛美半导体 2022 年为硅片和碳化硅衬底制造推出新型预清洗设备，2023 年推出了负压清洗平台，以满足芯粒和其他 3D 先进封装结构清除助焊剂的独特需求。芯原微电子于 2022 年实现集成电路前道芯片制程领域用单片式清洗机研发及产业化。

在量测设备领域，精测电子自主研发的光学膜厚测量设备可用于 28 纳米 FEOL 和 14 纳米 BEOL 节点制程，其 OCD 设备适用于 28 纳米节点及以上制程。另外，中科飞测已量产多款 28 纳米及以上量检测设备，2x 纳米套刻精度量测设备正在验证，1x 纳米无图形晶圆检测设备处于研发中。

七、人工智能

在人工智能辅助抗体设计领域，2023 年，清华大学计算机科学与技术系、智能产业研究院刘洋课题组与中国人民大学高瓴人工智能学院黄文炳提出了一种基于深度学习的抗体设计方法——多通道等变注意力网络（Multichannel Equivariant Attentive Network，以下简称 MEAN）。该方法巧妙地借鉴端到端神经机器翻译技术的核心思想，将抗体生成视为三维等变图翻译问题：给定抗体 - 抗原三维复合体，输出抗体 CDR 区域一维氨基酸序列和对接后的三维结构。MEAN 利用等变图神经网络直接在三维空间处理蛋白质结构，有效克服了传统方法仅能在预处理阶段纳入三维结构信息的局限性，通过充分利用目标抗原的完整信息和抗体的完整可变区域对抗体的复杂内外部物理作用进行建模，高效同步生成满足物理定律对称性的抗体 CDR 区域一维序列和三维结构，具有很强的抗原靶向特性、表达能力和泛化能力。在靶向抗原的抗体设计任务上，MEAN 的 CDR 氨基酸重构精度比麻省理工学院提出的 RefineGNN 提高了 23%。在亲和力优化任务上，MEAN 优化后抗体的吉布斯自由能差变化 ddG 比 RefineGNN 提高了 34%。

在人工智能芯片领域，清华大学魏少军、尹首一团队提出可兼顾能效、精度和灵活性的 AI 芯片新范式，即可重构数字存算一体架构。可重构数字存算一体架构将数字存算一体架构与可重构计算融合，从根源上解决模拟存算一体的固有缺陷：数字存算一体架构在存储器内实现纯数字逻辑，完全避免模拟计算导致的计算误差，兼顾能效和精度；可重构计算架构动态改变存内数据通路，提高 AI 芯片灵活性。基于此范式，设计出国际首款面向通用云端高算力场景的存算一体 AI 芯片 ReDCIM（Reconfigurable Digital CIM）。该芯片首次在存算一体架构上支持高精度浮点与整数计算，可满足数据中心级的云端 AI 推理和训练等各种需求。

在人工神经网络与计算领域，华为云计算技术有限公司田奇团队基于人工智能技术，提出了一种适配地球坐标系统的三维神经网络模型，称为盘古气象大模型。该模型能够有效处理天气数据中的复杂过程，并通过层次化时域聚合策略来有效减少迭代误差，成功实现了精准的中期天气预报。盘古气象大模型在某些气象要素的预报精度上超越了传统数值方法，且推理效率提高了上万倍。在全球高分辨率再分析数据上，盘古气象大模型在温度、气压、湿度、风速等重要天气要素上都取得了更准确的预测结果，将全球最先进的欧洲气象中心集成预报系统的预报时效提高了 0.6 天左右。盘古气象大模型也可用于极端天气预报。在 2023 年汛期，盘古气象大模型成功预测了玛娃、泰利、杜苏芮、苏拉等影响我国的强台风路径。

八、航空科学技术

近年来，在党和国家的坚强领导下，我国在航空科技领域取得了较大进展，显著加强了我国的国防实力和国际竞争力，为提升我国的大国地位贡献了重要力量。本文主要着眼于旋翼飞行器、发动机、超高声速动力技术、浮空器等方面的研究进展。

旋翼飞行器方面取得重要进展。民用直升机 AC 系列的 AC312E、AC332 和 AC313A 等直升机先后成功首飞，AC352 取得中国民用航空局颁发的型号合格证。新型军用通用运输直升机直 –20 完成研制并批量装备部队，实现了我国直升机从第三代向第四代的巨大跨越。eVTOL 电动垂直起降飞行器是旋翼类飞行器新的热门方向。2023 年，峰飞、沃飞长空、沃兰特、时的、御风未来和零重力航空等几家国内公司相继实现了全尺寸样机首飞；同年 10 月，中国民用航空局向亿航智能公司颁发 EH216–S 型无人驾驶航空器系统型号合格证，表明我国 eVTOL 技术及行业发展已经与世界先进国家同步。

军用航空发动机在技术和产品上取得了重大突破。“昆仑”发动机的研制成功结束了长期以来我国不能自行研制航空发动机的历史。以涡扇 –10“太行”系列发动机为代表的第三代军用涡扇发动机技术已较为成熟并大规模部署，实现了我国航空发动机从涡喷到涡扇、从中等推力到大推力、从第三代到第四代的历史性跨越。第四代军用涡扇发动机如涡扇 –15“峨眉”发动机和涡扇 –20 发动机的研制工作已到收尾阶段。自“两机专项”设立后，国产民用航空发动机已经成为我国发展的新重点。大型客机 C919 的配套国产发动机 CJ1000 研制进展顺利，已经取得了阶段性成果，为大型宽体客机 C929 配套的发动机 CJ2000 的研制也在顺利推进中。

超高声速动力技术方面，多座高超音速风洞建成并投入使用，为超燃冲压发动机试验研究提供良好基础。2022 年 7 月，西北工业大学“飞天一号”煤油燃料火箭冲压组合循环发动机完成飞行试验，验证了多模态平稳过渡和宽域综合能力。2022 年 1 月，清华大学“清航壹号”新型旋转爆震冲压发动机飞行演示验证试验成功，使我国在新型空天动力领域跻身世界前列。

浮空器方面，北京航空航天大学研制的“圆梦号”飞艇和中国科学院浮空器中心承担的鸿鹄先导专项“临近空间科学实验系统”任务中的平流层飞艇技术研究，均已持续开展相关的飞行验证，实现了跨昼夜长时控制飞行。多项关键技术已攻关突破，标志着关键技术成熟度已经达到较高水平，为临近空间平台早日进入工程应用打下坚实的技术基础，具有开创意义。

九、航天科学技术

航天运输系统加速向无毒、无污染、模块化、智慧化方向升级换代。长征系列运载火箭近地轨道运载能力达到 25 吨级，地球同步转移轨道运载能力达到 14 吨级，入轨精度处于国际先进水平。新一代长征系列运载火箭进入应用发射阶段，成功实施我国火星探测、月球取样、空间站建造等一系列重大工程任务。

固体运载火箭首次海上发射，可重复使用运载器成功首飞和复飞，首款固液混合动力运载火箭“长征六号甲”、固体运载火箭“力箭一号”成功首飞，“快舟一号甲”“双曲线一号”“谷神星一号”“天龙二号”等商业运载火箭成功发射，我国运载火箭谱系更加丰富，进入空间能力显著提升，综合指标达到国际先进水平。同时，大直径箭体结构技术、大推力氢氧发动机技术、发动机大范围推力调节、故障诊断与容错重构、着陆支撑等技术取得突破，对于推动空间科学发展具有重要意义。

通信、导航、遥感、科学试验卫星系统不断完善。通信卫星方面，“天链二号”第二代数据中继卫星 3 星组网完成，中国空间站上的航天员可以随时与地面进行视频和语音通信；中星 6C 卫星（2019 年 3 月 10 日发射）、中星 9B 卫星（2021 年 9 月 9 日发射）、中星 6D 卫星（2022 年 4 月 15 日发射）等系列通信卫星，成为我国广电安播工程业务的主力；

高通量卫星“中星 26”卫星（2023 年 2 月 23 日发射）的通信容量超过 100 Gbps，实现高速通联，覆盖范围持续扩大；“天通一号”卫星实现 3 星组网，用户只需要一部手机就可以实现互联互通，极大地提升了应急通信保障能力，具有里程碑意义。

导航卫星方面，“北斗三号”全球卫星导航系统自 2020 年 7 月 31 日正式开通，可提供定位导航授时、全球短报文通信、区域短报文通信、国际搜救、星基增强、地基增强、精密单点定位共七类服务，性能指标先进。定位导航授时服务方面，全球范围水平定位精度约 1.52 米，垂直定位精度约 2.64 米（B1C 信号单频、95% 置信度）；测速精度优于 0.1 米 / 秒，授时精度优于 20 纳秒，亚太区域精度更优。

遥感卫星方面，陆地、海洋、大气遥感卫星在探测精度、观测时效、探测手段等方面均显著提升，商业遥感卫星研制也蓬勃发展。“高分三号”海陆监视监测星座实现了高分辨率 SAR 图像在国民经济发展各行业的业务化应用，平均重访时间缩短至 4.8 小时；2023 年 4 月 16 日成功发射的“风云三号”G 星，是我国首颗主动降水测量卫星。

科学试验卫星方面，以“悟空”“慧眼”“羲和”“墨子”“张衡一号”“实践二十三号”“夸父一号”等为代表的空间科学探测卫星，为空间科学研究与发现提供了有力手段。其中，“夸父一号”（全称先进天基太阳天文台）综合性太阳探测专用卫星于 2022 年 10 月 9 日发射，首批太阳探测科学图像于同年 12 月 13 日发布。

中国空间站系统全面建成。2022 年年底，天宫空间站组装建造阶段 6 次飞行任务全部顺利完成，标志着中国空间站全面建成，我国载人航天工程“三步走”发展战略已从构想成为现实。中国空间站在空间机械臂、推进剂补加、在轨维修、货物装载、大型舱段空间转位、柔性太阳翼和驱动机构、大型组合体控制、元器件等方面的技术取得突破，实现了从搭载一人到多人升空、从舱内作业到太空行走、从短期遨游到中期驻留再到长期“出差”的历史性跨越。“天舟”货运飞船具备了船箭分离后 2 小时全自主交会对接的能力，“神舟”载人飞船实现了 5 圈快速返回。2023 年 5 月 30 日，“神舟十六号”飞船搭载三名航天员顺利进驻中国空间站，是我国载人航天工程进入空间站应用与发展阶段之后的首次载人飞行任务，开展了人因工程、航天医学、生命生态、生物技术、材料科学、流体物理、航天技术等多项空间科学实验，迈出了载人航天工程从建设向应用、从投入向产出转变的重要一步。

深空探测专业技术方面取得了系列突破。在月球探测方面，“嫦娥五号”月球探测器于 2020 年 11 月 24 日发射，于 2020 年 12 月 17 日从月球携带 1731 克月壤安全返回着陆，该探测器突破了月球探测轨道设计、月球采样封装、月面起飞上升等一系列深空探测关键技术，各项关键技术均处于国际先进水平。科学家通过月壤发现了“嫦娥石”，取得了丰硕的科研成果，对于研究月球的形成和演化具有重要的参考作用。

在火星探测方面，“天问一号”火星探测器于 2020 年 7 月 23 日发射，并于次年 5 月 15 日成功着陆于火星预选区域，成功实现了“绕、着、巡”的任务目标，我国成为世界上第二个成功着陆火星并开展巡视探测的国家。截至 2022 年 9 月 15 日，“天问一号”环绕器在轨运行 780 多天，“祝融号”火星车累计行驶 1921 米，完成了既定科学探测任务。该任务突破了火星制动捕获、火星进入着陆、火星表面巡视、长期自主管理、行星际测控通信等关键技术。2022 年，“祝融号”火星车在国际上首次在火星原位探测到含水矿物，对理解火星气候环境演化历史具有重要意义，并为未来载人火星探测的原位资源利用提供了可能。

十、采矿工程

在绿色采矿领域，研发了金属矿全尾砂短流程低能耗浓密与精准制备关键技术、矿山固废深部输送阻力特性与调控技术、全尾砂生态化处置与资源化利用关键技术、浓缩全尾砂地表无害化堆存关键技术，制定了金属矿膏体充填国家标准，建立了 1000 米以深金属矿绿色开采示范工程。研发了井下分选系统空间布置方法和优化设计技术、适应于井下条件的智能化模块化选煤装备、井下充填材料级配及制备技术及实用装备、井下工作面采煤与充填协调作业技术及装备，形成了深部煤矿井下分选及就地充填关键技术装备研究与示范。

在深地采矿领域，构建了深部岩体力学与开采理论，研究了深部岩体原位力学行为、采动岩体力学及渗流理论，研发了深部高应力岩体变形监测与动力灾害安全预警技术、深部高应力围岩大变形控制技术，提出了 2000 米以浅深部岩体力学与开采理论体系、安全监测预警技术体系。研发了深部矿产资源深竖井高效掘进技术及装备、深竖井大吨位提升与控制技术及装备，形成了 2000 米以浅深部矿产资源建井与提升技术装备能力。开展了深部金属矿高应力致裂破岩与控制爆破技术和深部金属矿床安全高效开采技术研究，开发了深部金属矿临界浓度充填采矿技术与装备，形成了 1500 米以深规模化采矿示范。提出了煤矿千米深井巷道围岩大变形机理新理论，形成了煤矿千米深井巷道围岩控制成套技术与装备、千米深井超长工作面安全高效智能开采成套技术及围岩稳定性监测预警平台。

在智能化采矿领域，研发了地下金属矿无人采矿工艺系统、采矿装备智能化控制技术、无人采矿多装备多系统集群控制技术、无人采矿系统增强现实与集控一体化平台，形成了多装备多系统无人驾驶集群生产作业系统的金属矿生产示范工程。开发了基于大数据的金属矿山装备智能管控及开采过程参数管理系统、矿山装备健康诊断与管控平台，研制了金属矿山充填智能化调控系统。揭示了特厚煤层自动化放煤机理和采放协调高效回采工艺方法，研制出煤矸识别装置，开发了智能放煤软件，研发了特厚煤层综放工作面无人操作关键技术与装备，建立了年产千万吨级特厚煤层智能化综放开采示范工作面。

十一、材料科学

二维半导体材料方面，北京大学彭练矛、张志勇团队制备了 10 纳米超短沟道弹道二维硒化铟晶体管，成就了世界上迄今速度最快、能耗最低的二维半导体晶体管。中国科学院物理所吕力、刘广同团队开展了高质量外尔半金属和伊辛超导体的合成，处在过渡金属硫族化合物低温量子输运领域的研究前沿。南京大学缪峰、梁世军团队在二维类脑器件领域，首次实现了读写对称的自旋存内计算器件。

超紧凑型太赫兹自由电子激光前沿技术主要用于大科学装置及国防高技术产业方面。中国科学院上海光机所李儒新、田野团队在超快光学领域，探究了太赫兹表面波驱动的兆伏电子加速，以“羲和”命名了目前全世界峰值功率最高的激光装置。长春光学精密机械与物理研究所刘春雨团队，西安光学精密机械研究所付玉喜团队在国际上首次突破了100mJ 量级的太瓦级超强中红外飞秒激光关键技术。

在超材料领域，2023 年清华大学周济团队基于超材料构建的局域磁电强耦合，提出一种在线性材料中产生高效非线性光学响应的普适性新方法，首次实现了室温硅基太赫兹二次谐波产生。东南大学崔铁军团队、电子科技大学邓龙江教授、哈尔滨工业大学宋清海/鹏城实验室余少华团队等在射频超表面的超透镜光谱学领域取得了重要进展，有力地增强先进的光学成像和光谱分析应用。

在信息功能材料领域，中国科学院上海微系统与信息技术研究所宋志棠、朱敏团队发明了一种基于单质碲和氮化钛电极界面效应的新型开关器件，充分发挥纳米尺度二维限定性结构中碲熔融结晶速度快、功耗低的独特优势，“开态”碲处于熔融状态是类金属，和氮化钛电极形成欧姆接触，提供强大的电流驱动能力，“关态”半导体单质碲和氮化钛电极形成肖特基势垒，彻底夹断电流。该晶 – 液态转变的新型开关器件，组分简单，可克服双向阈值开关（OTS）复杂组分导致成分偏析问题；工艺与 CMOS 兼容且可极度微缩，易实现海量三维集成；开关综合性能优异，驱动电流达到 11 兆安 / 平方厘米，疲劳寿命 > 108 次，开关速度 ~15 纳秒，尤其碲原子不丢失情况下开关寿命可大幅提升。该研究为发展海量存储和近存计算提供了新的技术方案。

十二、电力工业

在光伏领域，南京大学谭海仁团队的相关研究实现高效率的全钙钛矿叠层太阳能电池和组件。首先，通过设计钝化分子的极性，提升其在窄带隙钙钛矿晶粒表面缺陷位点上的吸附强度，显著增强缺陷钝化，大幅提升全钙钛矿叠层电池的效率。经国际权威检测机构日本电器安全环境研究所（JET）独立测试，叠层电池效率达 26.4%，创造了钙钛矿电池新的纪录并首次超越了单结钙钛矿电池，与市场主流的晶硅电池最高效率相当。其次，该团队开发出大面积叠层光伏组件的可量产化制备技术，使用致密半导体保形层来阻隔组件互连区域钙钛矿与金属背电极的接触，显著地提升了组件的光伏性能和稳定性，实现了国际认证效率 21.7% 的叠层组件（面积为 20 平方厘米）。

在空间太阳能电站（SSPS）方面，西安电子科技大学段宝岩团队完成了逐日工程—世界首个全链路、全系统 SSPS 地面验证系统落成启用。该系统远距离高功率微波无线传能效率（距离 55 米，发射 2081 瓦，波束收集效率 87.3%，DC-DC 传输效率 15.05%）与功质比等主要技术指标世界领先。逐日工程突破的远距离高功率微波无线传能技术，应用前景广阔。在太空，可助力构建空间能源网、空间充电桩，破解空间算力、星上信息处理、空间攻防及超远程探测的供电难题。在陆海空，可为空中飞艇、无人机群、海上移动平台、灾害及边远区域无线供电。

十三、能源科学技术

在核电技术领域，2023 年 12 月 6 日，我国具有完全自主知识产权的华能石岛湾高温气冷堆核电站示范工程商运投产，该核电站也是全球首个实现模块化第四代核电技术商业化运行的核电站，标志着我国在高温气冷堆核电技术领域实现了全球领先，对推动我国实现高水平科技自立自强、建设能源强国具有重要意义。该示范工程位于山东省荣成市，由中国华能集中产业链上下游优势资源，联合开展关键技术攻关和核心设备研制，研制出 2200 多套世界首台（套）设备，设备国产化率达 93.4%。

十四、城市科学

韧性城市是近年来城市科学十分重要的研究热点，其研究重点是城市在面对自然灾害、公共卫生事件等冲击时的适应能力和恢复能力。近年来，在韧性城市研究方面的代表性成果有：上海交通大学城市治理研究院重点关注城市社区韧性的问题，提出了基于社区参与和资源整合的社区韧性提升模式，为社区层面的韧性城市建设提供了实践指导。同济大学环境科学与工程学院提出了基于生态系统服务的城市生态韧性提升策略，为城市规划建设中的生态保护和恢复提供了科学依据。中国灾害防御协会标准与规划专委会、中国城市科学研究会韧性城市专业委员会等社会团体机构，深入研究社会主义市场经济条件下发展我国韧性城市建设的理论与政策，发挥桥梁与纽带作用，为促进我国韧性城市研究事业的发展作出贡献。

2023 年 3 月，首届海峡两岸暨港澳城市建设与持续发展科技论坛上，围绕以安全韧性城市建设推进公共治理现代化、城市全域三维韧性发展、城市地铁施工及运营安全综合快速检测技术及装备、澳门世界文化遗产保护及相关政策等议题进行了专题研讨，与会的城乡建设与公共安全领域专家学者共同提出了《韧性城市发展倡议》，从科技强韧、管理强韧、文化强韧 3 个方面为助推韧性城市建设提供了参考建议。对城市如何在日益增长且变幻莫测的风险和挑战中正常运行并保持韧性这一议题，有了更理性的思考。

十五、环境科学

在水环境领域，2022 年，清华大学胡洪营团队与沃顿科技股份有限公司等单位联合完成“再生水处理高效能反渗透膜制备与工艺绿色化关键技术”项目，该项目针对再生水利用过程中，反渗透处理工艺设计运行绿色技术缺位、高效能反渗透膜被国外垄断、浓缩水（以下简称浓水）高效低耗处理技术缺乏等突出问题，提出“水质特征识别、高效能膜研制、工艺绿色转型、浓水分盐脱毒”的总体技术攻关路线，拓展了再生水反渗透处理工艺进水水质设计指标体系，突破了再生水处理抗污堵反渗透膜及膜组件制备技术，开发出工艺高效能运行及浓水分盐和脱毒技术，形成了再生水绿色高效能反渗透处理成套工艺，并得到规模化应用。相关成果应用到北京、天津、山东等多个省（自治区、直辖市）60 多项再生水反渗透处理工程中，膜污堵防控效果达到国际领先水平。

2023 年，清华大学、中国科学院生态环境研究中心等单位联合完成“典型行业废水特征无机物转化控制关键技术及应用”项目。该项目以“污染物形态转化与高效控制”为创新思路，深刻揭示了非均相物理化学转化、微生物转化以及物化 – 生化过程协同的特征无机物转化控制原理，突破了多羟基微界面络合 – 吸附除氟、高盐废水微生物反硝化、高盐废水短程反硝化 – 厌氧氨氧化脱氮等关键技术，开发出高盐工业废水共存特征无机物转化去除的系列新材料 / 药剂、新菌剂和新装备，形成了可有效支撑典型行业高盐废水氟化物、无机氮等去除控制的新工艺并实现了规模化应用。相关成果在 40 多项高浓度氟化物和无机氮共存废水处理及受污染水体治理工程中得到成功应用，环境效益、经济效益和社会效益显著。

十六、制冷及低温工程

中国科学院理化技术研究所热声团队经过近 30 年的系统研究工作，创建了先进的热声分析和设计理论，提出了第四代热驱动热声制冷技术，研制了一种声场、温度场以及能量场多场协同型热驱动热声制冷机。实验结果表明，系统在 450℃的加热温度下，在标准空调工况下（7℃ /35℃）COP 最高达到 1.2，是目前公开报道的热驱动热声制冷机最高 COP 的近 3 倍，超过了当前主流的热驱动吸附式制冷技术和单效吸收式制冷技术，甚至完全媲美双效吸收式制冷系统。理论研究表明，当加热温度更高（例如采用燃气燃烧加热）时，其 COP 可望达到 1.5~2.0，将成为未来效率最高的热驱动制冷技术，具有广阔的应用前景和发展潜力。此外，该团队还率先在国内外开启了极寒温区、大温差应用的自由活塞热声斯特林热泵技术研究，研制了一台电驱动自由活塞斯特林空气源热泵样机。该热泵样机在冷端温度为 –40℃的工况下仍能正常运行且保持可观性能，显示出电驱动自由活塞斯特林热泵在寒冷地区供暖方面具有广阔的应用前景。

十七、基础农学

在农业信息学领域，新型农业专用传感器研发取得突破，部分实现国产化，“智嗅”系列农业气体传感器打破了欧美国家技术产品垄断。2021 年，北京农林科学院董大明带领团队研制的 SmartSoil 系列土壤成分快速检测系统是全球第一台可以在田间对多种土壤成分进行测量的仪器。首创了中国农产品监测预警系统 CAMES，支撑定期召开中国农业展望大会，发布 18 种主要农产品生产量、消费量、贸易量等的未来 10 年展望定量信息，实现了中国农产品预测性信息由国外主导到中国自主发布的历史性转变。以植物工厂为代表的设施工厂化生产得到广泛应用，我国在 LED 人工光源技术领域的研究取得重大突破，研究成果处于全球领先地位。

在农业资源环境学领域，产出了我国典型红壤酸化特征及防治关键技术构建与应用、黑土耕地质量提升“四位一体”技术、北方旱地农田抗旱适水种植技术与应用、典型农林废弃物快速热解创制腐殖酸环境材料及其应用、全过程全链条面源污染防控技术等一批重大成果，有力支撑了地力提升，保障了我国粮食安全，推进了绿色发展。

在农产品贮运与加工领域，近年来在玉米、花生等真菌毒素精准防控，淀粉、油脂、蛋白质等大宗组分精细化提取和加工，粮油、果蔬、畜禽水产副产物高值化开发与全组分利用等方面取得一系列进展。数字仓储、冷链物流等技术推动节粮减损和生鲜农产品高效储运，植物基食品和细胞培养肉开发、功能性配料生物合成等实现技术突破，传统食品工业化、标准化水平不断提升。超高压非热加工技术实现产业化并于 2022 年实施了我国首个超高压加工国家标准，蓝莓精深加工技术突破支撑了蓝莓花色苷于 2023 年成功获批为新食品原料。

在农药研发领域，2022 年 9 月，中国农业科学院植物保护研究所 / 农业基因组研究所杨青团队及其合作者通过前沿生物学技术包括冷冻电镜、扫描电镜、X 射线衍射等，解析了 4 个不同催化反应状态 PsChs1 的三维结构，揭示了几丁质合成酶实现几丁质生物合成的 3 个重要过程。这是第一次从原子水平上向人们展示了一个有方向性的、多步骤偶联的几丁质生物合成过程，标志着我国的农药研发水平提升到了基础理论原始创新的高度。

十八、种子学

在农作物种子产量品质研究方面，2022 年 11 月，中国科学院分子植物科学卓越创新中心巫永睿研究团队与上海师范大学王文琴研究团队合作从野生玉米中成功克隆到了首个控制玉米高蛋白含量的主效基因 *THP9*，能显著提高玉米整株及籽粒中蛋白质的含量，为粮食安全、生态环境保护和农业的可持续发展保驾护航。该研究成果发表在《自然》（*Nature*）杂志上。此外，我国首次从小麦近缘植物长穗偃麦草中克隆出抗赤霉病基

因 *Fhb7*，揭示其遗传和分子调控机理，为解决小麦赤霉病世界性难题找到了“金钥匙”，选育的新品种“山农 48”，已通过审定并大面积推广种植。

在植物基因组和遗传育种方面，中国农业科学院深圳农业基因组研究所黄三文团队取得多项成果：解析了栽培和野生马铃薯的遗传多样性和复杂演化史、通过图形泛基因组捕捉番茄缺失的遗传力为育种赋能等。

在种质资源保护体系建设方面，我国建成了国家野生稻种质资源圃，占地面积 180 亩，资源保存能力达 4 万份，年鉴定评价 1000 份以上，是全球最大、国际一流的野生稻种质资源保存中心。我国高度重视种质资源的收集和研发，已经建成了以 1 个国家种质资源库长期库、1 个复份库、10 个中期库、43 个种质圃、217 个原生境保护点和 1 个种质资源信息网为基础的国家农作物种质资源保护体系。当前我国保存的种质资源总量超过 52 万份，位居世界第二。

十九、植物保护学

近年来，植物保护学科在农业病虫害防控研究进展显著，在作物抗病机制、抗病品系构建、草地贪夜蛾、小菜蛾等防控技术、新型高效生物农药创制等方面均取得了重大突破。

小麦感条锈病机制领域取得重大突破。2022 年 7 月，西北农林科技大学植物免疫团队在《细胞》(*Cell*) 杂志在线发表了历经 18 年潜心研究的题为 *Inactivation of a Wheat Protein Kinase Gene Confers Broad-Spectrum Resistance to Rust Fungi* 的研究论文。该研究鉴定到小麦中第一个被锈菌效应子操纵的感病基因，对条锈菌和叶锈菌表现广谱持久抗性的感病基因编辑植株为小麦育种提供了极有价值的种质资源，具有重要的理论与应用价值，为利用感病基因改良作物抗病性提供了更坚实的理论与技术支撑，为作物抗病育种提供了新思路。

揭示植物受体抗病的“双重免疫”功能。2022 年 9 月，南京农业大学王源超团队和清华大学柴继杰团队合作在《自然》(*Nature*) 杂志上发表题为 *Plant receptor-like protein activation by a microbial glycoside hydrolase* 的研究论文，解析了细胞膜受体蛋白 RXEG1 识别病原菌核心致病因子 XEG1 激活植物免疫的作用机制，首次揭示了细胞膜受体蛋白具有“免疫识别受体”和“抑制子”的双重功能，这为人类认识复杂精密的植物与病原菌互作机制提供了新的认知，对改良作物广谱、持久抗病性具有重要指导意义，同时为开发绿色新型生物农药奠定核心理论基础。

揭示植物免疫受体监控病毒靶向激素受体诱导抗病新机制。2022 年 12 月，《自然》(*Nature*) 杂志在线发表南京农业大学植物保护学院陶小荣团队最新成果，该成果首次揭示病毒攻击植物激素受体有利自身侵染，植物则进化出了一种免疫受体模拟受攻击的激素受体，从而识别病毒、并激活免疫反应，该研究揭示了植物免疫受体监控病毒靶向激素受体诱导抗病的全新机制，提供了植物与病毒“军备竞赛”的新案例。

2023 年 6 月，《细胞》（*Cell*）杂志发表中国科学院遗传与发育生物学研究所陈宇航和周俭民合作团队研究成果，该团队克隆了广谱抗根肿病基因，以我国西汉著名将领“卫青”为其命名，并在植物中首次发现钙离子释放通道及其介导的免疫机制，阐明其作用原理。这项研究是我国科学家继发现植物抗病小体后在该领域取得的又一项重大理论突破，在十字花科作物抗根肿病育种中有良好应用前景。

二十、作物学

作物重要性状遗传解析和功能基因组研究取得新突破。华中农业大学严建兵团队联合中国农业大学李建生和杨小红团队研究发现，玉米 *KRN2* 和水稻 *OsKRN2* 基因均受趋同选择并通过相似的生化途径调控玉米和水稻的产量，揭示了玉米和水稻在演化过程中趋同选择的遗传规律。中国农业科学院作物科学研究所周文彬团队在水稻中发现一个关键基因（命名为 *OsDREB1C*），该基因通过协同调控光合作用、氮素利用、抽穗等重要生理途径，实现了水稻高产和早熟。中国科学院遗传与发育生物学研究所谢旗研究员科研团队与国内多家机构合作联合发现了耐碱基因 *AT1*，揭示了 *AT1* 通过调节细胞中的活性氧（ROS）水平来参与碱胁迫响应的分子机制。万建民院士领衔的中国农业科学院和南京农业大学的科研团队合作克隆了籼粳杂交不育性基因 *RHS12*，推动水稻亚种间超强杂种优势利用和高产新品种的培育，为利用亚种间杂种优势培育高产品种提供了理论和技术支撑。

作物育种关键技术与新品种培育取得重要进展。中国农业大学赖锦盛团队发现的拥有自主知识产权的基因编辑底盘酶 Cas12i 和 Cas12j，突破了国际基因编辑专利的制约。中国农业科学院作物科学研究所李立会团队突破“小麦 – 冰草”远缘杂交难题，创建了利用野生近缘植物基因改良小麦的远缘杂交技术体系。云南大学胡凤益团队利用长雄野生稻和亚洲栽培稻杂交，把长雄野生稻地下茎无性繁殖特性转移到栽培稻中，成功创制了多年生稻，入选《科学》（*Science*）杂志 2022 年度十大科技突破。我国玉米单倍体育种技术获得大规模工程化应用。北方稻区的圆粒广适水稻品种“龙粳 31”累计推广过亿亩，创制了聚合 *Yr30*、*Lr27*、*Sr2* 等 10 多个抗病基因的优异小麦新种质，“京农科 728”“裕丰 303”“东单 1331”等玉米品种推广面积持续增加，培育出耐密、高产油菜新品种“中油杂 501”。

作物丰产优质高效协调关键技术与机械化、智慧化栽培加速推进，综合产能显著提升。稻麦等作物丰产优质高效协调栽培技术通过种植制度重构、温光高效利用、水肥优化施用等关键技术的突破，促进了产量品质和效益的协同提升。稻麦等作物“无人化”栽培技术通过机艺智融合，基本实现机械化播栽、植保、施肥、收获等重点作业环节无人化，在江苏等地进行了示范，为全面走向大面积生产提供了技术支撑。大豆玉米带状复合种植，遴选出适应性强利于稳产增产的 4+2、3+2 等种植模式，创建了机械化工艺流程和装备体系，在全国 17 个省（市、区）推广应用，在大面积稳粮的同时提升了大豆的产能。水稻全程绿色智慧施肥技术，实现水稻长势的实时感知、基肥方案的精确设计、追肥方案的动态调控、肥料投入的精确作业这一技术链创新，促进了水稻全程施肥管理的绿色化和智能化。花生玉米机械化带状种植秸秆裹包混贮利用技术，通过玉米和花生的同步收获，进行玉米秸秆和花生秧的裹包混贮打捆，生产高值化的饲料，实现稳粮增油，种养结合，促进了农牧循环。

作物肥水资源高效利用与绿色栽培技术得到新发展。玉米密植高产滴灌水肥精准调控生产模式，实现了玉米生长全过程的精准调控，协同提高产量与资源利用，创造了一大批高产典型，推动产量水平大幅提升。大豆大垄密植浅埋滴灌栽培技术通过滴灌大豆实现适期播种，加宽垄体实现合理密植，破解东北地区春播期干旱频发、大豆出苗不全不齐等难题。创建了高产低碳减排稻作新模式，水稻增产 5%，稻田甲烷可减排 30% 以上。

二十一、生物学

在遗传学领域，中国科学院动物研究所刘光慧、曲静和中国科学院北京基因组研究所张维绮等利用多学科交叉手段，揭示人类基因组暗物质驱动衰老的机制。并据此提出古病毒复活介导衰老程序性及传染性的理论以及阻断古病毒复活或扩散以实现延缓衰老的多维干预策略。通过对人类基因组中蛋白编码区域的“逆老”基因进行系统排查，发现可重启人类干细胞、运动神经元和心肌细胞活力，逆转关节软骨、脊髓及心脏衰老的新型分子靶标，并构建一系列针对器官退行的创新干预体系。以上发现为衰老生物学和老年医学研究建立了新的理论框架，为衰老及老年慢病的科学干预和积极应对人口老龄化奠定了有益的基础。在基因组编辑方面，中国科学院遗传与发育生物学研究所高彩霞团队联合北京齐禾生科生物科技有限公司赵天萌团队利用人工智能辅助的大规模蛋白结构预测方法对基因组编辑新酶进行发掘。他们通过开展基因组编辑元件挖掘方法和技术体系创新，实现了单碱基到超大片段 DNA 精准操纵，为作物改良和基因治疗提供了重要支撑。

在生理学领域，军事科学院军事医学研究院生物医学分析中心李慧艳、张学敏等发现大脑视交叉上核（SCN）神经元的初级纤毛，这一初级纤毛可能作为机体中的“中央生物钟”的结构基础，参与生物钟内稳态的维持，而靶向 SCN 初级纤毛的 Shh 信号通路可能是治疗与昼夜节律紊乱相关的人类疾病的潜在治疗策略。该“有形”生物钟的发现，对于理解生物钟的构造以及分子层面与细胞层面生物钟的联系具有重要意义。中国科学技术大学薛天等揭示了光调控生物（小鼠和人）血糖代谢的神经机制。该研究发现了全新的“眼 – 脑 – 外周棕色脂肪”通路，回答了长久以来未知的光调节血糖代谢的生物学机理，拓展了光感受调控生命过程的新功能。这项工作发现的感光细胞、神经环路和外周靶器官，为防治光污染导致的糖代谢紊乱提供了理论依据与潜在的干预策略。

二十二、恶性肿瘤

近年来，我国在多个恶性肿瘤的理论研究和临床实践方面取得突破。

在肺癌方面，免疫治疗在肺癌的综合治疗中发挥了重要作用，为患者带来了更好的生存质量和预后。新型药物和靶向治疗手段的发展使得肺癌的治疗趋于个性化和精准化。此外，新型诊断方法如胸腔积液和心包积液 cfDNA 检测以及围手术期 MRD 预测等，提高了肺癌诊断和风险评估的准确性。

胃癌方面，不同化疗方案间的对比、在化疗基础上联合抗血管生成治疗或免疫治疗或抗 HER2+ 免疫治疗、双免疗法等的探索如火如荼，新的证据不断出炉。随着新型抗 HER2 药物的迅速发展，HER2 阳性胃癌的全程治疗格局也在发生改变。精准筛选免疫治疗获益人群及探索耐药机制成为优化结局的关键。

胰腺癌方面，诸如 KRASG12D 小分子抑制剂和新生抗原特异性 TCR-T 细胞这些曾经被认为很难实现的科研成果得到临床转化。

血液肿瘤方面，发展机制的深入研究、预后分层体系的不断细化、新型靶向治疗及免疫治疗药物的研发应用，血液肿瘤进入精准诊疗的新时代，针对患者实施有效的个体化治疗，有助于提高患者的缓解率，延长患者生存。

多原发和不明原发肿瘤方面，中国抗癌协会多原发和不明原发肿瘤专业委员会完成《中国肿瘤整合诊治指南：多原发和不明原发肿瘤》的编写和发布，通过全国巡讲的形式，增强相关医务工作者诊疗临床实践上的可及性、操作性和指导性。

妇科肿瘤方面，宫颈癌免疫治疗处于飞速发展阶段，越来越多的新靶点、新药涌现，大量的临床试验得以开展。宫颈癌的免疫治疗已经从后线提至一线。从单抗治疗到双抗治疗，从单免疫治疗到双免疫治疗，从单药到联合，免疫治疗正在革新宫颈癌的诊治模式。晚期子宫内膜癌，除传统治疗方法外，免疫单药及联合治疗方案取得较大进展。随着新药物的不断涌现，卵巢癌治疗将逐步慢病化。

在骨与软组织肉瘤方面，新型免疫联合治疗（免疫联合放疗、免疫联合化疗、免疫联合靶向药、双免联合）的肉瘤临床试验数据表现不俗。与单免疫治疗相比，免疫联合法治疗肉瘤不仅获得了较高的客观缓解率，部分研究还显示出了较明显的 PFS 获益。

结直肠癌方面，2023 年 3 月，中山大学肿瘤防治中心徐瑞华、王峰、陈功消化系统肿瘤团队发表了信迪利单抗用于错配修复缺陷的局部晚期直肠癌新辅助治疗的Ⅱ期临床研究结果。研究结果初步显示，在 dMMR/MSI-H 局部晚期直肠癌患者中，通过新辅助 PD-1 单抗免疫治疗，能让患者有机会达到临床完全缓解，从而避免放化疗及手术，实现器官功能保全，可能从根本上改变这一疾病的治疗方式。

二十三、产科学

在妊娠生理方面，2022 年 8 月，中国科学院刘默芳团队发现 RNA 结合蛋白 FXR1 通过液 - 液相分离参与后期精子细胞 mRNA 翻译激活和精子形成的分子机制。同年 9 月，山东大学陈子江团队发现了调控人类合子基因组激活和早期胚胎发育的关键转录因子，并首次报道了人类与小鼠在卵子向早期胚胎转变过程中，翻译水平动态变化存在物种差异。2023年1月，北京大学第三医院乔杰团队首次在单细胞水平上分析了人类植入前胚胎的蛋白质组，解答了阶段特异性蛋白质表达和蛋白质组学MZT模式的问题。7月，广州医科大学附属第三医院陈敦金团队建立了不同孕周、不同疾病模型的胎盘类器官，为开展妊娠期胎盘起源性疾病奠定了基础。

在产前诊断方面，2022年2月，首都医科大学附属北京妇产医院吴青青牵头中华医学会超声医学分会妇产超声学组、国家卫生健康委妇幼司全国产前诊断专家组医学影像组，发表了《超声产前筛查指南》，该指南为中国胎儿超声产前筛查指南的首次发布。2022年5月，河南省人民医院、北京协和医院等联合发表了全外显子组测序技术在产前诊断中应用的专家共识，对产前WES的适用群体、检测前咨询、取样及实验室检测、产前WES报告、检测后咨询、妊娠结局随访、产前WES复杂病例多学科会诊、产前WES样本与资料信息的保存等提出了建议。2022年10月，复旦大学黄荷凤团队研发了新一代NIPT，实现了对胎儿染色体非整倍体、染色体微缺失微重复和单基因遗传病的无创产前检测。

在妊娠合并和并发疾病诊治方面，制定或更新了产后出血预防与处理指南、妊娠期高血糖诊治指南、双胎妊娠期缺铁性贫血诊治与保健指南等多个疾病临床指南，妊娠期急性脂肪肝临床管理指南被指南STAR评级为2022年产科领域最佳指南。此外，北京大学第三医院赵扬玉团队发现孕晚期炎性细胞因子浓度与产后出血显著相关，并建立了预测模型。南京鼓楼医院胡娅莉团队发现了胎盘特异性高表达miR-155的子痫前期亚型，miR-155可能是该亚型预测和治疗靶点。北京大学第一医院杨慧霞团队从单细胞测序等基础研究着手，提出针对胎盘植入性疾病聚焦高危因素精细化管理策略。

在胎儿和新生儿健康方面，2022年4月，中国科学院赵方庆团队和北京大学第三医院赵扬玉、魏瑗团队联合发现遗传因素及宫内不良环境对生命早期菌群的塑造，揭示关键肠道微生物对生长受限新生儿的长期作用，团队还将微波消融创新性应用于复杂性单绒毛膜双胎的选择性减胎术。上海市第一妇婴保健院孙路明团队描绘了Ⅲ型选择性宫内生长受限的胎儿生长模式。2022年12月，重庆医科大学漆洪波团队开发了能够主动靶向改善炎症微环境实现治疗妊娠期血栓所致胎儿生长受限的纳米药物。

二十四、免疫学

2020年以来，新冠肺炎疫情极大地影响了全球的经济发展和人民生活。新冠病毒奥密克戎突变株及其变体持续涌现，及时地解析新冠突变株如何逃逸疫苗接种所建立的免疫屏障和病毒感染所产生的人体免疫力对于未来疫苗设计与疫情防控至关重要。2022年，北京大学、北京昌平实验室曹云龙、谢晓亮团队联合中国科学院生物物理研究所王祥喜团队率先揭示了新冠奥密克戎突变株及其新型亚类的体液免疫逃逸机制与突变进化特征，揭示奥密克戎BA.1中和抗体逃逸机制，及其与病毒刺突蛋白结构特征的联系；发现奥密克戎 BA. 4/BA. 5 变异可逃逸人体感染 BA. 1 后所产生的中和抗体，证明了难以通过奥密克戎感染实现群体免疫以阻断新冠传播；基于自主研发的高通量突变扫描技术，成功预测了新冠病毒受体结合域免疫逃逸突变位点，并前瞻性筛选出广谱新冠中和抗体。相关研究为广谱新冠疫苗和抗体药物研发提供了理论依据和设计指导，为全球新冠疫情防控提供了重要参考。

二十五、分子药理学

大多数动物（包括人类）均拥有一套主嗅觉系统来识别挥发性的气味分子。大量的嗅觉受体通过“组合编码”的气味识别方式，帮助动物识别数以万亿计的气味分子。嗅觉受体可以分为3个家族，第Ⅰ类是气味受体（OR）家族；第Ⅱ类是痕量胺相关受体（TAAR）家族，OR 和 TAAR 都属于 A 类 G 蛋白偶联受体（GPCR）家族；第Ⅲ类是非 GPCR 嗅觉受体。2023 年，山东大学孙金鹏团队和上海交通大学医学院李乾研究员团队合作阐明嗅觉感知分子机制。他们应用冷冻电镜技术解析了 TAAR 家族成员之一的小鼠 TAAR9（mTAAR9）受体在 4 种不同配体结合条件下与 Gs/Golf（嗅觉特异性 Gα）蛋白三聚体复合物的结构，进一步结合药理学分析揭示了 mTAAR9 感知配体后被激活的分子机制。同时，该研究也提出了嗅觉受体“组合编码”识别配体的结构机制，阐明了Ⅱ类嗅觉受体独特的激活方式。该研究为嗅觉受体家族识别配体奠定了理论基础，对开发靶向嗅觉受体的新药也有重要意义。相关研究成果同年 5 月 24 日发表于《自然》（*Nature*）杂志。

第三节　学科发展问题与挑战

针对现阶段我国学科总体发展现状，从学科总体布局、学科研究发展、学科成果转化、学科发展保障 4 个方面提出以下 6 个制约我国加快建设世界一流学科的问题与挑战。

一、学科布局有待优化，前瞻性系统性布局不足

当前我国学科布局规划存在学科设置滞后、顶层设计不足、学科发展特色不突出、学

科分类设置过细等问题。

（一）学科布局相对滞后，学科设置不完全遵循科学发展规律

构建学科布局及时调整机制，是学科建设的基本逻辑。学科设置作为科学研究、人才培养工作的基本管理手段之一，过于强调学科专业划分的管理功能、过于刚性，不利于发挥高校、科研院所的主动性，难以适应经济社会发展需要及科学发展要求。国际高校往往在开展教学、科研活动时，根据自身与社会需求自主设置、调整学科专业。但是，我国现有学科布局多为跟踪性布局，未及时挖掘研究前沿，实现前瞻性布局，存在一定的滞后性。这使得我国的学科体系在很长一段时间内相对固定、研究内容不连续，前沿科学思想、新的科技生长点以及处于两个学科边界的科学领域经常不被关注。另外，我国学科体系的形成和变化受到一定的外力影响，有时会因为短期的、迫在眉睫的需要而“一哄而上”。

（二）学科发展顶层设计不足导致研究无法形成巨大合力

当前我国的科学研究还存在科研力量较为分散，科研数据“孤岛化”、科研平台之间协同性不足等问题，这都折射出学科建设缺乏系统规划。导致基础研究相对薄弱，核心技术自主研发不足，缺少学科之间的交叉融合，且与产业结合不深入，迫切需要对学科发展和产业布局进行整体的规划和顶层设计，凝聚科研力量和资源。

（三）学科领域碎片化、分类设置过细

当前我国学科设置过于细化，2022 年印发的新版《研究生教育学科专业目录》共划分出 117 个一级学科，而美、英、日等国高等教育学科数量均低于 50 个。我国科学基金申请代码超过 2300 个，而其他国家均少于 500 个。国际通行、用于衡量科学研究绩效的 ESI 数据库仅将学科分为 22 个。学科专业过于细化可能导致学术活动内容支离破碎，形成“学科壁垒”，不适应学科之间、科学与技术之间、技术与工程、自然科学与社会科学之间的交叉融合发展。

（四）学科发展特色不突出，未能实现错位发展

我国对学科发展实行统一化管理，虽然不同定位和特色的高校所需的学者和学生、所营造的学术学科平台并不相同，但当前各高校并未据此优化配置资源、实现错位发展，导致“千校一面”。一些长期无法成为研究热点的学科逐渐无人问津，出现青黄不接、人才断档的局面。

二、学科交叉融合难，与现行成熟学科体制存在冲突

当今科学已经进入了以多学科交叉融合为主要特征的“大科学”时代，但是我国交叉

学科体系规划起步较晚，这导致我国学科交叉融合难，与现行的成熟的学科管理体制存在冲突。国务院学位委员会、教育部下发的《学位授予和人才培养学科目录（2011 年）》只规定了学科门类和一级学科，没有单独设置交叉学科。2020 年 8 月，教育部才增设交叉学科作为新的学科门类。2020 年 11 月，国家自然科学基金委成立交叉科学部，负责统筹国家自然科学基金交叉科学领域整体资助工作，旨在促进复杂科学技术问题的多学科协同攻关。不少研究型大学曾尝试重大科研项目横向管理与纵向校、院、系层级式人事管理体系交叉形成矩阵型的组织结构，以此解决大学学科交叉和科研合作上研究人员跨院系自由流动的问题，然而，目前我国高校院系设置已沿用很长时间，已形成相对僵化、固化的思维模式和管理体系，欠缺与学科交叉相适应的科研组织管理模式。

此外，在交叉学科研究中，常常出现“跨学科”以及“跨领域”的现象，需要多主体合作，但是现有考评体系带来的利益冲突导致发展交叉学科的激励机制不足。在目前的考评体系以及知识产权归属中，仍然主要认定为文章、课题的第一作者以及第一完成单位，这严重降低了合作者的积极性。适用于传统学科研究的制度体系使交叉学科的研究行为、研究成果面临不易评价、无法纳入考核晋升成果认定范围、在交叉学科团队协作中贡献难以计算等制度瓶颈，限制了交叉学科发展。从研究人员自身来看，学科交叉意识不足，长期在同一学科领域进行科学研究，形成了单一的学科思维习惯，无形中使得科研人员淡化了对其他学科的关注。同时，科研人员普遍认为在单一学科内进行考核评价相对容易，打破学科边界的意识和积极性不足，难以有效地开展学科交叉研究。

三、跟踪性研究偏多，原创性引领性研究少

党的十九届五中全会提出“坚持创新在现代化建设全局中的核心地位，把科技自立自强作为国家发展战略支撑”。这一重要表述是党中央基于我国现代化建设全局作出的重大战略判断。创新是引领发展的第一动力。当今世界，谁牵住了科技创新这个“牛鼻子”，谁就能走好发展的先手棋。而我国的学科创新体系还有较大提升空间，学科原始创新能力不足，基础研究和应用基础研究发展不足。

（一）基础研究重大原创性成果缺乏，对工程技术的支撑力度有待提升

近年来，我国基础研究虽取得了一批世界瞩目的创新性成果，如铁基超导材料保持国际最高转变温度，量子反常霍尔效应、多光子纠缠、中微子振荡、干细胞、利用体细胞克隆猕猴等取得重要原创性突破。然而，我国目前的研究成果仍然以跟踪研究为主，原创性和引领性研究少，重大原创成果仅呈现出点的突破，引领性研究和重大原创性研究的产出质量与发达国家相比仍有较大差距。此外，基础研究对工程技术的知识支持力度还有待提升。工程技术涉及国家生产生活的方方面面，是推动我国建成社会主义现代化强国的重要基础。现阶段我国工程技术领域仍存在许多基本科学原理、科学问题、科学思维和科研范

式没有取得明显进步和突破等问题，成为我国科技创新链的薄弱环节，归根到底是我国基础研究发展的不足。

（二）应用基础研究“量”大“质”弱

应用基础研究同样存在“量”大“质”弱的问题，表现为我国在许多学科领域仍然处于跟跑状态，未掌握新兴学科的发展先机，突破性理论及技术创新较少，解决“卡脖子”问题能力较弱。因而，我国若想实现关键核心技术的突破，必须在学习借鉴世界经验的基础上，靠自身努力，提升“从 0 到 1”的原始创新能力。

四、关键核技术发展不足，技术产业链亟待完善

习近平总书记在《求是》撰文指出：实践反复告诉我们，关键核心技术是要不来、买不来、讨不来的。只有把关键核心技术掌握在自己手中，才能从根本上保障国家经济安全、国防安全和其他安全。

改革开放以来，我国一些产业在相当长的一段时期里，通过引进成熟技术快速形成生产能力，粗放式发展而忽视了产业创新基础能力建设，没有形成内生的技术创新机制。由于国际大环境的变化，使得我国知识交流的渠道，以及过去主要依赖的“技术引进—消化吸收—再创新”的路径受到阻碍。传统的以先进技术替代落后技术的企业技术升级策略无法解决产业基础的系统性提升困境，技术知识体系和产业知识体系基础不牢，使得部分关键核心技术受制于人。目前，我国拥有全球最完整的产业体系，但是与欧美发达国家一百多年的工业化积累相比，我们在技术领域仍然有较大差距，研究短板依然突出，底层基础技术、基础工艺能力不足，在工业母机、高端芯片、基础软硬件、开发平台、基本算法、基础元器件、基础材料等方面瓶颈仍然存在。当外部环境变化时，我国很多技术领域都存在产业链完整度、自主权受制于人的情况。因此，推动关键核心技术自主可控，加强创新链产业链融合，通过自主创新破解“卡脖子”项目和前沿领域，补齐科技短板，提升国家创新体系整体效能，是我们目前亟待解决的问题。

五、产学研协同合作不充分，科技成果转化能力不强

我国产业界与学术界互惠互利的合作关系尚未形成，产业界不了解学术界科研攻关重点，未能有效借助学术界的研发资源带动产品性能与成本管理的突破。学术界不清楚产业界遇到的技术难题，无法有针对性地开展科研课题研究。如何建立长效的产研交流合作机制，实现产业界与学术界破圈重整，达到融洽自得、互惠互利的状态，成为推动我国学科建设实现突破式发展的关键难题。

当前，我国科技成果 46% 处于实验室研究阶段、40% 处于中试阶段，产业化阶段的

技术仅为 14%，成果转化应用比例低，与发达国家 80% 的水平差距较大。造成这一现象的原因主要有以下几点：首先，企业和学术界的目标不够一致。企业的目标是追求利润最大化，高校科研人员的目标则是多产出具有原创性的科技成果，呈现出追求国际学术发表的价值倾向。因此，企业缺乏大范围扩散技术的意愿，大学和科研院所缺乏提高技术成熟度的意愿。其次，产学界合作意向不够强烈。工业部门在 20 世纪 80 年代中期之后主要以跨国企业作为获取技术或者模仿和跟随的主要参照对象，而其与国内科研部门的合作则被边缘化。国内一流大学主要以国际期刊论文发表作为科研绩效考核依据，与产业界的合作并不是其优先考虑的目标。最后，产学研各部门缺乏联动机制无法有效推进产学研创新发展。支撑产学研创新发展的法律法规和政策之间也缺乏整体关联性，政出多门。法律法规和政策主要分散在国家发展改革委、工信部、科学技术部、财政部、人力资源和社会保障部、教育部、国家市场监督管理总局、国家知识产权局等多个行政管理部门。由于各政府部门职责不同、管理范围各异，政府对产学研创新工作机制、成效评价机制还不够完善，政府部门之间缺乏协同联动机制，容易造成沟通渠道不通畅、信息不对称问题，更有甚者造成资源浪费、重复投入现象。

六、人才发展机制体制不健全，科研生态环境有待优化

（一）复合型人才和顶尖人才数量不足

中国科学院院士、西湖大学校长施一公曾在 2023 中关村论坛全体会议上表示，尽管在研发人员数量上，中国已居全球首位，但是顶尖人才依然匮乏。获得诺贝尔奖可以视为一种衡量标准，自诺贝尔奖设立以来，从 1901 年到 2023 年超过一百年的时间里，只有 11 位华人获得过诺贝尔奖，其中 6 位是物理学奖、2 位是化学奖、2 位是文学奖、1 位是生理或医学奖，远远低于美国、英国和德国等发达国家。科睿唯安发布的 2023 年度“全球高被引科学家”名单表明，中国高被引科学家人数达 1275 人，所占比例从 2018 年的 7.9% 上升至 17.9%，位居全球第二，但远低于美国的 2669 人（37.5%）。除了顶尖人才数量不足，我国复合型人才也较为缺乏，以脑科学与类脑智能领域为例，从论文发表的通讯作者统计，截至 2020 年 12 月，我国类脑智能领域的 PI 人数仅占整个脑科学与类脑智能领域的约 1/10，显示出我国在该领域人才的不足。同时，在人工智能医疗等新兴前沿领域，跨学科人才，特别是具备生物学、医学与计算机科学专业背景的复合人才严重缺乏，这无疑限制了相关领域创新和发展步伐。

（二）人才培养体系和模式不健全

近年来，我国人才培养规模虽在持续扩大，但是人才培养的体系和模式还不够健全。首先，学科领域的分类过细，不利于大科学时代学科交叉型和复合型人才的培养。当前我国高校的学科专业“条块分割”，学科制度设计和院系组织壁垒阻碍了多学科交叉融合、

集成创新人才的培养。其次，人才培养存在“理科化”倾向，学生学业评价以考试成绩和研究论文为导向，对工程实践能力、系统思维的培养不足。虽然我国多措并举支持企业和高校加强人才联合培养，但实际上高校与产业界需求分化明显，人才培养与产业需求“两张皮”，导致高校的人才培养难以完全满足社会对人才的素质需求。

（三）人才评价仍存在“五唯”现象，落实分类评价力度不够

当前我国人才评价机制还存在分类评价不足、评价标准单一、评价手段趋同等问题，尚未形成有效的科研评价治理体系。以医学学科为例，基础生命科学是中国国际撤销论文“重灾区”，主要原因是对临床医生、药剂师、护士、医学检验师、医学影像师等各类卫生专业技术人员分类评价体系不完善，对他们生搬硬套基础生命科学的评价体系而催生的学术造假现象。对于基础研究来说，论文仍是体现研究成果最重要的载体，我们仍需要将论文作为一个重要的评价指标，但要摒除形式主义，不只看数量和发表期刊的级别，而应当更要看论文的影响力、贡献力和创新性；对于应用类学科而言，人才评价则应更多地考察科研成果与产品和市场能否有效对接、解决实际问题等。然而，“破五唯”虽提出 3 年多，人才评价制度仍未完善。

（四）科研诚信建设压力较大，仍需加强

科研诚信是促进科技事业健康发展的重要保障。当前，伴随我国科技创新快速发展，面临的科研诚信挑战也日益增多。一方面的表现为撤销论文数量较多，根据中国科学院文献情报中心 2023 年 12 月统计数据，Retraction Watch 数据库同期共收录全球撤销论文（含期刊论文和会议论文）43333 篇，其中中国参与署名了 11139 篇，占全球总量的 25.7%，在全球排在第一位，其中涉嫌失信行为的比例为 44.1%，这说明中国科研诚信建设任重而道远。另一方面，新的学术不端行为不断出现，带来新的学术规范问题。比如 ChatGPT 自 2022 年 11 月面世以来，生成式人工智能技术带来研究范式的变革，同时也对科研诚信建设提出新挑战，引发了学术界、教育界、出版界等各界人士思考。生成式人工智能等人工智能技术的实现基础和使用过程均缺乏透明度，其生成内容的真实性和可靠性让人担忧。生成式人工智能技术的滥用可能引发更为严重的科研失信行为，原有的相似度查重和图像识别等技术逐渐失效，加大了识别这些学术不端行为的难度。根据 Retraction Watch 的数据统计，自 2021 年到 2023 年 7 月 6 日共有 914 篇论文因“随机生成的内容”原因被撤稿，约占近三年来撤稿论文总数的 8.34%。这呼吁学术界要明确生成式人工智能这一新兴技术在科学研究过程中的合理使用边界。

第四节　学科发展启示与建议

针对第三节中提出的学科发展问题与挑战，本节为促进我国加快建设世界一流学科提出以下建议：加强学科顶层设计，建立学科动态调整机制；积极布局交叉学科，促进学科交叉融合发展；强化原始创新能力，实现高水平科技自立自强；集中力量攻克技术难点，提高产业链完整性和自主性；构建产学研协作合作机制，完善学科成果转化生态；创新人才培养和评价机制，夯实智力支撑基础。

一、加强学科顶层设计，建立学科动态调整机制

（一）加强学科发展顶层设计，完善发展机制

建设世界一流学科重在顶层设计，需要规划“学科 - 专业 - 人才 - 平台”的协调发展体系，明确学科建设目标，凝练学科发展方向，确定学科建设层次，组建学科团队，搭建学科研究平台，构建有效的学科建设机制。在建设什么层次的学科上，要按照地方、国家和国际一流的要求设立相应的建设标准，并以该标准为基础组建团队、搭建平台、构建机制。在建设什么学科的问题上，要明确发展目标，找好学科发展的方向。就我国大学而言，一方面，在学科上出现了同质化的倾向；另一方面，每个大学都想追求大而全，走综合大学的路子，这显然不符合集中力量建设、发展优势和特色学科的学科建设理念。在国家建设一流学科的大战略下，每个大学在学科建设和发展上以大而全或小而全为背景实现学科建设达到世界一流水平，实际上是一种难以实现的愿望。所以，各个大学为了建设世界一流学科所面临的根本问题是如何实现学科建设和发展的相对集中，做到这一点所要树立的根本理念就是要坚持有所不为才能有所为，要将已有的学科体系中那些与学校的优势和特色学科关联度不大、未来进一步发展缺乏基础的学科进行必要的舍弃，将学校的资源集中到优势和特色学科的建设和发展上[①]。

---

① 谢志华．建设一流学科重在顶层设计［J］．北京教育（高教），2018（6）：16-19.

（二）积极开展学科研判，构筑学科动态调整机制

进入 21 世纪以来，新兴学科和新兴技术高速发展，应据此开展学科发展研判，针对不同学科发展的特点，构筑学科动态调整机制，建立分层分类、多元化发展的体系，瞄准世界科学前沿和关键技术领域设置和优化学科布局，提升学科布局前瞻性。从政策诉求看，《统筹推进世界一流大学和一流学科建设总体方案》明确提出“引导和支持高等学校优化学科结构，凝练学科发展方向，突出学科建设重点”“强化绩效，动态支持”等具体要求。2023 年出台的《普通高等教育学科专业设置调整优化改革方案》[①] 也强调应建立专业动态调整机制，推动高校依据标准和人才培养实际动态完善人才培养方案。

建立学科动态调整机制，要注意政府、高校、市场三方的协同作用。政府、高校和市场在学科专业动态调整中具有不同的职能分工和责任。建立学科动态调整机制要避免过度依赖单一化的模式手段，要将行政干预、市场调节和学术自治充分整合，重构政府、高校和市场在学科专业动态调整中的关系生态。在具体的实践操作中，可进行自上而下与自下而上相结合的互动式调整，使得各方主体都能够基于自身的不同利益诉求参与到调整的过程之中。政府要加强国家整体宏观统筹及战略布局，从政策上指引高校学科动态调整方向，使高校有依据地“放开步子”进行学科动态调整。高校必须不断提高学科规划能力，能够形成较科学、清晰的学科结构，同时也要避免陷入“实用主义”陷阱的学科动态调整误区，要建立一种知识生成与学科发展的协同机制。市场要发挥好资源配置的决定性作用，鼓励企业与高校共同找准“真”问题，推进学科动态调整 [②]。无论是高校还是政府抑或是市场，在学科专业动态调整过程中的作用都是有限的，任何一方主体作用的有效发挥都会受到另外两方主体的牵制和束缚。重构学科专业动态调整机制需规范各方主体的权责范围和边界，以确立基于联合行动的合法性基础 [③]。

二、积极布局交叉学科，促进学科交叉融合发展

当前，新一轮科技革命和产业变革正在加速颠覆现有的研究范式、产业形态、分工和组织方式。为加快我国学科发展，需推进各学科交叉融合和多技术领域集成创新，厚植基础性学科，催生颠覆性创新。

---

① 中国政府网. 教育部等五部门关于印发《普通高等教育学科专业设置调整优化改革方案》的通知［EB/OL］.（2023-02-21）. https://www.gov.cn/zhengce/zhengceku/2023-04/04/content_5750018.htm.

② 楚旋. 近年来我国高校学科动态调整的特征分析与实践反思［J］. 黑龙江高教研究，2022，40（12）：95-101.

③ 田贤鹏. 高校学科专业动态调整：模式、困境与整合改进［J］. 高校教育管理，2018，12（6）：44-50.

（一）超前布局，推动交叉融合创新

针对各学科领域出现的新理论、新方法、新技术等，超前进行学科发展布局，例如设立重大科技专项、增加项目资助比例等，推动各学科之间的交叉融合创新，推动跨越式发展。前瞻布局未来可能产生变革性技术的基础科学领域，强化重大原创性研究和前沿交叉研究。

（二）突破学科界限，建立跨学科人才培养机制

人才储备是推动我国交叉学科发展的重要基础环节。政府相关机构需进一步完善学科设置与监管。政府赋予高校交叉学科专业设置更大空间的同时，应对高校交叉学科的专业设置和人才培养过程进行有效监管。高校应积极探索并实施与交叉学科发展相匹配的教学与科研模式，在教学过程中重视多学科知识与方法的传授，有计划地培养跨学科人才，引导人才培养、科学研究和社会需求有机融合、互相促进，在攻克"卡脖子"技术、破解经济社会发展重大难题中深化交叉学科人才培养。

（三）完善交叉学科评价和激励机制，鼓励科研人员合作交流

建立适合交叉学科发展的评价标准与方法，是交叉学科融合创新的重要制度保障。由于交叉学科研究的跨学科性，需建立与交叉学科研究相适应的评价指标体系，组建包含各个相关学科领域专家的交叉学科评审组，对开展交叉学科研究人员队伍的科研成果、业绩考核、项目申报及职称晋升等进行客观有效的评价。研究机构需要打破行政单位的考评体系壁垒，综合考虑参与交叉学科研究的科研人员的研究进展，适度考虑其在团队中的贡献与成果，制定更为合理、更为完善的绩效考评机制和奖励机制。同时，交叉学科研究需要不同学科经过长周期、深度融合的协同创新，实现远期目标。研究机构不应根据学科评估指标来制定交叉学科研究人员考评标准，要以实现长远目标为评价原则，在评价"论文、专利、获奖、项目、人才奖励"等近期显性成果的基础上，更要注重交叉学科研究过程中阶段性成果的认定与认可，提升交叉学科成果在考评中的占比当量，同时应延长交叉学科研究科研人员的考评周期。

（四）搭建跨学科共享平台

跨学科共享平台具有多学科交叉的优势，具有辐射面广、扩展性强的特点。科研机构应加强平台建设的顶层设计，根据交叉性的重点、重大研究的需求，凝练具有创新性的科学问题，着重建立相关跨学科共享平台。通过平台的建立，组织不同学科领域的研究者充分利用跨学科平台交流合作，联合开展交叉性研究，从而逐步培育具有学科交叉特色的研究者、研究小组、创新团队，孕育多学科交叉的学术创新成果，以跨学科共享平台为契机，带动交叉学科发展。

三、强化原始创新能力，实现高水平科技自立自强

继党的十九届五中全会提出要“把科技自立自强作为国家发展的战略支撑”之后，2021 年 5 月 28 日，习近平总书记在两院院士大会、中国科协第十次全国代表大会上提出，要“加快建设科技强国，实现高水平科技自立自强”。从“科技自立自强”到“高水平科技自立自强”，“高水平”这 3 个字既是要求、号召，也是鞭策、激励。加强原创性、引领性科技攻关，坚决打赢关键核心技术攻坚战，需要将学科研究探索与服务国家需求紧密融合。基础研究是科学之本、技术之源、创新之魂，要勇于探索、突出原创，拓展认识自然的边界，开辟新的认知疆域。科技攻关要坚持问题导向，奔着最紧急、最紧迫的问题去，从国家急迫需要和长远需求出发[①]。为强化原始创新能力，实现高水平科技自立自强，具体可从以下 4 个方面入手。

1）加强基础研究是学科发展的战略关键。把围绕国家战略需求和科学前沿重大问题的定向性、体系化基础研究作为主要任务，强化需求导向和问题导向的基础研究选题机制，推动基础与应用学科均衡协调发展。

2）围绕国家重大需求，加强基础前沿研究，加强对量子科学、脑科学、合成生物学、空间科学、深海科学等重大科学问题的超前部署。

3）围绕经济社会发展和国家安全的重大需求，突出关键共性技术、前沿引领性技术、现代工程技术、颠覆性技术创新。

4）围绕人民健康和促进可持续发展的迫切需求，加强资源环境、人口健康、新型城镇化、公共安全等领域科学研究。

四、集中力量攻克技术难点，提高产业链完整性和自主性

针对我国高技术产业链完整度和自主性有待提升的现状，我们需集中力量攻克技术难点，关注科学研究对工程和技术和实质性支撑，优化和稳定产业链、供应链，实现关键核心技术自主可控。具体可从以下 3 个方面入手[②]。

1）坚持自主可控、安全可靠的原则，分行业分阶段推进产业链供应链的优化与多元化，把关键核心技术牢牢掌握在自己手中，在重点领域形成产能备份，力争实现重要领域和关键节点的自主可控，打造以我为主的产业链、供应链。

2）针对高端芯片、基础软件、生物医药等重点领域，加快补齐在先进工艺、基础零部件、关键材料等方面的短板，着力攻克关键核心技术“卡脖子”问题，提升产业基础高

---

① 黄维．实现高水平科技自立自强［J］．上海企业，2021（6）：56.

② 邓子纲．增强产业链供应链自主可控能力［N］．人民日报，2021-03-17（09）.

级化和产业链现代化水平；对轨道交通、工程机械、航空航天、电子信息、新材料等已具备优势的领域，加紧实施产业基础再造和技术提升工程，以加强和巩固领先地位。

3）以智能化、数字化、物联网化为重点，加快推广应用新技术，加速产业数字化转型，确保相关产业发展始终站在全球数字产业链供应链前沿。

五、构建产学研协同合作机制，完善学科成果转化生态

为帮助产学研形成协作合作机制，可从 4 个方面构建产学研深度融合的学科发展生态圈。

1）真正发挥企业的创新主体作用，制定支持前沿探索、基础研究、应用攻关、产业转化全过程的相关政策，加大原始创新研发资金的投入力度，增强对相关实体行业创业的帮助和实体经济扶持，不断提升科技成果转化效率。

2）教育作为产学研链条中重要的一环，意义重大。各高校应担起培养新时代应用型高水平科研人才的使命，以市场实际需求和社会现状为导向进行人才培养，积极衔接高校实验室和企业双方平台，协同创新，打通创新链和产业链融合通道。针对高等教育环节：①要畅通产业高端人才进入高校的渠道，增加产业背景在校内教育资源调配上的话语权，改变高校科研目标只追求国际学术发表的倾向，将帮助企业追求技术革新和成熟度、实现更高利润纳入高校科研目标，增强产学界合作意向；②试点更加灵活的课程和专业更新机制，根据新兴产业领域发展快速调整专业和教材，试点放开专业和课程开设的审批权限；③形成实战型实习机制，在大学课程中设置系统化、周期化、定制化实习体系，从专业认知到实际操作，打通人才走向产业的通道，加快成熟人才供给；④支持企业与普通高校、职业院校共建现代产业学院、产教融合基地、高技能人才培训中心、卓越工程师培养基地等，建立适应产业发展需求的人才培养体系。

3）企业作为创新链上的重要环节，要加大对关键核心技术的各项投入，布局技术核心产业，突破并发掘产业战略支撑点。通过制定相关行业标准、提前谋划产业布局等，推进产业市场有序运行，发挥好市场配置资源的决定性作用。同时要积极拓展参与产学研融合的深度和广度，发挥科技领军企业“出题人”“答题人”“阅卷人”的作用，促进科研成果转化。

4）广大科技工作者要充分利用科学研究范式变革和学科交叉融合的机遇，助力产学研链条形成“基础研究 + 技术攻关 + 成果产业化 + 科技金融 + 人才支撑”全过程创新生态链[①]。在实现高水平科技突破的同时，加速研究成果走出实验室，加快科技成果孵化与转换，在企业中投入生产，尽快转化成社会科技力量。

---

① 黄维. 深入产学研协同 赋能创新发展［J］. 中国科技产业，2021（8）：5.

六、创新人才培养和评价机制，夯实智力支撑基础

现阶段，我国科技人才队伍建设与评价机制体制改革已进入深水区，需要根据学科方向和研究目标的不同，构建差异化的科技评价指标体系，创新科研人才评价机制，引导各类科技人才人尽其才、才尽其用、用有所成。

创新基础学科人才机制，具体可从以下 3 点入手：①重视高层次人才培养与引进，依托国家级平台和重大科技计划，实施创新人才推进计划，造就一批世界顶尖科学家和一流创新团队，积极开展国际科技合作，吸引国外高端人才参与，完善海外人才资助体系，打破人才壁垒；②加强中青年和交叉型人才培养，在各学科领域建立国际通行的访问学者制度，完善博士后制度，吸引国内外优秀青年博士在国内从事博士后研究，实施针对青年科研人员团队建设的专项行动，吸引国内外优秀青年人才加入科研人员团队，加强青年科研人员的国际化培养，优化国家留学基金管理办法，加大资助力度；③优化创新人才发展环境，拓展科研人员自由探索的空间以实现更多原创突破，完善人才评价机制，突出以人为本，丰富激励手段，促使人力资源不断发挥能动性，完善人才住房、就医、子女入学、配偶就业等保障服务，切实解决人才工作、生活中的困难。

针对高水平技术人才培养体系亟待完善的现状，需加强核心技术学科专业、实战、复合等高质量人才队伍建设，鼓励高校院所、企业、投资机构联手共同研发推动产业孵化和加强紧缺人才培养，具体可从以下 3 个方面入手：①依托高科技人才项目长期持续培养从基础理论、关键器件到核心技术、高端装备及超快应用的人才，提高关键核心器件研制人员的待遇，推进我国高端装备、关键技术自主化；②积极推动“走出去”和“引进来”的融合，积极举荐我国科学家在相关领域的国际组织担任重要职务，增强科技合作的主导性和引领性，加快引进海外高层次人才，助力核心技术领域科技创新进入全球领跑行列；③积极引导上下游联动，在技术链“卡脖子”节点上遴选培育选手，采用赛马机制等激发竞争活力。鼓励校企联手针对高技术人才开展联合资助、联合培养，开辟高科技人才成长快速通道。

针对应用类学科人才评价机制尚不完善的问题，应继续深化分类评价制度改革，深入落实《关于开展科技人才评价改革试点的工作方案》等政策要求。对应用研究项目评价应强调创新性、前沿性、应用性，对开发研究项目评价应强调创新性、推广性和持续性。对于应用研究和技术开发类人才，应以技术突破和产业贡献为导向，重点评价技术标准、技术解决方案、高质量专利、成果转化产业化、产学研深度融合成效等代表性成果，建立体现产学研和团队合作、技术创新与集成能力、成果的市场价值和应用实效、对经济社会发展贡献的评价指标。不得以是否发表论文、取得专利多少和申请国家项目经费数量为主要评价指标。探索构建专家重点评价技术水平、市场评价产业价值相结合，市场、用户、第三方深度参与的评价方式。

第二章

相关学科进展与趋势

第一节　数学

一、引言

数学是研究数与形的基础学科，是自然科学的基础，为科学研究提供精确的语言、严格的方法和新的研究范式。回顾科学的发展历史，不难发现几乎所有的重大发现无不与数学的发展与进步相关。数学也是重大技术创新的重要理论基础，我国许多“卡脖子”技术的“症结”归根结底是其中的数学原理没有解决。

本节中，将对我国当前数学学科发展情况开展综合研究，探讨学科发展现状，进行国际比较，思考问题挑战并提出对策建议。

二、本学科近年的最新研究进展

（一）数学作为知识和工具的重要性日益凸显

今天，数学科学几乎渗透到日常生活各个方面，并成为如互联网搜索、医疗成像、电脑动画、数值天气预报和其他计算机模拟、各类数字通信、商业、军事的优化以及金融风险分析等许多研究领域不可或缺的重要支撑。例如，广义相对论、黑洞的数学描述及其旋转中仍存在未解决的数学问题；希尔伯特空间算子为量子力学提供了自然框架；作为数据科学方法论的统计学研究加速了人类基因组测序的完成；天体物理学面临的重大挑战一部分依赖于科学计算与工程的发展；卫星集群和大型地基仪器提供的丰富数据资源需要结合统计学、科学与工程计算来进行整合分析。数学是自然科学的重要理论基础，数学和自然科学之间相互影响，并为促进自然科学发展提供了工具和知识。

数学作为最具有普遍性的工具和方法，改变着当今科学研究的方式方法，并通过与计算机科学的交叉，为科学研究提供了新的范式。例如，产生了计算流体力学、计算材料学、计量经济学等新兴交叉学科。科学与工程计算可以对无法进行解析求解且难以进行实验的物理过程进行模拟，如海啸、气候、核爆等。因此，模拟仿真被公认为实验和理论推理之后科学研究的新范式。大数据的分析与使用正在成为科学发现的重要手段，被称为继实验、理论与仿真之后的第四个研究范式，数学的发展推动了数据科学的产生，成为大数据分析必须依赖的理论基础和方法手段。

（二）新时代数学科研工作高质量发展

目前，我国形成了“政府－高校－科研机构－学会－期刊－奖项”多层级协同联动数学学科科研创新体系。与数学相关的全国学会、研究会主要有中国数学会、中国工业与应用数学学会、中国运筹学会、中国系统工程学会和中国现场统计研究会等；中国现阶段水平相对较高的一流数学期刊包括：《数学学报（英文版）》《中国科学：数学（英文版）》《数学物理学报》《中国数学年鉴》《北京数学杂志（英文）》《数学与统计通讯》《CSIAM 应用数学会刊》《计算数学（英文版）》《中国运筹学会会刊（英文）》等。

以论文情况来考察近年数学研究发展情况和产出规模，则可以聚焦数学四大刊，包括《数学年刊》（*Annals of Mathematics*）、《美国数学会杂志》（*Journal of the American Mathematical Society*）、《数学新进展》（*Inventiones Mathematicae*）和《数学学报》（*Acta Mathematica*）。它们在数学界具有较大影响力，虽然不能涵盖数学的全部重要领域，但在国际上被公认，很大程度上能够反映研究对象的数学研究水平。从四大刊的产出数据来看，在 2001—2005 年和 2006—2010 年，中国发文数分别为 13 篇和 21 篇，占四大刊总发文数比例均为 2%，2011—2015 年和 2016—2020 年，数学四大刊中国发文数增加到 44 篇和 39 篇，占四大刊总发文数比例均增长至 5%，说明中国近十年在数学顶级期刊发文数量比之前有了显著的提升。

（三）新时代数学科普工作高质量发展

近年来，党和政府、社会各界组织机构以不同方式推动数学科学普及、数学文化培育工作。以中国数学会为例，学会积极创新科普方式、传播数学文化、提升科学素质。为深入贯彻习近平总书记关于科技创新和科学普及的重要论述精神及《全民科学素质行动规划纲要（2021—2025）》，将科普工作深入到中国的各个地区，中国数学会年会期间当地大中小学通过多种形式邀请院士、大会报告人进校园作科普报告。多位院士在 2020 年学术年会期间，分别前往河北省的中学和高校作 4 场科普报告；2021 年学术年会期间前往云南省昆明市的中学和高校作 3 场科普报告；2022 年学术年会期间分别前往湖北省武汉市中学和高校作 8 场科普报告。“院士进校园”科普活动，大力弘扬了以爱国、求实、奉献、协同、育人为内核的科学家精神，涵养优良学风，激发了青少年数学梦想和数学志向，形成尊重知识、热爱科学、献身科学的浓厚氛围。

与此同时，中国科技馆也在数学科普方面作了很多工作，专门设置了数学展区——数学之魅展区不以知识的传授为主要目的，而是紧扣“探索”的主题进行展开，设置“探索中的数学”“生活中的数学”“思维中的数学”3 个分主题，着重体现数学诞生与发展过程中，人类的探索活动和取得的重大成果，以及人类在探索其他科学领域的过程中数学所起到的重要作用。

三、本学科国内外研究进展比较

党的十八大以来，我国数学研究水平稳步提升。根据科技文献数据库（WOS）统计，从 2019 年起中国数学学科发文总数已赶超美国，位居第一。但是，在全球数学界公认的四大数学顶级期刊上，美国发文量仍是第一，随后依次是法国、德国和英国，中国排名第五。从论文产出效率看，中国也排名第五，2011—2020 年产出效率[①] 为 4.21%。此外，中国数学学科规范化的引文影响力指标为 1.03，全球排名第十，超过平均水平。

考察国内外数学发展情况，还可以以国际数学家大会（International Congress of Mathematicians，ICM）为视角。国际数学家大会是由国际数学联盟（International Mathematical Union，IMU）主办的国际数学界规模最大也是最重要的会议。大会每四年举行一次，会议演讲分为 1 小时报告和 45 分钟报告，被大会邀请做 1 小时报告或 45 分钟报告是一个很高的荣誉。2010—2022 年的四届大会中分别有 195 人、211 人、238 人、226 人做了大会报告，其中 1 小时报告，分别有 20 人、20 人、22 人（21 场）、21 人，45 分钟报告分别有 175 人、191 人、216 人、205 人。对四届数学大会上做报告的科学家国籍进行统计，数量排名前十位的分别为：美国 301 人次、法国 124 人次、英国 58 人次、德国 50 人次、以色列 34 人次、中国 33 人次、加拿大 29 人次、俄罗斯 26 人次、日本 25 人次、瑞士 25 人次。其中美国以全球占比约 34.6% 绝对优势位居第一；中国位居第六，演讲者人数约为美国的 11.0%，约占全球的 3.8%。在四届大会的 1 小时演讲中，美国位居第一，演讲人数量有所波动，但基本持平；中国只在 2010 年和 2022 年分别有一位 1 小时演讲人。

总之，中国数学的发展已取得了诸多突破，但现阶段，中国数学距离世界一流水平还有一定的差距，面临着从数学大国向数学强国的转变。

四、本学科发展趋势和展望

展望我国数学学科未来，要以建设数学强国为目标，要培养国际级的数学大师，要提出具有较高国际关注度的数学问题。在数学领域，我国能否开辟出可以引领国际数学未来的新专题、新方向和新分支，能否形成中国学派，能否培养和打造具有较高国际声誉的数学奖项和比赛，这些值得思考和期待。

一是全球数学科学发展呈现明显的各分支学科之间相互交叉、多学科相互渗透融合的

① 重要期刊的产出效率指国家重要期刊发文数量占本国全部论文数量的份额。本书中重要期刊数据界定参考中国数学会数学科技期刊分级（2020 版）T1 中的国际期刊、综合学界认可的综合性高水平期刊构建。

态势。一方面，数学各分支之间交叉产生了许多新的知识生长点；另一方面，数学与自然科学、工程技术更加广泛的交叉融合，为航空航天、国家安全、生物医药等领域提供重要支撑。希望未来能够进一步提高中国的高水平数学论文数量和质量，不断发展和巩固具有国际影响力的中国数学期刊和数学科普品牌活动（如“国际数学日”），提升中国数学科研和科普的整体水平。

二是我国数学研究将在“十四五”期间仍然保持一个高速发展的态势。在基础理论领域，希望有一定数量的我国数学家成为数学研究领军人物，使我国成为引领国际数学发展的国家之一，继续培养和造就一些具有全球竞争力的青年数学家，希望本土数学家尽早在菲尔兹奖、沃尔夫奖等国际著名数学奖中获得突破；在实际应用领域，将继续鼓励数学家关心实际问题，承担和解决国家重大急需的问题，促进中国由数学大国向数学强国的快速转变，促成国防和民生科技的关键数学问题研究走在世界的前列。

三是期待更多支持数学发展的科技政策出台。党的十八大以来，党和国家高度重视我国数学科研教学工作，多次强调数学在基础研究中的重要地位，对加强数学科研教学工作提出了重要指示要求。直至今日，我国已从法律法规、政策举措和行业规范 3 个层面形成了自上而下的政策环境，在数学评价、学科布局、道德规范等多个方面不断完善体制机制，营造并巩固全社会尊重数学、重视数学研究的文化氛围。未来的数学发展一定是同国家整体科技发展相匹配、同国家科技政策、法律法规协同发展的战略性基础学科，既要杜绝急功近利的短视政策，又要有集中力量办大事的集中性举措。期待更多有利于数学发展、数学交叉应用的政策出台，落地实施。

第二节　高端科学仪器与集成电路先进装备

一、引言

仪器设备是科技社会里人们认识世界和改造世界的“工具”，它既是科技创新的成果，也是支撑科技进步的基础。高端仪器设备的理论原型通常源于物理学，经过与电子、自动化、软件等领域的交融，形成可应用于材料、化学、生物、集成电路制造等前沿领域的系统。高端科学仪器的需求日益增长，但市场主要由国外企业垄断，造成国内的科技“卡脖子”问题。特别是近年来，集成电路产业已上升为中美经贸争端的关键领域之一，高端仪器设备与集成电路产业息息相关，中国为发展自主可控的集成电路产业，加大了在国产仪器装备研发和产业化方面的投入力度与决心。与此同时，美国对出口的高端制造和高端测试设备采取了战略限制。这使两国在集成电路装备领域的竞争日益激烈，形成了一场科技与经贸的“冷战”。

高端科学仪器与集成电路先进装备相关学科的发展和产业进步是相互促进、相互依存的。在这个全球化的时代，先进的集成电路产品需要依赖先进的制造设备，掌握高端仪器设备技术已经成为一个国家综合实力和战略竞争力的重要体现。发展高端仪器装备技术离不开良好的科研环境和专业的技术人才，最终在这场竞争中，是人才的培养和科研技术能力的比拼，同高端科学仪器与集成电路先进装备相关学科的发展戚戚相关。解决当前本学科面临的人才短缺、基础研究薄弱、产学研脱节等问题，推动本学科的高质量发展，将大大增强我国集成电路产业的整体实力和国际竞争力。这不仅关乎国家安全和科技进步，也将促进全球科技合作，造福人类社会。

二、本学科近年的最新研究进展

根据集成电路制造设备的分类和高端仪器的技术进展情况，我们分别对前道设备和后道设备中的核心技术及相应研究进展进行简明阐述，分析相关的技术特点、发展趋势和核心的技术难点，指出相关技术国内外发展情况和我国面临的技术瓶颈。涉及前道设备主要有光刻机、刻蚀机、镀膜设备、离子注入设备、清洗设备、热处理设备、化学机械抛光（CMP）设备和量测设备，涉及的后道设备主要有减薄设备、引线键合设备、倒装焊设备、电镀设备、晶圆键合设备、分选设备、划片设备、测试机和探针台等。

从全球发展态势以及竞争态势来看，集成电路设备领域总体上保持较高的增长态势，尤其是自动化、智能化、高效化的设备，为集成电路产业在摩尔定律驱动下的快速成长提供了基础支撑。然而，集成电路设备的消费与生产“倒挂”，我国集成电路产业的发展在设备环节的“短板”极为显著，成为影响产业安全和产业链稳定的重要隐患。从全球集成电路设备的领先企业来看，顶端巨头拥有行业内的绝对控制能力，并成为行业发展的技术路线主导者，但由于集成电路制造过程的多流程、高精度、高可靠性要求，在专业化分工的驱动下，中小企业在一些细分领域获取一定的竞争力，这也为处于后发追赶阶段的我国集成电路产业发展提供了“机会窗口”。

科学的研究离不开高端仪器设备相关学科的支持，在集成电路相关的科研方面，我们主要分析了用于制造纳米器件，探索新材料和新现象的仪器设备，涉及的主要有基于光、电、磁、力等众多物理原理设计的相关仪器和技术发展趋势，相关设备有电子束曝光设备、磁性测量设备、原子探针显微镜、高频电子仪器、X 射线设备、脉冲激光沉积（PLD）和分子束外延（MBE）等。对可能在集成电路产业中实现应用的科研仪器展开论述，分析科研仪器相关学科技术在集成电路应用中的未来发展趋势。

科研设备在集成电路领域的应用是促进产业升级的一个重要因素，是推动集成电路设备发展的"母技术"，因此是分析设备行业发展和技术研究的重要因素。传统集成电路装备与科研仪器之间在功能、设计、性能和应用领域等方面存在明显的差异。在功能和应用领域方面，传统集成电路装备主要用于大规模制造过程，如光刻机、气相沉积设备等，而科研仪器则用于科学研究和实验室环境中，如原子力显微镜、拉曼光谱仪等，用于探索新材料和现象。在设计和制造方面，传统集成电路装备注重工程化需求，以满足大规模生产，而科研仪器更加灵活，以适应多样的研究目标。在性能和精确度方面，集成电路装备需要保障一致的产品质量，而科研仪器则需提供高分辨率的测量和分析。至于自动化程度，传统集成电路装备在生产过程中高度自动化，而科研仪器通常更注重研究人员的灵活操作。

从集成电路装备技术的发展历程上看，传统集成电路装备的发展离不开科研设备的影响和启发。随着信息技术的飞速发展，集成电路作为现代电子产品的核心，其制造和研发日益复杂化。先进科学仪器的运用为集成电路行业提供了更精确的测量、分析和控制手段。尤其在下一代集成电路研发方面，科研设备有力推动了行业的技术进步，有助于保障集成电路的性能和质量，推动技术的创新。这些科学仪器在集成电路制造过程中的潜在应用涵盖了工艺研究、质量控制和故障分析等领域，对于提高生产效率、降低成本以及优化产品性能具有至关重要的作用。虽然科研仪器行业体量不大，但在整个集成电路产业发展中的作用却十分重要，20 世纪 90 年代初，美国商业部国家标准局一份报告中提到：仪器仪表工业总产值只占工业总产值的 4%，但它对国民经济的影响达到 66%。随着技术的不断进步，科学仪器在集成电路行业的应用将持续演化，为行业的发展带来新的机遇和挑战。

三、本学科国内外研究进展比较

自 1958 年世界第一块集成电路发明以来，集成电路已从实现电路小型化的方法演变为信息系统的核心。在人类进入信息时代后，集成电路成为引领新一轮科技革命和产业变革的关键力量，是国家综合实力的重要标志，是大国竞争的战略制高点。近年来，集成电路成为全球战略博弈的主战场，我国在集成电路领域所面临的核心技术不可控、行业领军人才缺乏、专业人才培养投入不足等问题日益凸显。习近平总书记多次强调要重点发展集成电路产业，推进学科交叉融合。2020 年 12 月 30 日，国务院学位委员会正式批复在交叉学科门类设置"集成电路科学与工程"一级学科，推进集成电路相关行业人才培养。

集成电路是多学科、多领域交叉融合的基础研究和前沿工程技术的结晶。随着集成电路科技和产业加速变革，一些重要科学问题和关键核心技术已经呈现革命性突破的先兆，新的学科分支和新增长点不断涌现，学科深度交叉融合势不可挡，经济社会发展对高层次创新型、复合型、应用型人才的需求更为迫切。以欧美和日韩为代表的国家集成电路学科发展较早、学术理论水平先进、相关工业基础深厚，相关学科一般设置于电子电气工程（EE）学科下，专业实力较强的院校有麻省理工学院、斯坦福大学、加州大学伯克利分校、普渡大学等。国内在2019年之前长期参照国外相关学科设置，将集成电路学科设置为电子科学与技术的二级学科，实力较强的院校有清华大学、北京大学、复旦大学、电子科技大学、西安电子科技大学、上海交通大学、南京大学、东南大学、浙江大学、华中科技大学等。国内集成电路学科发展较为落后，学科实力在国际上排名不高。教育部为了推动集成电路学科的发展，在2015年发布《关于支持有关高校建设示范性微电子学院的通知》，支持北京大学、清华大学等9所高校建设示范性微电子学院，北京航空航天大学等17所高校筹备建设示范性微电子学院；2020年7月30日，国务院学位委员会会议投票通过“集成电路科学与工程”成为一级学科。2020年12月30日，教育部正式发文设立“集成电路科学与工程”一级学科。2021年10月26日教育部正式发文，18所高校入选新增“集成电路科学与工程”一级学科博士学位授权点名单。近两年来，全国有十多所双一流高校相继成立了集成电路（科学与工程）学院。集成电路科学与工程学科迎来了新的发展阶段。

集成电路科学与工程一级学科建设对于我国是新生事物，特别是针对集成电路先进装备的学科布局在今天非常稀缺。如前所述，因其与高端科学仪器有很大重合度，国内高校中涉及光学工程、仪器科学与技术、控制科学与工程学科支撑了相关领域，但集成电路先进装备并没有对口的学科，高端科学仪器相关的科研和人才培养也散落在各学科中，没有建制化的发展力量。

四、本学科发展趋势和展望

我国面临着高端仪器装备领域的人才短缺问题，特别是高端人才的培养需要时间和投入。国内高校的人才培养质量和数量与国外存在明显差距，学科之间割裂问题严重，特别是基础学科方面不足。高端仪器行业的成果产出周期较长、人才吸引力不强，这也影响了仪器研发人才队伍的规模和结构。因此，亟须加强相关学科的发展，加快培养紧缺高层次人才，以支持中国成为制造和科技强国。因此，要鼓励高校院所、企业、投资机构联手共同研发推动产业孵化和加强紧缺人才培养。

高端仪器装备对于科技创新和国家发展至关重要，我国在这一领域面临着挑战，但也机会巨大。观察集成电路产业的发展，各国政府牵头成立的技术联盟在不同阶段的集成电路产业地缘竞争中几乎起到了扭转乾坤的作用，例如日本VLSI计划、美国SEMATECH

等技术开发联盟。我国在这方面应加强顶层规划，组建自己的技术联盟，培育自己的人才库。

我国正在持续不断推动自主可控的高端仪器设备方面布局和本学科的发展。集成电路产业链具有投资大、回报周期长的特点，需要重点突出、持续地投入，要补链强链，将自主可控放在追求商业回报之前。集成电路设备的进步要比集成电路制造提前 3~5 年，要注意提前布局。EUV 光刻机前期研发用了 20 年，真正走向量产又用了 20 年，这期间需要持续的投入支撑。同样，本学科要以前沿应用为牵引，打通产学研“障碍墙”，将最先进科研成果应用于实际产业界中，提升学科的全球竞争力，注重自主知识产权体系建设。

回顾近年来我国在高端科学仪器与集成电路先进装备发展历程和技术发展趋势，中国的科技创新实力虽然在不断提高，但高端科学仪器和集成电路先进装备市场目前仍主要由国外企业掌握，这导致了国内科技工作的“卡脖子”风险，进而影响我国高端科学仪器与集成电路先进装备相关学科的发展。因此，我们必须认清高端科学仪器在科技创新和国民经济发展中的战略地位，注重人才培育，优化科研体系导向和政策扶持，加速科学仪器相关学科的发展，相信经过 5~10 年培育，能够将本学科提升至世界先进水平，实现国内高端科学仪器与集成电路先进装备的自立自强。

第三节　化学

一、引言

习近平总书记在关于科研工作的论述中指出，科研工作要坚持“四个面向”，把世界科技前沿同国家重大战略需求和经济社会发展目标结合起来。化学作为一门研究物质的性质、组成、结构以及变化规律的基础学科，是人类用以认识和改造物质世界的主要方法和手段之一，是解决资源、信息、环境和生命健康等关乎可持续发展问题的关键学科。近四年来，我国化学工作者在党中央和国务院的指挥部署下，克服新冠疫情带来的困难，在化学学科的各专业领域取得了极大的进展。本节总结了近四年来我国化学工作者在科研和教育方面的成果，涵盖了物理化学、分析化学、无机化学、有机化学、高分子化学和化学教育学6个主要分支学科，以及核化学与放射化学、化学生物学、公共安全化学、环境化学、有机固体化学、分子医学、能源化学和燃烧化学等交叉学科的最新进展、国内外比较和未来展望。

二、本学科近年的最新研究进展

（一）物理化学

物理化学研究的前沿集中在发展先进的原位表征技术，深入理解化学体系的结构、过程和机制，在原子及分子水平上实现化学反应的理性设计。2020年以来，我国科学家建立了离子液体离子率的定量测定方法、以Flash DSC表征微尺度材料热导率和界面热阻的新方法，以及极端条件下测量化学过程热效应的技术；揭示了氢相关的重要系列基元反应的量子效应；利用极紫外自由电子激光光源－串联高分辨质谱的新方法，为解析蛋白质结构提供了新技术；实现了能源材料的理性设计和性能优化；在利用神经网络算法构建的高精度势能面方面处于国际领先地位；在催化研究领域解决了高选择性和高催化活性的“跷跷板”难题；将化学过程与生物工程耦合，从二氧化碳出发实现了淀粉等高附加值物质的催化制备；发现纳米尺度曲率匹配对于手性传递的重要性；实现了分子自组装由溶液分散体系向宏观连续材料的跨越；证明了化学吸附水的覆盖度随电位的变化会对电化学界面造成一个负电容的贡献，阐明了双电层内的氢键网络可能是电化学析氢反应pH效应的来源；设计出维生素K3生产的绿色新路径；揭示了多种生物分子在生命过程的物理化学机制；自主开发了高性能分子动力学模拟软件SPONGE，为生物物理化学领域的研究提供了理论研究工具和平台；成功开发和集成移动机器人、化学工作站、智能操作系统、科学数据库，研制出数据智能驱动的全流程机器化学家。

（二）分析化学

分析化学是研究物质的化学组成、含量、结构和形态的科学。2020—2023年，我国分析化学在单分子和单细胞分析、活体生物分析、基于功能性核酸的生物分析、纳米生物分析等领域的研究不断发展；在色谱研究，包括样品预处理、色谱固定相及柱技术、多维和集成化以及在复杂样品（蛋白质组、代谢组、中药组等）分离分析等诸多方面取得了显著进展。

（三）无机化学

无机化学的分支包括配位化学、主族元素化学、晶体化学、分子筛、无机纳米材料、原子簇化学及生物无机化学等。针对COF/MOF/HOF材料的研究，我国学者着重探讨晶态化合物的构效关系和产业化应用，在高效率/高选择性的吸附/分离、高效率和高选择性催化、高效发光和高灵敏光电响应、燃料电池 / 氢能 / 储能等领域涌现许多高质量成果；在配位聚合物、无机非金属化合物等晶态化合物的探索、低维晶体材料的制备、晶态材料的功能基团序构等方面也有众多成果涌现；新型分子筛拓扑结构的合成、分析表征技术的开发以及在催化及吸附分离等领域的应用拓展，我国学者均取得了一系列重要突破性进展。

（四）有机化学

近年来，我国有机化学学科取得了长足进步，实现了系列有机化学反应的高效、高选择性转化，在一些研究方向上达到国际领先水平。天然产物化学研究领域逐渐向化学生物学领域融合和转向小分子生物学功能的发掘。

（五）高分子化学

过去 4 年里，我国高分子科学相关研究工作水平不断提高。在生物医用高分子和光电功能高分子领域的研究始终保持强劲的发展势头，在重大疾病治疗相关高分子材料及有机光伏材料方面取得突破性进展。烯烃配位（共）聚合、开环（共）聚合、含硫聚合物合成、可闭合回收高分子等方面也取得一批原创性高水平研究成果。

（六）化学教育学

2020—2023 年，我国化学教育工作者在推进基础化学教育课程改革、培养高水平的中学化学教师、提升高等化学教育教学质量等方面，取得了重要的成果。修订了普通高中化学课程标准、义务教育化学课程标准及 4 个版本的高中教材。我国学者深入探索教材的知识结构、教学目标及学习内容，系统性分析教材设计的原则及方法，评估教学策略、学习活动及组织结构，进一步深化了化学学科能力素养的结构与内涵。借助实施手持技术数字化实验、传感器、信息技术、摄影技术、可视化技术等现代技术手段开展中学化学实验研究。

（七）交叉学科

核化学与放射化学领域，揭示了 Cf–C 键合性质并精确模拟了锎化合物的 UV–vis–NIR 光谱，通过诱导分解—重组策略制备了目前已知的最大的钍系 – 银团簇 [Th_9Ag_{12}]。创新性地设计了可精准匹配六价镅配位构型的保护性基团，实现了六价镅在水溶液和固体中的长时间稳定，取得了至今国际上报道的六价镅和三价镧系之间的最好分离效果。

化学生物学研究经历了跨越式的发展：生物大分子和生命机器实现高效合成、构筑；利用小分子探针、生物正交化学反应等工具，实现生物大分子及生命过程的在体调控；生物大分子动态化学修饰及其参与生命过程的标记与探测；生物大分子机器及生命过程的机制解析；基于化学生物学策略的靶标验证与靶向化合物开发。

公共安全化学方面的毒品代谢机理研究取得重要进展；危险物质检测技术快速化、精准化、智能化等特点凸显；单根毛发微分段毒品检测、污水中毒品监测、爆炸物安全检测等新技术成果得到初步应用。

在环境化学领域开展了大量国际前沿的研究工作，在基于非靶标分析技术发现新污染物、基于稳定同位素的污染物溯源、大气污染的发生机制、水污染治理的新材料研发与制备、土壤污染修复与农产品质量安全、污染物环境毒理与健康等方面的研究工作获得了国内外广泛关注。

分子医学研究在基于核酸适体测序的单细胞蛋白图谱、分子诊断与治疗技术、纳米生物医学、活体生物分析等方向涌现诸多前沿创新成果。

在能源化学基础利学和工程应用领域取得系列突破性进展，纤维锂离子电池、二氧化碳到淀粉的人工合成和海水原位直接电解制氢等重大创新性成果入选了年度中国科学十大进展。

燃烧化学方面，在碳氢燃料、燃烧动力学仿真、燃烧机理和航空动力领域的工程计算软件和数据库方面都取得了长足的进步。

三、本学科国内外研究进展比较

化学是一门涵盖面极其广泛、应用范围几乎包括科技前沿和国计民生的各个方面的学科，本部分按学科分类，对近年来国内外化学学科领域的研究进展进行比较。

（一）物理化学

在化学反应的量子效应研究方面，国内理论和实验研究水平均处于国际领先的状态，但我国的离子—分子反应动力学研究刚刚起步。随着我国自行建造的极紫外自由电子激光投入使用，小分子的光化学动力学的研究得到快速发展，中性团簇红外光谱研究占据了国际气相团簇科学研究的制高点。国际星际化学随着建模发展和一流望远镜的投入使用，新的发现不断涌现。国内由于全冬晖等团队的努力，也发展很快，在此方面在国际处于第一梯队。

目前，国际主要依赖脉冲宽度为纳秒尺度的 193/157 纳米真空紫外激光解离 – 串联质谱技术研究生物大分子及其复合物动态结构和相互作用，在对超级蛋白质复合物等在解离时间和效率上仍然受到限制。极紫外自由电子激光的投入使用实现了对生物大分子的高效解离和序列结构表征，有望实现更大分子量生物体系更高时间分辨率的动态结构和相互作用探测。

在真实体系表面结构及其动力学方面，国际上合频光谱技术发展较快，为表面光电催化领域的应用提供了新的研究手段。我国科学家利用高分辨合频振动光谱研究了界面超分子手性自组装分子机理、复杂生物界面分子结构及分子间相互作用，也取得了重要成果。

在液相结构和动力学研究方面，兆电子伏的液相超快电子衍射（MeV-LUED）是一种研究液相超快动力学的尖端科学仪器，于2020年在美国SLAC国家实验室诞生。2022年，清华大学成功研制出世界第二台MeV-LUED仪器，使我国在该领域跻身世界前列。

在表面反应动力学方面，近年来，国际广为采用分子束—表面散射技术，为人们在原子尺度上理解异相催化过程的微观机理提供了新方法。我国在这一领域尚未开展研究。

（二）分析化学

近年来，随着大脑成像、脑机交互、大数据处理等新技术的不断涌现，脑科学与计算机技术、人工智能等学科交叉融合，类脑研究上升为西方发达国家的科技战略重点之一。我国在分子层面的脑化学研究进展相对缓慢。

未来实现类脑信息处理，需要摒弃0和1的僵化计算语言，构筑新的神经形态器件。流体忆阻器应运而生，通过移动的离子来携带和储存信息。2023年我国和法国学者在《科学》（*Science*）期刊上同期发表相关研究成果，分别描述了纳流体仿神经器件能在接近生物系统的电压和功耗下工作，并实现神经可塑性多功能模拟。

（三）无机化学

在分子筛研究领域，我国科学家产出的科研成果在国际逐渐形成了具有学科引领之势的蓬勃发展势头。在稀土材料科学方面取得了具有创新性的系统成果，实现了利用稀土配合物的结构调控来开发新型光电器件、诊疗探针、高效催化剂和太阳能电池的目标。分子铁电体研究进入化学设计时代，我国学者提出并建立了分子铁电体设计思想，引领了国际上该领域的发展。

（四）有机化学

近年来，我国有机化学学科取得了长足进步，已接近和达到国际先进行列，部分领域得出了引领性的研究成果。一批原创性配体被成功开发出来，实现了系列有机化学反应的高效、高选择性转化，并被国内外学者和企业频繁应用于学术研究和工业生产中。周氏催化剂、冯氏配体和Ullmann-Ma反应，都是其中的杰出代表。

（五）高分子化学

2020年，世界范围的高分子学者都在讨论高分子需要关注的研究前沿，重要的共识体现在3个方面：发展新的高分子合成方法、发现高分子的高级性质和功能、实现高分子材料可持续性发展。高分子化学呈现通过合成化学的发展创造“双碳”时代所必需的可循环或可回收高分子材料的发展模式。我国学者也关注到这一重要领域，在高性能可闭合回收聚合物，包括聚酯、聚碳硫酯、聚硫酯、聚氨酯及交联聚合物等方面取得一批原创性成果，处于和国际同行并行竞争的行列。

（六）交叉学科·其他学科

1. 化学生物学

生物正交反应是指可以在生物体系中进行且不会与天然生物化学过程相互干扰的一类化学反应，是化学生物学领域发展的一个重要方向。我国在生物正交反应方面的研究、开发与应用方面具有扎实的研究基础。目前处在生物正交反应由活细胞向活体动物升级的关键时期，抓住机遇发展用于活体动物层面的生物正交反应，并用于在活体动物中研究生物学问题、揭示疾病发生发展机制等，将进一步巩固我国在该领域的领先优势。

2. 分子医学

近年来，基于功能核酸的分子医学已成为许多国外研究机构和制药公司研发的重点和热点，被称为“化学家的抗体”的核酸适体具有广阔的应用前景。中国科学家在基于功能核酸的分子医学领域处于国际领先地位，率先发展了以活细胞为靶标的功能核酸 Cell-SELEX 筛选新技术，针对多种癌细胞筛选出了多种功能核酸组，解决了如何在疾病标志物未知的条件下，获得特异识别病变细胞及其他复杂生命体系的分子探针的关键科学问题。研发的骨硬化素核酸适体获得美国 FDA 孤儿药认定，用于治疗成骨不全症，是核酸适体药物领域的一项重要突破。

四、本学科发展趋势和展望

随着我国国民经济发展进入新常态，中国的经济已经从对量的追求转为对质的提升，增强经济发展的科技内涵，处理好发展与环境的关系。对于化学学科来讲，就是要面向国际科学前沿，面向国民经济主战场，面向新常态经济下在资源、环境、健康等方面提出的重大课题，在国际化学领域实现从“跟跑者”向“并行者”和“领跑者”的转变。

（一）物理化学

将人工智能、理论与实验相结合，揭示、预测所研究体系的过程和机制、发展新的物理化学概念、原理、理论和技术，指导实验和生产，是物理化学各个分支共同的发展趋势和战略需求。未来 5 年，我国在物理化学领域的重点发展方向包括：①建立通用的高效电子结构计算与分子模拟平台，发展适用于复杂体系和过程的多尺度模拟方法；②发展新型催化、分子组装、生物及仿生材料，解决学科自身发展和国家能源、环境、医药以及国防需求；③发展具有超高时空分辨力的原位表征技术和适用于原子、分子、纳米层次的热力学表征技术；④人工智能指导下的研究范式改变，为实验研究提供新的理论洞察力和实践基础。

（二）分析化学

基于纳米孔、等离激元纳米颗粒、荧光纳米探针的分析应用以及纳米传感器等纳米分析的研究是分析化学的热点领域，如何在复杂的机体中实现对低浓度物质的灵敏、时空精确的成像是今后的研究重点。活体分析化学已经成为分析化学重要的前沿领域之一。流体忆阻器为化学和神经形态器件的深度交叉融合带来了全新的机遇，二维材料、纳米管、凝胶等新材料为基于流体忆阻器的神经形态器件的多功能性提供了无限可能。

在色谱理论和方法上，应继续对色谱保留值、色谱峰形的影响因素进行研究，并发展多元混合物分离新理论优化特定任务和目的的色谱分离条件。如何面对生命健康，特别是多组学分析领域不同生物分子分离的需求，以及面对公共应急事件及时发展基于色谱的分析新方法，也是需要注意的问题。

近年来，我国分析化学基础研究已经取得了长足的进步，然而我国在相关研究领域还有相当大的提升空间，未来还必须加强分析仪器装置研究的原始创新性工作，摆脱目前高精尖分析仪器过度依赖进口的局面。

（三）无机化学

分子筛是低碳催化及催化剂领域关键核心材料，未来五年面向国家在能源、化工、环境等领域对新型分子筛催化材料创制的重大需求，分子筛学科的重点发展方向是实现系列新型分子筛催化材料的创制、拓展分子筛催化材料在能源、环境、光电、生物质转化和其他领域的高效应用。

晶体化学与材料领域亟须深入探索如何实现 COFs/MOFs/HOFs、簇合物 / 功能配合物、超分子晶态化合物、块体功能晶体材料、低维晶态材料等体系的晶体结构和电子结构的精确解析，理解材料从微观原子到宏观块材的演变是晶体化学研究的未来重点发展方向。

结合人工智能和机器学习技术，探索并发展出高效 / 高选择性的催化材料、新颖光电磁功能材料、智能材料、新药物、医用材料等，在国民经济、国防安全等领域获得广泛应用。

（四）有机化学

在有机合成方法学方面，通过设计合成新型配体 / 催化剂，发展更高效的反应和合成策略，构建绿色可持续、精准可控的高效催化体系，耦合光、电和磁等外场调控，加强与生物技术的融合，实现重要功能分子的精准创制，为重要精细化学品的生产提供理论基础与技术支撑。以人工智能为代表的新兴技术涌现为有机合成方法学和天然产物化学的研究范式的变革提供了强有力的基础。我国有机化学工作者要主动积极投入“基于人工智能的合成化学”研究中。

（五）高分子化学

今后 5 年，高分子科学的发展趋势包括：实现有明确分子量、分散性、组成、序列、立体及拓扑结构等精密聚合物的高效定制合成；开展基于物理和数据驱动的计算方法在预测设计新的高分子结构和功能以及阐明新的高分子物理学方面的应用研究；全面和预测性地了解高分子复合体系在所有相关尺度上的行为，为多组分高性能高分子材料的制造和加工奠定基础。

（六）化学教育学

加强化学教育领域与国际化学教育及科学教育的对话，提高研究者的国际交流及发表经历，提高研究者的国际化培养经历；加强化学教育领域与国家科技发展战略的对接，形成全维的化学教育学科体系；加强化学教育领域与国家教育改革战略的对接使得化学教育学科建设能够在国家教育改革、国家教育研究中真正发挥作用；加强大数据及信息技术在化学教育中的应用，为教师提供针对性的指导及反馈，进一步提高学生的学习成果。

（七）交叉学科及其他学科

未来 5 年内，核化学和放射化学学科的发展将迎来爆发，预期在配位环境对锕系元素光谱的影响相关理论化学、三价锕镧及镅锔分离化学、放射性废物处理及处置化学、支撑乏燃料后处理和高放废液综合利用的核燃料循环化学、支撑放射性药物研发的核化学及标记化合物化学等方面都将取得显著进展。

研究生物大分子动态修饰已成为生物化学中备受关注的前沿方向之一，发展原位标记和时空探测技术对于此领域具有重要的意义，利用化学工具来探索免疫识别过程、调控免疫识别信号等方面具有重要的意义。这些方向代表了化学生物学领域未来发展的重要趋势。

环境化学方面，随着一些化学物质的环境和人类健康危害被认识而被限制使用，一些新的化学替代物出现，加强新型化学替代物的识别、监测和健康风险评估研究具有重要意义。人工智能的发展为环境计算毒理学的建模带来了机器学习技术，为揭示复杂非线性关系带来了契机，需进一步促进学术界、政府部分以及企业界间的数据共享，从顶层设计构建数据库，进一步规范数据的收集、注释和存档，平衡阴性与阳性数据数量，为机器学习准备高质量数据，提升模型预测效率。

分子医学专业方面，发展核酸适体等功能核酸高效筛选方法，构建核酸分子探针，突破生物医学研究中的瓶颈问题；将分子组装行为与新兴纳米技术、纳米材料有机结合，为生物传感器、分子器件、高效药物载运、智能医用材料等领域的发展提供重要基础；利用新型分子识别探针，发展单分子水平的高灵敏生物医学分析新方法，研发高度特异性识别癌症、病毒感染、病原菌感染等相关疾病标志基因的工具酶，构建精准、无创体外诊断新体系：针对以分子靶向治疗及免疫治疗为代表的精准治疗临床医学需求，发展起多种动、植物天然药物分子的分离鉴定和新型药物有机合成新方法，为生物医药领域提供新技术方法。

未来 5 年，能源化学的研究将继续围绕化石资源清洁高效利用与耦合替代、清洁能源多能互补与规模应用、能源化学前沿科学三条主线开展科学研究布局，为推动能源革命和实现“双碳”目标奠定坚实基础。

第四节　水环境

一、引言

水环境是地球表层系统各圈层水体所处环境的总称。水环境学科是研究水环境系统的特点及其演变规律、机制及其调控理论、方法和技术的学科。根据研究对象不同，水环境学科可分为湖泊环境科学、河流环境科学、地下水环境科学和城市水环境科学等；根据研究目的不同，可分为水环境监测、水环境评价、水污染治理、水环境管理和水生态保护等。近年来，我国水生态环境质量持续改善，水生态环境保护工作取得明显成效，水环境科学在理论、方法、技术等方面呈现新的发展趋势。理论上，可持续发展理论成为研究水环境的指导性思想，水文学、水动力学、水化学、环境水力学、水污染处理等水环境传统学科与生态学、经济学等学科深入交叉。方法上，水环境科学将多种水环境要素整合，面向流域及区域系统开展研究。技术上，多学科交叉、已有专业技术有机集成，高新技术不断应用于水环境专业领域，人工智能、大数据等新兴技术在水环境领域中不断得到应用，极大地提高了水环境科学的研究深度与广度。

二、本学科近年的最新研究进展

我国水环境污染问题依然严峻，成为生态文明建设的突出短板。治理水污染、保护水环境，关系人民福祉和国家未来。以习近平同志为核心的党中央把水生态环境保护摆在生态文明建设的重要位置，把解决突出水生态环境问题作为民生优先领域，把打好碧水保卫战列为污染防治攻坚战的三大保卫战之一。“十四五”时期，我国水生态环境保护事业进入新阶段，水生态环境保护由污染治理为主向水资源、水生态、水环境协同治理、统筹推进转变。

我国水环境问题的特殊性、复杂性和解决水环境问题的紧迫性，成为水环境学科快速发展的强大动力。水环境学科在我国推进生态文明、建设美丽中国进程中的地位愈发凸显，已逐渐成为支撑水生态环境高水平保护、推动高质量发展、创造高品质生活不可或缺的战略性关键学科。水环境学科近年的最新研究进展主要体现在以下方面。

（一）控制对象更加关注高风险新污染物识别控制和复合污染物的协同控制

水环境污染物种类不断增加，新污染物不断涌现，复合型污染特征日益凸显。关注常规单一污染物在水环境中的污染形成机制与控制原理，已不能满足日益复杂的水环境污染治理需求。因此，高风险新污染物识别、微观转化机制与控制原理、水质标准制定基础理论、复合污染物协同控制理论与技术等成为水环境学科的主要发展方向。

（二）控制目标更加关注水生态保护修复和水生态健康保障

我国水环境治理的目标已经从常规污染物排放量削减，向水环境质量改善、水生态保护修复和水生态健康保障发展。因此，水生态修复与安全保障理论和技术体系、水生态健康评价理论、方法和技术已成为新的发展需求。

（三）控制手段更加关注减污降碳和资源循环利用新理论和新技术

低效高耗的水环境污染物末端治理模式，已不符合我国社会经济高质量发展的新需求。水污染物全流程防控、减污降碳协同增效理论与技术、水生态循环利用以及资源能源高效回收利用理论与技术的突破越来越受到重视，实现资源循环利用、全流程控制与精细化管理是水环境学科未来发展的必然要求。

（四）研究方法向微观解析和宏观模拟发展

水污染物迁移转化研究方法向电子转移跟踪、超微结构解析、微纳米界面观测发展，水污染物生态效应研究方法向分子、细胞、微生物群落方向发展，水环境系统模拟方法向区域模拟、流域模拟和全球尺度模拟发展，大数据信息化手段的应用将会更加广泛和深入。

（五）理论创新更加关注与新兴学科的深入交叉与深度融合

水环境学科基础理论与前沿技术的发展与现代生物技术、信息技术、生物技术、新能源技术、新材料和先进制造技术等的交叉融合越来越深入和紧密。多元化的学科交叉和大数据、人工智能等新兴技术的引入，为水环境学科基础理论的原始创新、颠覆性技术的突破和多尺度、跨流域、跨区域水生态环境问题综合解决方案的制定提供了强劲动力。

三、本学科国内外研究进展比较

从水环境学科整体研究进展看，污水资源能源回收、水系统低碳优化设计运行、人工智能智慧化手段应用等发展方向既是国际研究前沿，也是国内的研究热点。同时，针对“十四五”时期我国水生态环境保护事业的新需求，水环境学科的研究重点也逐渐由污染治理为主向水资源、水生态、水环境协同治理、统筹推进转变。

在水质水生态评价与环境基准标准制定理论领域，面向复杂环境的综合性指标、区域差异化基准标准和系统性评价研究是发展的主要趋势。当前，国内外传统的水质指标侧重物理和化学性质评价，对水的组分特征及其效应关注不够。基于水环境复杂体系的非线性变化特征和水处理过程中的质变特性，我国学者提出了“水征”的概念，为认识水质水生态状况提供了更加全面的评价视角。

发达国家已经建立了相对完善的水质基准标准技术方法和体系，我国对水质基准标准的研究起步相对较晚，但发展迅速。自“十一五”以来，我国生态环境基准工作取得了突破性进展，基于我国区域差异化特征陆续发布了一系列水质基准及其制定技术指南；经过多年的发展和修订完善，我国水质标准也已逐渐形成了符合我国国情的成套体系。我国水质基准标准的研制水平基本达到甚至超过美国、欧盟等发达国家或地区。生态系统健康评价的研究始于20世纪80年代，但国际上对“生态系统健康”迄今尚无统一观点。在我国，经过不断深入研究，对生态系统健康评价的研究由河流扩展到湖泊、水库等多种水环境类型，逐渐形成了自己的体系，可为制定流域生态修复目标、评估生态修复效果以及环境立法与执法提供数据支撑。

在水处理理论与技术领域，国际水处理领域呈现出减污降碳协同增效的发展趋势。随着经济社会发展，污染物种类日趋复杂，同时公众环境意识增强，对水环境质量要求不断提高。因此，一些发达国家的污水处理厂正在由生物脱氮除磷向强化脱氮除磷方向发展。同时，高级氧化、纳滤、反渗透等深度处理技术的应用也越来越广泛，以达到对环境内分泌干扰物、药物和个人护理品等新兴污染物的去除，满足更加健康安全的水环境质量需求。低碳处理和能源开发、气候变化问题和能源危机要求城市污水处理实现低碳化，在处理过程中实现节能降耗，提高能源自给率。欧美发达国家已针对各自国情，就再生水利用、污水生物质能回收、氮磷回收等领域展开各有侧重的研究。国内同样开展了以城市污水资源概念厂为代表的系列研发工作，以期推动水处理行业的转型升级。

在湖泊治理理论与技术领域，湖泊生态系统健康和生态功能全面恢复正在成为新的目标和要求，因此湖泊治理正在由单纯的水质治理到水质与水生态协同改善转变。如何在流域社会经济快速发展的同时，实现湖泊更高质量保护治理目标，是目前国内外需要重点解决的问题。近年来，国内湖泊治理的研究及实践，结合传统水质治理理论与技术体系，借鉴发达国家湖泊治理的路径，也在不断地探索湖泊生态恢复的新机理和新方法，以满足现在及未来治理的需求，湖泊保护修复的研究与实践水平基本达到甚至超过发达国家或地区。一方面，湖泊流域环境治理技术水平切实提升，在通过对传统技术进行优化以获得最佳治理效能同时不断融入最新理论与技术，寻找湖泊治理的技术增量，提高湖泊治理科技支撑能力。如新型钝化剂，新型湿地、生物栖息地修复构造技术和基于物联网、卫星遥感、人工智能、eDNA 的湖泊环境监测新技术等都在湖泊治理过程中取得了较好的应用效果。另一方面，湖泊流域协同治理的理念不断强化，优化调整流域产业结构，加强水资源科学调度和实施湖泊生态修复工程，以此推进湖泊治理相关立法、政策、规划、标准等管理措施与治理技术的有机结合，逐渐形成系统的治理体系。

四、本学科发展趋势和展望

未来，水环境学科将聚焦国家生态文明建设重大战略需求和国际学术前沿，针对我国城镇化进程和社会发展中显现的水环境问题，贯彻绿色低碳和可持续发展理念，推动前瞻性、原创性、颠覆性理论创新和技术突破，不断丰富和完善水环境学科基础理论、方法体系和技术体系，形成创新思想丰富、创新人才集聚、原始创新成果辈出、复杂水环境问题解决能力提升的学科发展新局面。“十四五”乃至更长一段时期，水环境学科的发展应注重长远布局、合理规划和重点突破。为促进我国水环境学科的未来发展，提出如下建议。

1）基础性研究：水环境学科的研究方法、水环境标准制定方法、水环境测定标准物质、基本概念、术语定义和基础数据等，在水环境领域具有基础性地位，对学科发展具有重要奠基作用和推动作用，但我国在这方面的研究和积累仍然不够，对学科发展的基础性贡献有待提高。

2）系统性研究：截至 2023 年，我国的水环境学科研究，大多聚焦在水污染控制技术原理方面，“技术孤岛”现象突出，对水污染控制技术集成理论、组合工艺设计与控制理论、饮用水全流程安全保障理论、城乡一体化供排水系统建设理论、重点流域水环境治理系统整体优化理论、区域再生水循环利用系统建设理论等方面的研究仍然不足，水环境学科研究的系统性亟待提高。

3）发现型研究：发现新的水环境问题和新的污染产生机制，对保障水环境安全、促进学科发展具有突破性或颠覆性意义，我国在解决已有水环境问题方面具有较为明显的优势，但是发现的国际上公认的新问题、新现象和新机制还十分有限，面向国际学术前沿以及我国重大需求，亟须提出和实施解决水环境问题的中国思路，引领未来发展。

第五节　制冷及低温工程

一、引言

制冷及低温技术是指利用人工方法获得低于环境温度的技术。其学科基本任务是研究获得并保持不同于自然界温湿度环境的原理、技术和设备，并将这些技术用于不同场景。其范畴包含狭义的冷量输出以维持低温度，也包括除湿、环境参数调节和热泵等技术。制冷及低温技术几乎与国民经济所有部门密切相关，与人民生活紧密联系。此外，低温技术是物理学前沿研究不可或缺的手段，在医疗卫生、工业技术、资源环境、空间技术等领域也有着广泛的应用。

在环境问题凸显和各国科技竞争加剧的背景下，众多重大全球性议题和国家政策的实施以及科学问题的解决依赖于制冷低温科学的发展，因此制冷及低温技术越发受到各国的重视。当前，在政策和市场驱动下，传统制冷技术面临强制性更新换代，新型制冷和低温技术迅猛发展。及时根据国家发展战略调整学科重点发展方向，同时跟进国际新兴研究方向保持完善的技术储备，对于保证学科的高速发展以及学科对国民经济发展和国家战略实施的可持续支撑至关重要。

二、本学科近年的最新研究进展

近年来，制冷及低温工程学科围绕国际科技前沿和国家重大需求，从学科研究、应用和发展的角度出发，在制冷和低温技术等方面都涌现了一批科学意义重大且社会经济效益显著的科技成果，为城乡建设、清洁供暖、余热回收、新能源利用、物流、新能源汽车、空分、天然气液化等众多领域的发展起到了强有力的支撑作用，尤其助力我国先端物理、量子技术、超导等前沿科技领域快速发展。

在制冷技术方面，制冷工质替代是研究热点，现阶段已进入第四代制冷工质替代的关键时期；而蒸气压缩制冷技术在循环的设计与应用、压缩机与换热器性能的改进和智能调节方面也取得了一定进展。在热驱动制冷及热泵技术方面，吸收式制冷与热泵技术主要发展了两类吸收式换热器并推进了新型热泵与制冷系统的工程应用；吸附式制冷与热泵技术则开发了多卤化物等高性能复合吸附材料并构建了多级复合循环系统；而热声制冷与热泵技术侧重于研制双行波式、气－液/固耦合式热声制冷系统。在固态制冷技术方面，镁基热电材料、微型高热流密度热电制冷器件和柔性制冷器件得到了快速发展；使用增材制造技术可制备形、性可控的弹热工质，长寿命、低场强电卡材料取得了突破，无流体的紧凑式电卡制冷器件性能得到了大幅度提升；磁制冷机在室温和低温均做出了产业化尝试；弹热制冷系统正向千瓦级制冷量稳步发展。

制冷技术应用方面的研究在近年快速发展。在空调系统制冷技术中，室内热湿环境控制发挥着关键作用，包括温湿度独立控制、系统流程创新、新型干燥剂等；空调系统的关键部件研究也尤为重要，包括磁悬浮、气悬浮等冷水机组产品升级以及换热器的优化等；在产品层面上，低环境温度空气源热泵、多联机系统发展迅速，而线性能测量、大数据技术的发展降低了系统能耗和维护成本。在冷链装备技术方面，主要进展包括基于流态冰的差压预冷装备、移动式压差预冷装备、智能立体冻结隧道等。高温热泵领域最新研究进展主要包括采用双螺杆压缩机和喷水冷却的高效水蒸气高温热泵、兼顾低温热源利用和高温输出的大温升复叠压缩式热泵循环、大温升吸收－压缩耦合式热泵循环以及开式闭式结合的高温热泵直接蒸汽供应技术。在车用热泵领域，近几年涌现了车辆热管理无霜运行控制策略等热管理新技术，部分车型推广应用了电池直冷直热技术，而热管理回路高度集成模块产品的使用也愈发广泛。

低温技术的发展同样令人瞩目。在空分技术方面，近年的研究主要聚焦于高精度的低温混合工质物性预测、高效的空分部机研制以及灵活可靠的大型空分优化控制等；以杭氧、川空、开封空分为代表的民族企业已实现大型、特大型空分设备成套技术国产化。对于液化天然气技术，全球范围内的大型天然气液化工厂流程选择高度集中，我国在小型和非常规天然气液化领域有丰富实践，用于海上气田的浮式液化天然气近年来发展迅速，大型化和海上浮式储存与再气化装置（FSRU）是液化天然气储运领域的关注重点。在大型低温系统方面，我国先后完成百瓦、千瓦液氢和液氦温区制冷机的研制工作。在回热式制冷机方面，我国在微型化、深低温和大冷量制冷方面均已实现了产品化，国产 4 开 G-M 制冷机开始服务于国家高科技前沿领域，单级斯特林脉管制冷机已实现空间在轨应用；在 1 开以下极低温制冷方面，目前吸附制冷和磁制冷技术还处于研发阶段；而稀释制冷机最低温能达到 10 毫开以下，并开始了产品化推广。在低温冷冻治疗及保存领域，多模态冷冻治疗和仿生绿色化冷冻保存是低温生物医学领域的研究热点。多模态冷冻治疗旨在联合低温、纳米、生物医学等技术提高抗癌疗效；仿生绿色保存旨在优化保存过程，扩大保存维度，同时积极发展生物样本库建设。

三、本学科国内外研究进展比较

近年来，我国在制冷及低温学科持续快速发展，与国外先进水平的差距不断缩小，在部分领域处于优势地位。但在基础制冷技术方面缺乏原始创新，在关键技术领域仍存在“卡脖子”问题。

在制冷技术方面，我国在替代技术、标准制订等领域都取得了卓越的成果。欧盟、美国和日本在制冷工质替代方面走在世界前列，制定了大量的政策、法规和配套的行动措施。在热驱动制冷和热泵方面，国内在吸收式热泵与压缩式热泵耦合系统的工程应用以及吸收式制冷机的小型化方面领先于国外；在吸附制冷技术中吸附剂开发和高效循环及系统构建方面的发展快于国外；而国内在传统热声制冷系统的研究处于主导地位，新型湿式热声制冷循环正成为国际研究热点。对于固态制冷技术，尽管这类研究始于国外，但国内在各固态制冷技术的材料和器件研究方向达到了与国际同行并驾齐驱的水平。

制冷技术服务于国民经济的众多领域。我国是世界上最大的空调系统生产制造国，但仍需补齐短板。在冷链装备技术方面，冷加工领域内的智能立体冻结隧道等少数设备性能已优于国外主流产品，但大多还存在差距；冷冻冷藏领域的氨制冷剂应用技术以及冷藏运输领域的信息化与国外相比仍存在差距；而冷藏销售领域整体上处于国际先进水平。在高温热泵方面，国外对天然工质和低 GWP 工质的高温热泵以及兆瓦级工业余热源高温热泵进行了研究；国内总体起步较晚，但也出现了空气源高温热泵的特色技术。在车用热泵方面，随着欧洲 P-Fas 法案的出台，R-1234yf 等 HFO 技术路线热度衰减，国内主流企业主导二氧化碳作为制冷剂，而国外企业也有了二氧化碳热管理量产产品，并开始针对 R-290 开展研究。

低温技术为工业领域和先端科学提供重要支撑。在空分方面，民族企业已实现大型、特大型空分设备成套技术国产化，并成功抢占外资企业的国内市场；总体来讲，我国低温空分技术已达到与外资企业同台竞争的水平。在液化天然气技术方面，国内在小型（包括撬装式）和非常规天然气液化装置方面已进行了大量工程实践探索，但迄今尚无大型装置实践；国外对浮式液化天然气的关注较多，国内尚无具体案例；国内在大型液化天然气储罐和运输船建设方面正在接近世界先进水平。在大型低温系统方面，欧美国家已经具备成熟的大型低温系统研制经验；我国在这一领域尽管起步较晚，但已具备 2500W@4.5K 制冷机和 500W@2K 制冷机的自主研制能力。在微小型低温制冷系统方面，回热式制冷机在微型化、深低温、大冷量方面基本有与国外水平相当的样机和产品，但在产品化方面，G-M 脉管制冷机、稀释制冷机等与国外还存在一定差距。在冷冻医疗和保存方面，冷冻治疗的应用规模与国外仍有差距，但关键技术研发居领跑地位；冷冻保存在仿生材料开发、小尺度样本保存方面居并跑甚至领跑状态，但对大尺度样本保存研究仍需追赶；生物样本库建设迅速，但仍需向国际先进水平学习。

四、本学科发展趋势和展望

整体来讲，学科内交叉渗透融合趋势日益明显，不断涌现高水平原创性基础研究成果，关键核心技术与国外先进水平差距越来越小。未来，将进一步增强学科融合，增进原始创新，发展关键器件和基础技术，摆脱对国外的依赖，更好地服务国家战略和经济建设。

对于制冷技术，为了助力我国“双碳”目标的完成，制冷空调行业应当进一步探索全链条制冷工质减排技术方案，同时加快天然 / 替代制冷工质循环优化和制冷关键部件的开发，并且推动人工智能和信息化全价值链应用技术的发展。对于热驱动制冷和热泵技术，吸收式热泵技术应重点发展热量灵活变换的新流程和系统；吸附式制冷及热泵技术则注重变热源和环境温度下的高适应性工质对和多效循环系统的研发；而热声制冷系统应提升声场调控水平，开发高性能交变流动换热器以及强化传质的新型热声制冷循环。对于固态制冷技术，热电制冷有望在高热流密度、高控温精度、柔性热管理等场景得到进一步发展；三种固态相变制冷技术的材料和系统性能预计将得到大幅提升，应用场景仍有待探索。

在制冷技术的应用方面，大众最为熟知的是应用于住宅的空调制冷技术，未来研究的热点是降低住宅建筑空调系统的碳排放。在冷链装备技术方面，未来的发展趋势主要包括绿色低碳、健康安全、精准环控、信息化与智能化。在高温热泵方面，发展趋势主要包含基于高温热泵的高温热能和蒸汽供应、面向工业余热就地消纳利用的热泵技术。在车用热泵方面，未来的发展趋势主要包括绿色高效化、功能一体化、结构模块化和控制智能化。

关于低温技术，在空分技术方面，为实现低温空分设备的大型化、节能化、智能化，亟须开发新流程、新部机设计以及新运行控制策略。在液化天然气方面，海上天然气液化、液化天然气领域碳减排、非常规天然气液化和液化流程模拟优化算法可能成为研究热点；对大型气化器国产化研究仍需持续发力。在大型低温系统方面，重点突破氢氦压缩、系统流程与集成调试、透平膨胀机与冷压缩机、低温储运技术和技术标准建设，研制万瓦级液氦温区制冷机和千瓦级超流氦温区制冷机。在微小型低温制冷系统方面，微型化、深低温、大冷量是回热式制冷机发展的主要方向；此外，需要进一步提高 4 开 J–T 制冷机的有阀线性压缩机性能，以及 1 开以下极低温区稀释制冷机的研发水平和产品化水平。在冷冻医疗和保存方面，未来将推进精准低温生物医学建设，实现能量与物质的精准调节与靶向可控，同时在现有基础上优化样本库建设水平。

第六节 人工智能

一、引言

人工智能是一门利用计算技术对世界上的智能行为进行模拟的科学，其利用知识计算、机器学习、深度学习与大模型、自然语言处理、计算机视觉、语音处理、信息检索、多智能体系、具身智能、对抗技术等技术，让计算机帮助或替代人类高效完成部分智能任务，为人类社会进入智能时代发展提供重要技术与理论支撑。

二、本学科近年的最新研究进展

由于人工智能学科持续且高速的发展，新的研究方向与课题层出不穷，乃至学科本身都处于更新之中。因此，下面从目前发展热门的较大学科分支来展示整个人工智能学科的最新发展。

（一）知识计算

随着人工智能的蓬勃发展，知识表示从传统符号知识模型进一步扩展到隐知识模型，近年来隐知识模型地位有了极大的提升。传统符号知识是显式的符号表示，而隐知识通常隐藏在非符号化非结构化的数据之中。近期研究进展主要集中于神经网络的知识表示、大语言模型、暗知识的提取与表达、知识图谱的搜集与构建等方面。

（二）机器学习

近年来，机器学习在工业界的实际应用为机器学习的发展注入了新鲜的血液。以自监督学习为代表的无监督训练方式使得模型对数据的利用和处理能力得到了极大提升，从而引导机器学习算法从实际应用中不断取得突破。在蛋白质结构预测等领域，机器学习模型显示出巨大的优势，强化学习、终身学习等方法也让机器学习算法的适应性显著提升。

（三）深度学习与大模型

深度学习技术通过模拟人脑神经网络结构，从大量数据中自动学习特征并进行复杂任务建模。大模型则具有庞大的参数量，在处理大规模、复杂任务方面具有显著优势。二者的结合为人工智能领域带来了诸多创新和突破。在计算机视觉领域，深度学习技术已经广泛应用于物体识别、目标检测、语义分割等任务，使得计算机能够像人类一样对图像和视频进行理解和分析。

（四）自然语言处理

自然语言处理在预训练语言模型、情感分析、机器翻译和文本生成等方面取得了显著成果。BERT、GPT 等模型在大规模语料上的训练，为自然语言处理任务提供了高质量的表示方式，使得模型在下游任务上取得了更好的性能。情感分析是自然语言处理领域的一个重要应用。深度学习技术已经能够自动地从文本中提取情感信息。机器翻译则是自然语言处理领域的另一个重要方向。

（五）计算机视觉

计算机视觉是利用计算机模拟人类视觉功能的技术，旨在从图像或视频中获取信息并进行处理和分析。在目标检测和识别方面，基于深度学习的算法不断刷新准确率记录。计算机视觉在图像生成、增强和超分辨率方面也取得了重要突破。生成对抗网络（GAN）的提出为图像生成提供了新的思路，能够生成高质量、多样化的图片。同时，图像增强技术也得到了发展，通过调整图像的属性来改善图像质量，为后续的计算机视觉任务提供了更好的基础。

（六）语音处理

语音处理通过计算机技术对人类语音进行采集、处理、分析和理解。深度学习算法的引入和应用，使语音识别的准确率得到显著提高。此外，语音合成技术也取得重要进展，通过生成对抗网络等技术，能够生成更加自然、逼真的语音，为语音交互提供了更加真实、流畅的体验。

（七）信息检索

近年来，信息检索在基于深度学习的信息检索技术、跨语言信息检索、多模态信息检索等方面取得了较大进展。深度学习模型被应用于文本表示学习，从而提高信息检索的准确性和效率。随着跨语言信息检索的需求日益增加，通过机器翻译、双语词典等技术手段，实现了不同语言之间的信息检索，提高了信息检索的适用范围。多模态信息检索通过将不同类型的信息进行融合，实现了多模态信息检索，提高了信息检索的丰富性和多样性。

（八）多智能体系统

多智能体系统是一种分布式人工智能系统，由多个独立决策的智能体组成，通过协作与通信解决复杂问题。在协同控制方面，取得了如一致性算法的收敛性分析、针对时延和不确定性的鲁棒控制策略等成果。在智能体建模方面，基于大型语言模型的智能体展现出强大的感知、决策和行动能力，为人工通用智能的实现提供了新方向。

（九）具身智能

具身智能正不断突破涵盖了感知与动作的集成、认知建模的深化以及智能体设计与制造的创新。智能体的感知与动作融合技术已经能够实现高效的环境感知和精确控制，而在认知建模方面的进展使得智能体能够模拟人类的适应和学习机制，以应对复杂多变的环境。在设计与制造领域，利用三维打印和软机器人技术，智能体能够拥有更加灵活和适应性强的物理形态。

（十）AI 安全

近年来，AI 安全技术在对抗攻击防御、隐私保护以及模型安全性评估等方面取得了显著的研究进展。在对抗攻击防御方面，已开发出多种策略来识别和抵御欺骗 AI 系统的恶意输入。在数据隐私的维护方面，差分隐私和同态加密等技术已被广泛应用，以确保数据在存储和传输过程中的安全。模型安全性评估方面也取得了突破性进展，新的评估框架和工具能够从多个维度全面评测 AI 模型的安全风险。

三、本学科国内外研究进展比较

人工智能作为引领新一轮科技革命和产业变革的战略性新兴技术，在推动科技进步、产业升级、经济发展等方面发挥着重要作用，正成为各国竞相发展的重点领域。在国内外，人工智能的应用领域和发展水平都有很大差异。

（一）知识计算

我国在知识计算领域的研究虽起步较晚，但已取得显著成就。在知识表示与建模方面，我国逐渐形成了具有特色的知识表示方法，不断提升知识的表达与建模能力。在知识获取与推理技术上，我国紧跟大数据时代的步伐，取得了实质性进展，尤其在数据规模和处理速度上表现出明显优势。

（二）机器学习

跟国外相比，我国在算法优化和传统机器学习应用方面取得了一定进展，特别是在大数据处理和模式识别等领域表现出诸多优势。然而，与国外相比，我国在机器学习理论创新和前沿技术探索上仍存在差距，尤其是在深度学习、强化学习等领域的研究深度和广度上。

（三）深度学习与大模型

我国在算法优化、模型压缩和应用落地方面展现出了明显的优势，特别是在计算机视觉等领域的商业化应用方面取得了突破。然而，与国外相比，我国在大模型的原创性研究、模型训练的规模和效率以及模型的泛化能力和鲁棒性方面仍存在不足。尽管国内在预训练模型方面取得了一定进展，但在模型的国际影响力、通用性以及引领全球技术发展趋势方面，仍有较大差距。

（四）自然语言处理

与国外相比，我国在自然语言处理的理论深度、技术创新和跨语言处理能力上仍有差距。在应用技术方面，我国已成功地将自然语言处理技术应用于诸多领域，如机器翻译、情感分析、信息提取等，展现出较强的市场适应能力，但在自然语言理解的深度和广度以及自然语言生成的原创性和多样性方面，仍有较大的提升空间。

（五）计算机视觉

我国在视觉特征表示、稀疏编码、视觉注意力机制等方面开展了深入研究，取得了一系列重要成果。在视觉 Transformer 架构的探索与应用上，我国虽起步较晚，但已在多模态融合、视频理解等领域取得了一定突破。目前，在基础理论研究与部分尖端技术领域，如视觉推理、三维视觉、跨模态视觉理解等方面，我国仍待进一步加强。

（六）语音处理

在语音识别、语音合成、说话人识别等关键技术方面，我国已达到国际领先水平，尤其在中文语音识别和语音合成方面优势明显。然而，与国外相比，我国在语音处理的基础理论研究、原创性算法创新以及跨语种、多场景的适应性等方面仍存在一定差距。此外，国内在语音识别的准确性、鲁棒性以及自然语言理解等复杂问题上，仍有较大的提升空间。

（七）信息检索

与国外相比，我国在中文分词、词性标注、命名实体识别等技术上具有明显优势，在学术搜索引擎、知识图谱、问答系统等应用领域也取得了一系列重要成果。然而，我国在信息检索的基础理论研究、原创性算法创新以及跨语言、跨领域的适应性方面仍存在一定差距。此外，国内在检索结果的准确性、鲁棒性以及实时性等复杂问题上，相较于国外仍有较大差距。

（八）多智能体系统

我国在多智能体系统领域的技术研发进展迅速，尤其在多智能体强化学习算法优化及复杂环境下的协同控制方面具有一定优势。我国在通信协议设计与信息共享策略上取得创新，有利于提升多智能体系统内的协作效率和稳定性。然而，我国在多智能体系统的基础理论研究及实验设施建设上尚存不足，如部分尖端技术如分布式优化理论、多智能体博弈论等领域的研究深度还有待加强。

（九）具身智能

我国在具身智能领域展现出积极态势，尤其在机器人感知与控制、虚拟现实与增强现实的交互应用上形成了自身特色与优势。在机器人技术层面，我国在力觉传感、视觉识别以及运动控制算法等方面取得重要突破。在虚拟现实和增强现实技术中融合 AI 具身智能，成功实现了高精度的人体动作捕捉与自然交互体验。然而，我国在仿生材料与结构设计、脑机接口技术以及智能体在复杂环境下的长期自主学习和适应性优化等方面仍有待提升。

（十）AI 安全

在 AI 安全领域，我国的技术研究取得了一定的进展。在模型安全与鲁棒性方面，我国研发了一系列针对深度学习模型的对抗样本防御技术，有效增强了模型对噪声干扰和恶意攻击的抵抗力。同时，我国在多方安全计算、差分隐私保护等关键技术上取得了显著成果。然而，与国外相比，我国在 AI 安全基础理论体系的构建、跨领域的安全攻防研究以及标准化制定等方面，亟须进一步加强。

四、本学科发展趋势和展望

人工智能多学科交叉融合特征明显，且技术综合性强，而随着技术的不断进步，人工智能将在更多领域发挥重要作用。

（一）多模态模型促进图像、文本、视频深度融合

目前已有的生成式人工智能技术融合应用的形式还较为单一，多数仍是文字生成图片的变相应用。而多模态模型使文本、图片、视频等多种内容形式的综合理解，以及多种内容的结合输出成为可能。从视频生视频到文生视频、图生视频，多模态的发展重视用更少的用户输入信息量实现更丰富的 AI 生成结果。

（二）高质量数据成为大模型价值跃迁的法宝

随着人工智能模型迭代发展，高质量数据集的需求将进一步增长。从自然数据源收集

取得的原料数据并不能直接用于有监督的深度学习算法训练，必须经过专业化的采集、加工，形成相应的工程化训练数据集后才能供深度学习算法等训练使用。目前，带有监督学习的算法对于训练数据的需求远大于现有的标注效率和投入预算，基础数据服务将持续释放其对于算法模型的基础支撑价值，而高质量的数据处理、数据标注服务以及完善的数据收集和评估体系的价值将进一步凸显。

（三）具身智能成为 AI 发展新形态

具身智能是可以和物理世界进行感知交互，并具有自主决策和行动能力的人工智能系统。这种智能形式使主体可以以主角的角度感知物质世界，通过与外界互动后与自身学习相融合，从而获得对客观世界的理解与改造能力。业内专家认为，大模型并不是 AI 最终形态，下一个 AI 浪潮将是 10 年后的多模态“具身智能”，即将智能算法与机器人相结合，使机器能更智能地执行任务。

（四）AI 可解释性和透明度亟待增强

可解释性 AI 是人工智能发展的关键点之一，它不仅支撑了可信任 AI、负责任的 AI、可问责的 AI，还能够提高公众对于 AI 的理解和信任。AI 可解释性与算法治理、算法安全息息相关，缺乏可解释性的 AI 带来的负面影响能跨越所有经济阶层。目前，算法的可解释性仍处于早期发展阶段，而如何让人们理解和信任人工智能系统，已经成为人工智能未来发展的重要方向之一。

第七节 工业互联网

一、引言

工业互联网是新一代信息通信技术与工业经济深度融合的新型基础设施、应用模式和工业生态，通过对人、机、物、系统等的全面连接，构建起覆盖全产业链、全价值链的全新制造和服务体系，为工业乃至产业数字化、网络化、智能化发展提供了实现途径，是第四次工业革命的重要基石。

工业互联网学科是围绕工业互联网相关理论、技术、应用和管理方法的知识体系，涉及物联网、云计算、大数据、人工智能、5G 等多个前沿技术，并融合了制造工程、通信技术、计算机科学等多个学科领域的知识和方法。旨在培养具备工业互联网技术、应用能力的工程技术人才和管理人才，推动工业互联网相关技术的研究和应用，促进工业制造业的转型升级。本报告的定位是全面反映工业互联网学科的研究现状和趋势，主要任务是对工业互联网学科的发展历程、研究现状、应用场景、技术体系等方面进行全面系统的梳理和分析，为工业互联网的发展提供有力的学术支撑和智力支持。

二、本学科近年的最新研究进展

（一）工业互联网观点

“工业互联网”概念在通用电气公司发布的《工业互联网：打破智慧与机器的边界》中被系统性地提出，认为工业互联网的目的是提高工业生产效率，提升产品和服务的市场竞争力。德国政府认为工业 4.0 将生产方法和最先进的信息和通信技术相结合，以更灵活、定制和可持续的方式生产，从而带来智能生产流程和新的商业模式。美国将先进制造定义为现有产品制造的改进创新，以及由先进技术实现新产品的生产。思科是国际知名的数字通信技术公司，其认为工业物联网（IIoT）是一个由设备、传感器、应用程序和相关网络设备组成的生态系统，它们协同工作以收集、监控和分析来自工业运营的数据。《工业互联网创新发展行动计划（2021—2023 年）》中指出工业互联网是新一代信息通信技术与工业经济深度融合的全新工业生态、关键基础设施和新型应用模式。中国工程院院士邬贺铨认为工业互联网是工业数字化、网络化和智能化发展的基础，需要满足企业应用的高安全性、超可靠、低时延、大连接、个性化以及信息技术跟操作技术兼容的要求，需要开发对工业互联网优化的 ICT 技术。中国工程院院士李伯虎提出我国工业互联网的发展已经进入开启万物智联的“智慧工业互联网系统——工业互联网 2.0”新阶段。中国工程院院士刘韵洁预测到 2030 年，工业互联网对生产率提升可以使中国经济规模比目前增加 5.6 万亿美元，承担着升级改造传统行业重任，将推动互联网发展由“消费型”向“生产型”转变。

（二）工业互联网新成果

工业互联网专利情况方面，专利在工业互联网技术的研发、保护、创新和合作等方面具有重要的作用。国内专利在国家知识产权局的专利检索名称中包含“工业互联网”，申请日在 2022 年 12 月 31 日前，共收集到 1735 条工业互联网专利信息。专利申请数量总体呈现上升的趋势，其中中国信息通信研究院、深圳玄羽科技、山东浪潮申请工业互联网专利数量最多。通过对工业互联网相关专利的类别进行分析，报告发现在工业互联网相关专利中发明专利的占比最多，高达七成。国外专利在德温特专利情报数据库中检索篇名包括“IIoT”“Industria 4.0”“Industry 4.0”，2022 年 12 月 31 日之前的专利，共收集 67 项国外工业互联网专利信息。国外专利数量上总体呈现上升趋势，其中工程、计算机科学是热门主题。

工业互联网论文情况方面，国内论文以知网为检索平台，搜索题目包含“工业互联网”和“产业互联网”的高质量论文，时间范围限定在 2018 年 1 月 1 日—2022 年 12 月 31 日，共计 258 篇工业互联网论文，呈现论文数量不断上升趋势，研究领域主要集中在信息经济和工业经济两大学科。国外论文以 Web of Science 核心合集为检索平台，搜索题目中包含“IIoT”“Industria 4.0”“Industry 4.0”的论文，时间范围限定在 2018 年 1 月 1 日至 2022 年 12 月 31 日之间，共计 5321 篇相关论文，论文数量同样呈现上升趋势，且国外工业互联网论文研究领域广泛，主要集中在计算机科学和工程两大研究方向。

（三）工业互联网新方法和新技术

5G 方面，2022 年 8 月 25 日，工业和信息化部办公厅印发《5G 全连接工厂建设指南》，指导各地区各行业积极开展 5G 全连接工厂建设，带动 5G 技术产业发展壮大，进一步加快“5G+ 工业互联网”新技术、新场景、新模式向工业生产各领域、各环节深度拓展，推进传统产业提质、降本、增效、绿色、安全发展。2019 年，张云勇介绍了 5G 面向产业发展的网络特性，指出 5G 将不断夯实万物互联基础，全面使能工业互联网，加速推动多产业互联网化融合。时间敏感型网络的发展与 5G 密不可分，2020 年张强等学者深入研究 5G 技术在工业互联网中的应用，分析了 5G 与时间敏感型网络技术融合面临的挑战，以及在工业互联网中的应用。

云计算和边缘计算方面，云计算和边缘计算解决了工业互联网在数据处理上的困难。2020 年，罗军舟团队重点调研和归纳了云计算和边缘计算结合领域所涉及的几个重要方向的研究现状和面临的挑战，并提出一个云端融合的工业互联网新型体系结构及相关关键技术的解决方案。2021 年，李辉团队论述了工业互联网边缘计算架构及推动工业互联网边缘计算发展的核心技术，并阐述了当下工业互联网边缘计算的现状与挑战

标识解析方面，标识解析技术是产业链上下游数据互联互通的重要技术之一，该技术贯穿工业生产生命周期，可以实现制造实体、过程、产品等相关数据信息互通。2019 年，任语铮等学者讨论了工业互联网标识解析体系设计原则和关键支撑技术，关键支撑技术包括标识方案、标识分配机制、注册机制、解析机制、数据管理机制与安全防护方案等。Ren Yuzheng 等学者讨论了 IIoT 中身份解析系统的重要性，提供了关于可能用于 IIoT 的身份解析系统的综合调查，并从是否满足 IIoT 要求和技术选择的角度进行了比较。

人工智能方面，工业大模型是人工智能技术的关键最新和应用，是工业互联网领域内的应用中一项重要而具有前瞻性的技术，它通过整合大量的数据、模拟和分析，提供对工业系统的全面、实时了解。2022 年，任磊等学者认为基于工业互联网形成的大规模智能

制造产业链供应链系统的分析预测与优化管控问题，涉及具有复杂关联关系的跨行业、跨企业、跨流程的多层级工业制造系统，对此类复杂系统内在运行机理与模式规律的学习，将面对超大规模多源异质工业数据驱动的、大规模子问题深度学习的大集成问题，当前已有的适用于特定领域单项子问题的工业智能“小模型”技术难以应对，迫切需要探索适用于此类大型复杂工业系统问题的新型工业智能技术。

数字孪生方面，数字孪生技术允许企业在数字世界中创建、模拟和测试其实体设备、系统和流程的虚拟副本。数字孪生技术的功能主要有以下几种：①数字孪生技术允许制造商模拟和优化生产流程，减少生产中的浪费和不必要的成本；②通过虚拟测试，可以提前发现问题并进行改进，从而提高生产效率；③数字孪生技术可以创建物理设备的虚拟副本，实时监测其状态，并预测潜在的故障，这有助于降低设备维护成本，减少了因设备故障导致的停机时间；④数字孪生技术使产品设计团队能够在虚拟环境中测试和优化产品原型，加速新产品的开发周期，提高产品质量；⑤数字孪生技术可以整合各种传感器和设备数据，提供实时的数据分析和预测能力，帮助企业更好地理解其运营状况并做出智能决策。

三、本学科国内外研究进展比较

工业互联网是近年来国际上十分受关注的一个领域，各国纷纷将其列为国家战略并加大投入，较为著名的有德国工业 4.0，美国的“Advanced Manufacturing”等。目前，中国、美国、德国是世界上工业互联网发展最快的 3 个国家。对于国内的研究情况，将从中国国家自然科学基金委员会资助的工业互联网相关研究计划和项目，以及知网中工业互联网相关的论文来描述国内工业互联网的研究情况；对于国外的研究情况，将根据德国研究基金会和美国国家科学基金会入手并与国内研究情况进行比较。

在工业互联网项目的数量上，中国目前与美国、德国保持统一步伐，都在近五年内不断发展。一方面是因为各国政策的大力支持；另一方面是因为信息技术等快速发展，与制造的及时结合，促进了制造业的更新迭代。

在工业互联网项目的资助金额上，中国和美国在工业互联网的发展上投入很大，相对而言美国对工业互联网项目的资助的规模和平均金额要比中国高一些，这一方面是因为美国国家科学基金会整体体量比中国国家自然科学基金委员会要大；另一方面是因为美国国家科学基金会除了资助科研项目还会对教育项目和技术商业转化项目进行资助。纵向对比来看，中国的资助规模整体呈上升的趋势。

在工业互联网项目的关键词上，不同国家有共性点也有各自侧重点。数据作为工业互联网的核心要素，中国、美国、德国 3 个国家都十分看重。各国也有各自侧重的方面，中国更加重视网络方面的建设。

在工业互联网项目的学部背景上，中国和美国的学部划分比较类似，且都更加注重计算机和信息科学，工程科学和管理科学为辅，其他学部也有所涉及。而德国则是以机械和工业工程部为主，计算科学部为辅，其他科学部作为工业互联网的应用领域也有所涉及。我国与德国的差别主要原因可能是我国工业互联网更加注重技术层面，而德国更加注重实践方面，这和我国与德国对工业互联网架构的设计特点相对应。

四、本学科发展趋势和展望

随着信息化和工业化融合的趋势不断加强，工业互联网的学科发展的战略重要性正不断凸显。

一是工业互联网学科发展是工业互联网战略的重要支撑。全球各国在工业互联网领域的合作不断加强，竞争愈发激烈。工业互联网学科发展将为工业互联网战略在人才端和技术端不断提供支撑。

二是工业互联网学科发展是促进产业和企业数字化转型的重要基础。工业互联网学科发展将为产业和企业数字化转型提供复合型专业型人才，是产业和企业数字化转型的重要基础。

三是工业互联网学科发展是新型工业化时代人才培养的重要领域。工业互联网学科发展是新型工业化时代人才培养的重要一环，为新一代学生成为具有工业互联网应用能力和创新精神的高素质人才提供平台。

工业互联网学科未来五年的发展趋势主要体现在以下 3 个方面。

一是工业互联网学科特点愈加鲜明、学科融合持续深入广泛。一方面学科呈现基础化、综合化、实践化、个性化特点愈加鲜明；另一方面学科与计算机、工程等学科，与能源、航空、化工等领域交叉融合更深入、更广泛。

二是多元化、国际化的“数字工匠”培育基地布局更广泛。体现在多元化人才培养体系持续升级；国际化人才培养模式稳步落地；“数字工匠”培育基地加速形成；中国特色、世界一流的人才培养高地加快建设等方面，并借鉴德国双元制人才培养模式，吸收美国麻省理工学院工业互联网领域的前沿技术、方法。

三是助力三大产业协同发展，推进新型工业化转型。体现在继续巩固、提升工业互联网在制造业的优势地位；推动制造业高端化、智能化、绿色化发展；突破产业发展中的主要瓶颈——安全可靠性问题等方面。另外，随着新型工业化在相关行业领域蓬勃发展，农业领域发展前景广阔，进一步提高农业生产效率和质量，促进农业产业升级转型。

第八节　航空科学技术

一、引言

航空制造业是战略性高技术产业，具有知识与技术密集、多学科集成、产业链长的特点，处于制造业的高端，是我国重点发展领域。航空科学技术的发展水平，是国家科技水平、国防实力、工业水平和综合国力的集中体现和重要标志。我国航空科技近几年得到较快发展，本节选择其中的浮空器、复合材料、发动机、维修工程、可靠性工程、旋翼飞行器及其系统、无人飞行器及其系统等专业领域进行研究，以反映最新进展。

二、本学科近年的最新研究进展

近年来，在党和国家的坚强领导和持续稳定投入下，我国航空科技取得了较大进展。军用航空方面，歼-20 正式列装空军，并换上国产发动机，进入了自主发展的新阶段；运-20 列装空军，并在疫情期间为国际社会提供了重要援助；直-20 开始批产。在民用飞机方面，万众瞩目的 C919 取得型号合格证并投入商业运营，顺利运营突破 1000 小时，进入产业化阶段，现有订单超过 1000 架；ARJ21 飞机目前已累计交付 116 架，运送旅客超过 960 万人次，已成为我国支线航空市场的主力机型；ARJ21 系列化发展，首批 2 架货机成功交付，公务机、医疗机完成取证；AG600 水陆两栖民用飞机灭火型 AG600M 完成首飞，并进行了应急灭火演练演示；直-15 取得适航证走向市场。这些成果的取得，显著加强了我国的国防实力和国际竞争力，为提升我国的大国地位贡献了重要力量。

（一）浮空器

我国高空气球技术及其应用已经比较成熟，气球产品系列化、体积从几百立方米到一百万立方米、最大升空高度 43 千米、最大载重 4 吨，总体技术水平处于世界前列。固定式系留气球我国有成熟产品，体积 12000 立方米的固定式大型系留气球系统，载重能力

1 吨多。“云中漫步”系列载人观光系留气球取得了我国特种设备制造许可证。3500 立方米民用载人对流层飞艇项目，已进入适航取证试飞阶段。对流层无人飞艇国内研制了多型产品，执行多种任务。平流层飞艇工作高度在 20 千米左右，我国多家机构开展了相关关键技术攻关，目前完成了跨昼夜长时控制飞行，验证了相关关键技术。

（二）复合材料

1. 航空复合材料应用

复合材料在国产大型客机 C919 上的用量达到 12%。C919 复合材料机翼研制典型盒段静力和损伤容限试验顺利完成。远程宽体客机复合材料机身曲面加筋壁板工艺验证件完成研制。宽体客机复合材料前机身攻关全尺寸筒段顺利总装下线。另外，山河 SA160L、ZA800 轻型飞机等多型全复合材料通用飞机研制成功。

我国航空功能复合材料方面，结构吸波复合材料先在三代飞机隐身集成考核验证后，实现了在四代机的大量应用；结构透波复合材料已发展到当前的低吸湿、低介电、低密度和高综合力学性能，扩大了应用，支撑了航空装备探测能力的提升。

2. 复合材料原材料发展现状

（1）纤维材料

国产碳纤维在性能、工程化和应用研究等方面发展迅速，核心技术不断突破，产品品质逐步提高。攻克了国产 T300 级、国产 T700 级、国产 T800 级碳纤维和国产 M40J、M55J 石墨纤维的工程化和应用问题；突破了国产 T1000 级、T1100 级碳纤维和国产 M65J、M40X 石墨纤维的关键制备技术，实现了工程化生产。在芳纶纤维方面，我国已突破对位芳纶和杂环芳纶关键技术，建成多套工业化装置。国内陆续突破连续碳化硅纤维及其复合材料的工程化制备技术，开展了工程应用验证，同时也初步建立了氧化铝纤维、氮化硅纤维、系列陶瓷前驱体等验证试验线。

（2）树脂基体

国内研制成功新一代高韧性环氧树脂基复合材料 AC531，达到了第三代韧性复合材料的水平。国内已形成第一代、第二代和第三代双马来酰亚胺树脂基复合材料。目前，高韧性环氧和双马复合材料在四代机、大型运输机、武装直升机等结构大量应用。

3. 复合材料制造技术发展现状

目前，航空复合材料成型工艺依然以预浸料热压罐工艺为主。我国在热压罐复合材料成型工艺技术方面已经比较成熟。自动铺带和铺丝工艺技术目前在国内航空装备中实现装机飞行考核和小批量生产，而且在大型民机机翼机身结构实现了工艺验证。

（三）发动机

军用航空发动机在技术和产品上都已经取得了重大突破。“昆仑”发动机的研制成功结束了长期以来我国不能自行研制航空发动机的历史。“太行”系列发动机则实现了我国航空发动机从涡喷到涡扇、从中等推力到大推力、从第三代到第四代的历史性跨越。自“两机专项”设立后，国产民用发动机已经成为我国发展的新重点。大型客机 C919 的配套国产发动机 CJ1000 研制进展顺利，已经取得了阶段性成果。为大型宽体客机配套的发动机 CJ2000 的研制也在顺利推进中。

超高声速动力技术方面，多座高超音速风洞建成并投入使用，为超燃冲压发动机试验研究提供良好基础。“飞天一号”火箭冲压组合动力成功发射，突破了多项关键技术，飞行试验圆满成功。腾云工程圆满完成了我国首次液体火箭冲压组合发动机模态转换首次飞行验证，实现了空天飞行组合动力技术重大突破。

（四）维修工程

目前，国内民用航空维修基本上是按照西方发达国家的模式进行维修管理和维修生产，总体上保证了民航飞行安全和经济有效运营。国内军事航空维修领域的主要进展体现在以下几个方面：在基于状态维修（CBM）、推行视情维修领域开展了大量的探索应用，维修技术取得新进展，维修数字化开展了新探索。

（五）可靠性工程

近年来，运用系统工程理论方法，在可靠性工程基础上进一步发展出可靠性系统工程，制定了一系列针对航空装备的标准和技术规范，构建了贯穿航空装备产品全寿命周期的故障防、预、诊、治的可靠性技术途径，已形成可靠性综合集成的基础理论，以基于模型的可靠性系统工程、故障诊断与预测等为代表的应用技术，并持续在运-20、歼-20、直-20、C919 大飞机、新型军用 / 民用无人机为代表的航空装备产品应用实践。

（六）旋翼飞行器及其系统

近年来，我国旋翼飞行器的发展重要体现在以下几个方面：一是民用直升机 AC 系列的研制取得重要进展，AC312E、AC332 和 AC313A 等直升机先后成功首飞，AC352 取得中国民用航空局颁发的型号合格证；二是我国新型军用通用运输直升机直-20 完成研制并批量装备部队，实现了我国直升机从第三代向第四代的巨大跨越；三是无人直升机热度不减，社会力量聚集，型号研制踊跃；四是关键技术取得长足进步。

（七）无人飞行器及其系统

由于无人机自身的特殊优势以及在军民用领域的迫切需求和广泛应用，国内研发无人机的热情高涨，军工集团、国有企业和民营资本纷纷成立专业无人机公司。目前，国内有数千家无人机企业，通过激烈竞争，无人机技术取得长足进步，已形成较为丰富的无人机整备体系，各种类型、各种功能无人机投入使用。

三、本学科国内外研究进展比较

（一）世界航空科技的主要进展

1. 浮空器

欧美有许多世界知名的浮空器制造企业，研制和使用经验丰富，技术体系完整，电池、电机、航电等关键配套先进成熟可靠。近年仍在进行大型载重飞艇、各种新概念飞艇等的研制工作，各种关键技术不断突破，新技术不断提升。

2. 复合材料

复合材料已经成为航空飞行器最主要的结构材料。碳纤维复合材料在小型商务飞机和直升机上的使用量已占70%~80%，在军用飞机上占30%~45%，在大型客机上占35%~50%，在无人机上占90%以上。

美国、日本和欧洲等国家和地区在高性能纤维及其复合材料领域已形成先发优势，并形成其产业生态圈。俄罗斯等传统东欧国家有机纤维、黏胶基碳纤维及复合材料的技术水平较高，各种热加工设备实用可靠，可基本满足其国防工业需求。

复合材料自动铺放工艺发展最为成功并得到了不断推广应用。低成本液体成型复合材料形成了系统的技术体系。基于干纤维铺放的液体成型（VARI）复合材料及其成型工艺实现了自动化，大大提高了液体成型复合材料的抗冲击损伤性能。总体来说，国外复合材料自动化制造工艺技术日益成熟并得到广泛应用，已经占据了复合材料成型工艺的50%以上。

3. 发动机

目前，具备独立研制航空发动机能力并形成产业规模的国家有美、俄、英、法、中等少数国家，具有技术和商业优势的仅有美英两国。

涡扇发动机是目前应用最多的航空发动机。民用市场主要由美国GE、英国罗罗、美国普惠三大发动机巨头及其参与的合资公司瓜分；军用方面，除上述三家巨头外，欧洲喷气动力公司、法国斯奈克玛、俄罗斯联合发动机等也占有一定市场份额。

传统的航空发动机正在向齿轮传动发动机、变循环发动机、多电发动机、间冷回热发动机和开式转子发动机发展，非传统的脉冲爆震发动机、超燃冲压发动机、涡轮基组合发动机，以及太阳能动力和燃料电池动力等也在不断成熟。

4. 维修工程

近年来，以美国为代表的发达国家空军和民航公司围绕提升飞机完好水平、提高飞行安全、降低全寿命维修成本开展了大量工程实践工作，并取得进展。主要表现在：①新维修理论得到深化和应用；②维修工程数字化取得重大进展；③智能测试、增材制造等维修新技术得到了广泛应用；④装备维修工程向信息化深度推进，智能维修技术、远程维修技术及虚拟维修技术等均取得了较大进展并得到应用。

5. 可靠性工程

以美国为代表的国外先进航空工业推动可靠性技术逐渐向模型化、仿真化、智能化发展。美国国防部正在推进《数字工程战略》（*Digital Engineering Strategy*）。欧洲防务局在促进欧洲国家在装备研发和可靠性方面进行合作，联合研发项目使得欧洲各国可以实现装备的数据共享和交换，收集装备的可靠性相关数据，以提高装备的性能和可靠性。同时，国外学者也大力发展了基于数据驱动的可靠性技术。

6. 旋翼飞行器及其系统

近年来，载人旋翼类飞行器重点是研制高速旋翼类飞行器，复合式、共轴双旋翼与倾转旋翼构型成为发展的热点。无人旋翼飞行器的新型能源化、自主智能化和集群协同化成为新发展方向，新机型不断涌现，其中电动垂直起降飞行器 eVTOL 成为热门方向。

7. 无人飞行器及其系统

目前，美国和以色列的无人机技术水平最高、发展最快、应用最广，紧随其后的是俄罗斯、英国、法国、德国、意大利等欧洲国家。

无人机性能不断提升。大 / 中 / 小 / 微型、远 / 中 / 近程、高 / 中 / 低空等各种类型无人机全谱系全面发展。在役无人机的任务半径实现了近至视距、远至 5000 千米以上的无缝覆盖，既可用于阵地作战支持，又可满足远程甚至洲际作业任务需要。低空可至近地，伴随地面部队前进，高空已达临近空间低界。

无人机功能不断扩展。已覆盖情报 / 监视 / 侦察、探潜、电子对抗、通信中继、察打一体、隐身对空对地攻击作战、战场运输等任务领域。

（二）航空科技的国内外差距

近年来，我国航空科技和航空产业实现了快速发展，但与世界先进水平相比，还有一定的差距。

1. 浮空器

浮空器属于“小众”研究领域，我国大力发展的时间也就十几年，基础理论、基础技术研究和试验研究积累少，难题瓶颈依然存在。囊体材料、薄膜太阳能电池、风机等关键材料和器件，与国外比还有差距；国内搭载的航电系统产品小部分是自主研发，大部分依赖进口。我国近年加大技术研究，目前，国内外浮空器结构和制造工艺基本处于同一水平；国内储能电池发展水平与国外相当，比能量超过 500 瓦 · 时 / 千克；高可靠电机、螺旋桨等多项技术国内相关单位已达到世界先进水平，产品已投入使用。部分技术实现了超越，平流层飞艇试飞技术走在世界前列。

2. 复合材料

与国外相比，我国高端纤维及其复合材料仍存在代差。国外航空航天领域已经大规模应用以 T800 级碳纤维为主要增强体的第二代先进复合材料，而我国总体上仍处在第一代先进复合材料扩大应用、第二代先进复合材料考核验证阶段，落后一代以上，而且高强高模、超高模量碳纤维尚未建立有效的自主保障能力。在高性能有机纤维、陶瓷纤维等领域，同样存在高端产品缺乏、质量一致性差等问题。产业技术成熟度不够，大规模高效低成本的成套工艺与装备技术仍未完全突破。

3. 发动机

我国航空发动机产业基本具备了军用大中小型涡喷、涡扇、涡桨、涡轴等各类型航空发动机的研制生产能力，军用涡扇发动机已步入以自主研发为主的良性发展道路。民用航空发动机仍受制于人，国内巨大市场基本被国外产品垄断；长江–1000 发动机取得了重大阶段性成果，长江–2000 发动机研发进展顺利。军民用领域，据先进国家差距均在 20 年以上。

在高超声速动力方面，美、俄、日、韩等国家加速推动技术发展。我国也加大研究，目前试验技术处于世界先进水平。

4. 维修工程

我国的差距主要表现在：国际先进理念和先进理论的学习和实践不充分；先进维修技术如复合材料结构损伤修复、装备原位修复和智能修复、零件现场增材修复平台技术等方面尚有关键技术需要突破；维修数字化发展仍在起步阶段，维修对象数字化、维修产线数字化、维修业务数字化和支撑环境数字化等方面与先进水平存在较大差距。

5. 可靠性工程

可靠性技术逐渐向模型化、仿真化、智能化发展和升级，国外的航空可靠性工程技术发展注重理论创新、新技术开发和法规建设，国内差距主要表现在：国内技术水平存在差距，相关法律法规不健全，相关标准规范亟待更新。

6. 旋翼飞行器及其系统

国内差距主要表现在以下 2 个方面。

1）型号少，技术和性能相对落后，市场竞争力差。当前我国自产服役的直升机全部为单旋翼带尾桨的常规构型直升机，尚缺重型直升机、高速直升机、非常规构型直升机等。大部分直升机型号相对老旧，技术相对落后，寿命短、可靠性差和维修性差，市场竞争力弱。

2）先进技术少，众多技术亟待突破。我国独立研发的直升机都有参考样机，尚未具备完整的独立自主设计能力。技术储备不足，许多技术亟待提高和突破，动力及部分民机配套产品受制于人，离世界强国有 10~20 年以上的差距。

7. 无人飞行器及系统

我国无人机发展与世界先进水平存在明显差距。高空长航时无人机方面，我国无人机飞行高度不高、飞行时间较短；隐身长航时无人机方面发展严重不足；中空长航时无人机方面，留空时间较短、载重低；集群无人机、有人 – 无人协同无人机、垂直起降固定翼长航时无人机、大型隐身加油无人机、大型预警无人机、新能源无人机等方面，发展尚处于初级阶段，离实用化、实战化需求还有较大距离。

四、本学科发展趋势和展望

（一）浮空器

高空气球技术的发展趋势主要是大型化、小型化、长航时、轨迹控制、组网应用几个方面。系留气球未来的发展方向主要是完善型谱、提高功能性能，拓展应用领域。对流层飞艇将向着大航程和长航时、多功能一体化、大承载能力和高安全性能（抗打击、抗天气、抗干扰等）方向发展。平流层飞艇向着长航时、大载重、快速响应和低成本的方向发展。

（二）航空复合材料

复合材料制造技术将继续向整体化、自动化、数字化、智能化和低成本化发展。在航空级纤维方面，需从提高纤维强度、模量、改善工艺以及降低成本几个方面进行加强，树脂的发展方向主要是提升韧性、耐热性、改善工艺以及增加可回收性等。航空预浸料将向超薄的方向发展。

（三）发动机

未来的发动机将实现更高的效率 / 经济型、更长的寿命和更好的环保性能，同时也将变得更加智能化和高度自动化。发动机制造商将不断探索新的材料和技术，开发出更加轻量化、可靠和安全的发动机。

新型航空发动机技术将得到快速发展，如齿轮传动发动机、变循环发动机、多电发动机、间冷回热发动机和开式转子发动机。非传统的脉冲爆震发动机、超燃冲压发动机、涡轮基组合发动机，以及太阳能动力和燃料电池动力等国外也在不断成熟。这些发动机的发展将使未来的航空器更快、更高、更远、更经济、更可靠、更环保，并将使高超声速航空器、跨大气层飞行器和可重复使用的天地往返运输成为现实。

（四）维修工程

国内民用航空维修按照国际先进模式，不断更新理念、学习先进理论和技术，保证民航机队的飞行安全和经济有效运营。同时，创新技术和建立适合国产民机的民机维修体制和规章标准，保障国产民机安全高效运营，促进国产民机产品改进完善和民机产业健康发展。

未来军机维修工程，应借鉴西方先进理念和模式，推动其工程应用，加大维修新技术攻关和应用，推进航空装备数字化维修转型，完善相关法规和标准规章。

（五）可靠性工程

针对新时代航空装备质量建设需求，重点发展跨尺度通用质量特性综合设计技术、可靠性数字孪生技术、故障诊断与健康管理辅助决策技术、面向能力的装备维修性正向设计技术，加快可靠性综合集成平台及生态建设和质量信息交换平台建设，优化资源配置，提高产品的可靠性。

（六）旋翼飞行器及其系统

直升机技术的发展趋势是高速化、智能化、信息化、轻量化和绿色化。电驱动旋翼飞行器、跨介质变体旋翼飞行器、模块化组合旋翼飞行器等各类新构型、新概念旋翼飞行器将得到快速发展。eVTOL 技术发展方向将主要集中在更加高效的电池、更稳定的控制系统以及更优秀的材料等技术创新领域。多旋翼飞行器将变得更加智能化和自主化，将逐渐成为各行业中的常规工具。

（七）无人飞行器及其系统

无人机在军民用领域的地位和作用越来越重要。无人机从战场辅助角色逐步变为信息化作战时代的主战武器，并不断改变战争方式。在经济建设领域，无人机的应用越来越多，将成为国民经济发展新的增长极。

无人机技术发展方向呈现多元化：向高空、高速、长航时大型无人机方向发展，提供持久作战能力；向临近空间方向发展，开辟空天一体化战场；向无人空战发展，开辟无人空战时代；向舰载方向发展，对未来海上作战产生重大影响；向微型化发展，助力复杂环境下的特种作战和小分队作战；新型部署与回收技术不断发展，提高作战快速响应能力；自主性和智能化程度迅速提高，体系化、网络化协同能力迅速提高；向低成本、作战可消耗方向发展，提高战争中费用不对称优势；同时，作战无人机的对抗性、隐身性能和战场生存力持续提升。

第九节　航天科学技术

一、引言

近五年来，我国航天事业取得了一系列辉煌成就，航天学科迈入了创新发展的“快车道”：一是航天运输系统加速升级换代，带动了系列关键专业技术发展；二是通信、导航、遥感、科学试验卫星系统不断完善，小卫星及星座系统蓬勃发展；三是空间站系统全面建成，载人航天迈入空间站时代；四是探月“绕落回”、探火“绕着巡”，不断创造我国深空探测的新高度；五是航天推进、空间能源、航天制导导航控制、航天探测与导引、航天智能探测与识别、航天遥测遥控、空间遥感、航天先进材料、空间生物与医学载荷、航天器回收着陆等专业基础技术领域均取得突出进展。

本节针对五年来航天学科发展情况，按照重大标志性、关键紧迫性、技术可行性等原则，遴选出 2022—2023 年航天科学技术学科专业 15 项，重点阐述了最新进展，分析了国内外研究进展比较，展望了未来发展。

二、本学科近年的最新研究进展

（一）航天核心关键技术取得突破性进展

1. 航天运载器技术

我国长征系列运载火箭加速向无毒、无污染、模块化、智慧化方向升级换代，运载能力持续增强。长征系列运载火箭近地轨道运载能力达到 25 吨级、地球同步转移轨道运载能力达到 14 吨级，入轨精度处于国际先进水平。长征十一号海上商业化应用，快舟一号甲、双曲线一号、谷神星一号、天龙二号等商业运载火箭成功发射，不断增强、丰富我国运载火箭多样化发射服务能力。多项可重复使用运载器飞行演示验证试验取得成功。

2. 卫星系统技术

通信、导航、遥感、科学试验卫星系统支撑和构建起多种功能、多种轨道的国家空间基础设施。通信卫星领域，实现天地相“链”，担起“安播”卫士，加速技术融合；导航卫星领域，北斗三号全球卫星导航系统全面建成，全球组网，服务全球；遥感卫星领域，高分辨率对地观测系统重大专项空间段建设圆满完成，遥感卫星定量化、业务化、产业化发展的态势已经形成；空间科学与技术试验卫星领域，发射了一系列科学卫星，推动了空间科学、太阳物理、地球科学等发展。

3. 小卫星技术与应用

小卫星平台与载荷功能和性能不断提高，商业小卫星蓬勃发展，特别是低轨巨型星座在天基全球通信、遥感等一系列领域的应用。以高景一号为代表的首个 0.5 米级高分辨率遥感卫星星座，增强了中国商业遥感数据服务能力；“灵巧通信”卫星在轨验证了基于通信小卫星一体化设计、智能天线等技术的低轨移动通信；国内多家企业成功发射低轨卫星增强北斗卫星导航系统的性能，实现高精度位置服务。

4. 载人航天器技术

我国空间站系统全面建成，载人航天事业实现历史性跨越。全面突破了空间站设计、建造和运行的一系列关键专业技术，主要包括空间机械臂、推进剂补加、再生生保、出舱活动、在轨维修、货物装载、大型舱段空间转位、柔性太阳翼和驱动机构、超快速交会对接、载人天地往返快速返回等。

5. 深空探测器技术

探月“绕落回”、探火“绕着巡”，突破了一系列深空探测专业技术。嫦娥五号月球探测任务突破并掌握了月球探测轨道设计、月球采样封装、月面起飞上升等一系列深空探测关键技术。天问一号火星探测任务突破了火星制动捕获、行星际测控通信等一系列关键技术，实现对火星的科学探测。

（二）航天专业基础技术实现整体跃升

1. 航天推进技术

液体发动机方面，确立了我国第三代运载火箭型谱，完成了从有毒到无毒、小推力到大推力、开式到闭式循环的技术进步。商业航天液体动力取得长足进步，天鹊 -12 成为全球首款成功入轨的液氧甲烷发动机。固体发动机方面，整体式固体发动机综合性能达到国际同类发动机领先水平。霍尔电推进系统实现小批量飞行应用，形成型谱产品；电推进、核推进完成了实验验证。

2. 空间能源技术

在发电单元方面实现技术突破，中国空间站实现了我国大型柔性太阳电池阵的首次成功应用；同位素温差电池首次应用于空间型号——嫦娥四号月球着陆器。在储能单元方面，国内继续保持领跑态势。在能源控制单元方面，卫星导航、低轨遥感和深空探测器上采用了国产高性能、长寿命电源控制器。在元器件、原材料方面，国内已实现了方钴矿材料的全流程自主研发；部分元器件实现国产化。

3. 航天制导导航控制技术

突破了一系列关键技术，并广泛应用于我国空间站和深空探测等领域。主要包括高精度自适应返回的制导 / 导航 / 控制技术，大尺度柔性航天器动力学行为辨识与分布式控制技术，自主交会对接精确控制方法，深空软着陆与表面起飞上升智能自主的制导 / 导航 / 控制技术，航天器超精超稳超敏捷复合控制技术等。

4. 航天探测与导引技术

协同探测技术方面，我国嫦娥四号任务发射的“鹊桥”中继星，成为世界首颗地球轨道外专用中继通信卫星，实现协同探测。探通一体化技术方面，“行云工程”计划建设中国首个低轨窄带通信卫星星座，将实现小规模组网与探通一体化建设。边缘计算与先进计算技术方面，已经应用于我国北斗卫星系统与车联网结合等。

5. 航天智能探测与识别技术

神经形态器件技术方面，围绕模拟器件在结构模型、突触功能层、随机神经元电路等方面开展了研究。在神经形态成像传感器技术方面提出了一些新型事件相机。类脑计算芯片技术方面，在以节能方式运行跨计算范式的多种 AI 算法、首台类脑电脑等方面取得进展。

6. 航天遥测遥控技术

在测控技术的理论与标准规范方面，提出空间和星内一体化网络协议体系结构。在测量与控制技术方面，遥测码率从 2 兆字节 / 秒加快到 10 兆字节 / 秒，遥测频段从主用的 S 频段向 Ka 频段发展；已建成 C 频段测控网以及 S 频段测控系统网络；测控设备由传统单测控设备向测运控 / 数传一体化综合设备发展。在跟踪与数据中继技术方面，天链组网具有全球覆盖能力。在激光测控通信技术方面，在轨验证了多项关键技术。

7. 空间遥感技术

光学成像技术方面，发展了一体化高精度目标定位、星上全频段微振动抑制与隔离技术。计算成像技术方面，超衍射极限成像、无透镜成像、大视场高分辨率成像及透过散射介质清晰成像成为可能。光谱探测技术方面，开发了国际首台专门设计用于探测太阳诱导植被荧光的载荷。我国在激光探测技术、微波遥感技术等方面逐步走向国际前列。

8. 航天先进材料

在金属材料与工艺技术方面，开展了第四代 600 兆帕级以上铝锂合金材料研制及应用研究；多种合金材料研发取得新进展；为重型运载火箭研制提供了技术支撑。在非金属材料及工艺技术方面，特种密封材料设计与应用技术取得进展。在先进功能复合材料及工艺技术方面，突破了大尺寸 C/SiC 陶瓷基热结构材料设计制备关键技术；在航天器高效隔热材料领域形成了隔热瓦、隔热毡和纳米隔热材料三大体系。在表面工程材料技术方面，首次实现加工成型硅橡胶体系防热涂层的工程化应用；突破了多项关键技术。

9. 空间生物与医学载荷技术

在细胞 / 组织研究方面，问天实验舱的生物技术实验柜，可以开展细胞培养和组织构建。在微生物 / 植物 / 动物研究方面，我国空间站配置了微生物载荷、空间辐射计量及生物损伤评估技术装置。在核酸 / 蛋白质研究方面，梦天实验舱提供了特定微生物的快速检测研究条件。在空间芯片载荷方面，微流控芯片技术已经应用于中国空间站和“天问一号”火星探测器的空间生命科学实验研究。

10. 航天器回收着陆技术

基础与前沿探索方面，初步构建进入减速与着陆系统可重复使用性能评估方法；开展了充气高速再入系统概念研究、方案论证、气动设计、防热设计演示验证试验。技术研制与应用方面，对可重复使用关键技术进行初步验证；突破了接近第二宇宙速度跳跃式再入返回的减速与着陆技术等。

三、本学科国内外研究进展比较

五年来，我国航天事业取得了一系列辉煌成就，使我国加速迈入航天强国。航天运载器突破了高频率、常态化发射瓶颈，并加速向无毒、无污染、模块化、智慧化方向发展；卫星系统技术水平稳步提升，支撑和构建起多种功能、多种轨道的国家空间基础设施；空间站系统全面突破了空间站设计、建造和运行的一系列关键专业技术；深空探测领域突破了一系列关键技术；众多专业基础技术领域均取得突出进展。但总体而言，与世界航天强国相比，我国在航天核心关键技术和专业技术领域仍存在不少差距。

（一）国内外航天核心关键技术领域主要差距

在航天运载器领域，运载系数等核心性能指标与国际主流水平相比仍有差距；我国新一代火箭运载能力达到大型运载火箭水平，与美国德尔塔Ⅳ相当，无法支撑载人登月、大型空间基础设备建设对运载能力的需求；运载火箭尚不具备重复使用能力。

在卫星与应用领域，我国通信卫星载荷重量的系统容量与国际在轨和在研存在差距，以中星 26 卫星为例，其通信容量为 100 千兆比特 / 秒，单位重量系统容量为每 100 千克 12 千兆比特 / 秒，而“卫讯-3”的系统容量已经突破 1 太比特 / 秒[①]，其单位重量系统容量为每 100 千克 50 千兆比特 / 秒；遥感卫星定量化探测能力存在较大差距，探测要素还有一定欠缺，例如，对于 $PM_{2.5}$、臭氧等气态的探测手段尚属空白。

与美国太空探索技术公司（SpaceX）相比，我国小卫星技术迭代与突破、系统降本增效、批量发射部署[②]等方面能力不足。

---

① Viasat. Exceptional Capacity：1 Terabit Per Second［EB/OL］.（2023-12-24）. https://www.viasatprovider.com/viasat-3.

② Shaengchart Y，Kraiwanit T. The Spacex Starlink Satellite Project：Business Strategies and Perspectives［J］. Corporate & Business Strategy Review，2024，5（1）：30-37.

（二）国内外航天专业基础技术领域主要差距

在航天推进技术领域，我国固体动力技术在性能和规模上存在代差；在核推进、绳系推进等前沿推进技术方面，我国还处于初期阶段。在空间能源技术领域，我国发电、储能、能源控制的产品集成化发展、批量化制备等方面与国外差距较大。在航天智能探测与识别技术领域，我国在神经元模型等类脑理论研究、类脑芯片与神经形态传感器等硬件研制方面存在较大差距。在航天遥测遥控技术领域，我国在遥测遥控理论与标准规范存在短板；全域覆盖、泛在互联的能力不足。在航天先进材料领域，我国金属材料与非金属材料体系尚不健全；先进功能复合材料领域的基础研究相对薄弱。

四、本学科发展趋势和展望

航天运载器技术将在国家战略需求的牵引和市场经济的推动下持续提升能力，并向可重复使用、新模式研制、重型化、多样化应用等方向发展。卫星系统技术在提升通信、导航、遥感等卫星性能的同时，进一步向体系化、业务化、一体化方向发展。小卫星技术与应用的主要发展方向将是星地一体、“通 – 导 – 遥”综合应用等。载人航天器技术将向可重复、低成本、长期在轨运行、载人深空探测等方向发展。深空探测器技术将通过系列重大任务，推动一系列核心关键技术突破，催生一系列空间科学成果。相关的材料、制导导航与控制、动力、能源、空间生物医学等基础技术将进一步获得突破。同时智能化、信息化、空天地多域一体化协同等前沿技术在航天工程中的应用日益深入，其他学科领域在智能化、信息化等方面的前沿学术成果与航天的交叉融合将成为重要方向。

第十节　现代先进毁伤技术及效应评估

一、引言

毁伤是军事打击链路的最终环节，影响战局进程、决定战争胜负。毁伤的科学本质是“向目标释放、传递能量的过程”，当作用于目标的能量密度超过一定阈值时，促使目标材料、结构破坏或功能丧失，达到毁伤目标的作战目的。常规武器毁伤目标的能量源于火炸药爆炸释放的化学能，能量十分有限，难以做到“一击即毁”，具有较大的随机性和显著的概率性，称之为“常规毁伤的概略性”。因此，本学科的基础科学问题是“高密度能量储存、控制、高效利用与评价”，是融合了力学、工程学、材料科学、化学、爆炸科学、凝聚态物理、计算科学等多个学科的前沿交叉领域，具有典型的跨尺度特征。现代先进毁伤技术及效应评估学科发展研究报告结合近年来毁伤技术的快速发展，总结提炼了现代先进毁伤技术发展的成果和取得的技术突破，分别从能量富集与创制技术、能量释放与控制技术、能量高效利用技术、能量利用效应评估技术等 4 个方面论述了现代先进毁伤技术及效应评估技术的进展，通过对比分析国内外发展现状，预测了未来的发展趋势并提出了对策。

二、本学科近年的最新研究进展

（一）能量富集与创制技术

能量富集与创制技术重点解决毁伤能量来源问题，以含能材料为载体实现能量富集，研究能量储存和稳定途径，掌握能量激发和转化机制，拓展含能材料的能谱空间。目前，二、三、新一代含能材料研究十分活跃，呈现三大发展趋势：一是二、三代含能材料成体系化发展，相关技术研究不断深化，主要体现在产品微纳米化、表征方法升级、表面修饰改进、钝感技术革新、制造工艺改进、工程应用创新等方面；二是由传统的单一硝基储能单元为主向多种储能单元相结合的方向发展，以氮氮单 / 双键为主的全氮化合物能量密度超过梯恩梯当量数倍，打破了传统碳氢氧氮系含能化合物能量的“天花板效应”（密度不超过 2.1 克 / 立方厘米，爆速小于 10000 米 / 秒），若成功合成将引发高能量功率物质和爆轰物理的重大变革；三是由化学键储能为主向高张力键与化学键储能相结合方向发展。

（二）能量释放与控制技术

能量释放与控制技术重点解决高密度能量利用问题，传统碳氢氧氮系含能材料已接近硝基的能量极限，通过热力学调控组分 – 产物能态来提升能量水平的空间有限，更多学者将目光放在火炸药爆轰反应区上，通过动力学调控来提升能量水平，主要进展为：一是能量释放与控制技术取得突破性进展，提出了基于不同物理机制的反应区测量方法，包括自由面速度法、电磁粒子速度计法、电导率法、激光干涉测速法等，以激光干涉法的物理机制最为明确，且时间分辨率最高；二是能量释放与控制技术出现颠覆性变革，通过多相反应的高密度能量贮存、释放及高效率转化的热力学和动力学规律，将高能物质蕴含的物理能、化学能或物理、化学作用耦合于目标结构及功能，从而大幅提升对目标的破坏效果。

（三）能量高效利用技术

能量高效利用技术重点解决高密度能量利用效率的问题，针对常规爆炸“近场能量过剩、中远场能量不足”的制约瓶颈，利用装药爆轰热力作用强化、多域能量耦合叠加、毁伤效应调控等威力场精准控制的新思路，通过装药精密爆轰波形起爆控制、毁伤元与炸药能量结构匹配、组分体系内外能量耦合的技术途径，实现对高能炸药的有效利用。具体表现在：一是能量释放的组合化和一体化发展。采用同轴双元或多元装药，通过不同能量输出结构的内、外层装药组合，提高毁伤能量的多模式和多任务适应性；炸药与活性材料组合，产生的活性破片打击目标时的破孔直径是惰性弹丸或破片的数倍，破坏作用明显增强。二是能量输出结构的多样化和精细化发展。通过精确控制起爆时间，使炸药装药部分燃烧、部分起爆，使输出能量与目标相匹配，实现毁伤能量威力可调可控；发展了先进的定向起爆网络，可通过选择轴向、偏心方向或者邻位、间位等多种起爆方式来调整爆炸能量输出结构，大幅提高能量利用率。三是能量利用方式向多样化、异形化和灵巧化发展，先进增材制造技术为多层、异形、微装药的制造提供了一条全新的途径，为能量释放时序上的高精度控制和空间上的精细化分布提供了新方法。

（四）能量利用效应评估技术

能量利用效应评估技术重点解决高密度能量运用效能评价问题，构建了基于毁伤的目标体系、目标易损特性分析及毁伤标准、典型目标毁伤准则及等效靶设计方法，在威力场及目标毁伤效应评价基础上，综合考虑目标易损特性、威力场分布和环境等因素，对能量利用效率进行评价。具体进展为：一是从目标系统构成、结构特点、防护能力等角度，构建了常见军事目标体系，在目标易损特性分析基础上，确定了各类型目标关重构件（分系统、或部位）及其毁伤准则、等效靶设计方法等。二是构建了能量利用的威力评估模型。基于理论分析与试验研究，得到了工程计算模型、数值计算仿真模型和基于人工智能大数据的知识图谱模型；利用三维图像显示技术，实现了威力场三维重构。三是构建了能量利用的毁伤效应模型。基于长期工程实践，形成了适应不同目标和环境的破片模型、冲击波超压模型、侵彻模型、内爆模型、靶后效应模型、热辐射模型等，开发了具有自主知识产权的仿真软件和毁伤效应数据库。四是发展了多种类目标的能量利用效应评价技术，基于多种能量利用效应评价方法，掌握了人工智能图像识别、激光高速摄像、脉冲 X 射线闪光摄像、嵌入式水下爆炸冲击波测试、动态云雾浓度测试等多种能量利用效应测试技术。

三、本学科国内外研究进展比较

（一）高密度能量创制技术

国外在高密度能量创制技术领域探索活跃，美国已利用超高压技术得到了 N5 负离子金属盐、聚合氮、金属氢、聚合 CO 等样品，其得到的世界首个 CO 聚合物样品，理论爆速超过 10000 米 / 秒，爆压是梯恩梯的 5 倍。美国试图在 2040 年前抢先突破具有超过 10 倍梯恩梯毁伤效能的新技术，将工作重点放在亚稳态纳米材料、高能储氢材料、高张力键能释放材料等颠覆性含能材料发展上，力图将常规武器毁伤提升到 6 倍、15 倍、100 倍梯恩梯当量。与国外相比，我国掌握了二代、三代含能材料高品质、低成本制备技术，应用水平持续提升，新一代含能材料研究进展显著加快，已赶上国际发展步伐，部分研究成果居世界前列。金属氢、全氮材料以及高张力键能材料合成技术取得重大突破。

（二）高密度能量控制技术

国外在高密度能量创制技术领域不断提升，深化爆炸反应动力学研究，掌握了剪切点火、热点增长、点火抑制等机制，突破了铝粉反应动力学、爆轰反应动力学等技术。研制了 LX-19、PBXC-19、PAX-29 等十余种 CL-20 基高能炸药配方，部分进入装备应用；开发了新型含能黏合剂和增塑剂，发射药火药力提升至 1200 焦 / 克以上；发展了 HMX、CL-20、ADN 基高能战术推进剂新配方，标准理论比冲接近 280 秒。近年来，我国通过自主创新，打破了配方设计方法单一、能量输出结构不成体系等技术瓶颈，相关武器技术指标与国外同类武器性能相当。掌握了多种爆炸能量耦合叠加的设计方法，开发了种类丰富的高能炸药；发射药能量性能、炮口动能、射程等技术指标达到国际同类产品的先进水平；突破了新型含能材料在固体推进剂中应用的关键技术，形成了高能、低特征、高燃速、不敏感等多种类高性能战术固体推进剂。

（三）高密度能量利用技术

国外在高密度能量利用方式实现多样化，在反装甲、反坚固目标、反舰船、反空中和空间目标等方面均进展迅速，法国的多用途串联破甲技术和美国的 13.6 吨巨型钻地弹的毁伤技术和威力显著提升。相比于国外，我国掌握了非理想炸药爆轰反应机理与反应动力学、环境匹配及目标耦合关系、能量输出规律与控制方法等基础理论与技术，建立了威力可调 / 效应可选择、低附带、强光、电磁脉冲等能量利用新原理、新方法，毁伤性能达到欧美同类装备的技术水平。

（四）高密度能量安全应用技术

美欧等国的高密度能量安全运用评估体系逐步完善，根据弹药在典型的勤务处理或作战使用场景中所面临的威胁类型，建立并形成了针对 7 类典型刺激因素，以 MIL-STD-2105D、STANAG-4439 为代表的一系列弹药安全性试验及评估方法标准。与之相比，我国合成的多种热稳定性优于 TATB 的耐热化合物实现了工程应用，突破了力热及复合刺激下炸药点火机制、热刺激下装药反应烈度控制、阻燃隔热材料制备与结构设计等关键技术，掌握了反应烈度量化测试表征技术和弹药安全性评价模型与方法，形成了弹药安全性试验与评估规范。

（五）高密度能量利用效应评估技术

美国积累了大量目标易损性试验数据，形成了通用化评估方法和软件，建立了完善的目标易损性评估方法和数据库，毁伤评估方法正从基于模型向基于评估应用系统过渡，技术数据采样速度向百兆和千兆级发展，PXI 总线测量仪器开始大量应用于毁伤效应测量中。相比于国外，我国构建了火箭撬、大口径平衡炮等大型动态威力试验系统，建立了常见军事目标的易损特性评估方法及毁伤等效靶设计方法，获取了大量试验数据，形成了能量运用与评价数据库，能量运用效能评价技术研究将重点转向复杂条件下的动态威力分布规律研究和多毁伤效应评估技术攻关。

（六）高密度能量利用全流程模拟仿真

美、欧等国家在量子化学、分子动力学、反应动力学、材料模拟与人工智能辅助设计系统等领域形成了多种专业化、系列化仿真软件，如 LS-DYNA、AUTODYN、Cheetah 等软件，具备毁伤试验的全数字化仿真能力，可实现快速设计和性能预测，大幅缩减研究周期，极大提高研发和评估效率。相比于国外日趋成熟的模拟仿真能力，我国已建立了基础材料性能数据库和试验数据库，形成了可描述高密度能量释放的物理模型和能量利用的毁伤效应工程模型，掌握了多种有限元和流体动力学求解算法，形成了覆盖高密度能量富集、输出、转化、利用的全流程精确分析和仿真模拟手段。

四、本学科发展趋势和展望

新型技术的应用使未来战场逐渐趋于智能化、全域化、精确化、多元化和自主化，现代先进毁伤技术及效应评估需要聚焦新质能量融合和利用、毁伤精准控制和表征、毁伤效能预示和验证、目标损伤特性等技术，掌握火炸药增能和能量调控、能量激发和转化机制、能量与目标耦合作用规律、爆炸燃烧作用的全域精准控制、毁伤能量融合等新方法，突破多域能量耦合、新质毁伤等核心关键技术，逐步构建技术全面、水平先进、谱系完整的先进毁伤技术体系，实现武器威力可调、动力随控，提升先进武器装备的应用水平和打击效率，牵引毁伤科技发展。

第十一节　采矿工程

一、引言

采矿工程学科是支撑我国国民经济持续发展的关键学科之一，是以矿物资源的安全、高效、环境友好地开采为目的的应用性基础学科，是矿业工程学科的二级学科。该学科历史悠久，改革开放后，我国矿业取得了举世瞩目的成就，采矿工程学科也进入了新的蓬勃发展时期，保障了我国国民经济发展对矿物资源日益增长的需求。

近年来，我国采矿工程学科在绿色开发、深部开采、智能采矿方面取得了显著的进展。我国10种有色金属产量连续多年位居世界第一，2022年全国煤炭产量达到了45.6亿吨，超过全球煤炭产量的54.7%。但是，与国际相比，在高质量发展、开采深度、开采智能化程度方面还存在较大差异。未来，负碳高效充填、矿产与地热共采、深部固体资源原位流态化开采、深空资源探测与开采、深海资源开发等是我国采矿工程学科发展的重要趋势。

二、本学科近年的最新研究进展

（一）金属矿开采

1. 金属矿膏体充填理论与关键技术装备

膏体充填技术是实现绿色开采的关键技术。深入研究了全尾砂深度浓密流变行为、膏体搅拌流变行为、膏体输送流变行为和充填体流变行为，构建了金属矿膏体流变框架体系。研发了膏体充填成套技术与装备：①全尾砂膏体深度浓密技术与装备；②膏体均质化搅拌技术与装备；③膏体长距离稳态输送技术与装备；④膏体充填全流程智能化控制技术与装备。制定了国家标准《全尾砂膏体充填技术规范》与《全尾砂膏体制备与堆存技术规范》。

2. 复杂环境矿山边坡灾害防控理论与技术

探明了岩石破裂诱发机制，构建了岩石强度准则。创新了边坡设计安全系数的不确定性问题的确定性解决方案，制定了《露天矿山岩质边坡工程设计规范》。研发了完全自主知识产权的边坡变形监测设备——S-SAR 合成孔径雷达，首次提出了基于牛顿力变化测量的滑坡“双体灾变力学理论”，形成了滑坡“加固 - 监测 - 预警”一体化防控技术，首次提出了单一地连墙结构和锚拉式地连墙结构两种地连墙止水固坡结构。

3. 金属矿山采动灾害监测预警云平台

提出了一种亿级自由度结构化六面体网格智能划分算法、弹脆性损伤本构模型、微震数据驱动的岩体力学参数时空变异性动态修正算法，形成了监测数据驱动下动态模拟云服务，形成了监测和模拟相结合的灾害预测预警方法。提出了基于案例挖掘的预警指标体系构建算法，最终形成地质灾害案例库 + 现场监测大数据挖掘 + 模拟云计算分析 + 专家系统评判“四位一体”矿山灾害风险动态评价方法，并搭建了金属矿山地质灾害监测预警云平台。

4. 地下金属矿生产作业链全过程高效智能协同技术

研究和攻克了矿山生产作业数据持续精准快速获取和多源异质数据集成管理、井下无轨装备自动驾驶、井下环境和作业工况可视化集成管控以及全作业链生产过程实时调度等核心难题，实现了地下金属矿生产作业链全过程高效智能协同，实现矿山智能化回采。革新了地下金属矿山作业模式，实现了地下金属矿生产作业链全过程高效智能协同，建成了典型金属智能矿山示范。

5. 金属露天矿智能管控关键技术

建成了一种全方位的新型金属露天矿无人采矿智能生产管控系统，攻克了智能管控关键技术：①无人驾驶车辆开放式体系结构设计与集成控制技术；②露天矿区复杂环境下的无人驾驶自主运行及避障技术；③多金属多目标露天矿全要素智能精细化配矿技术；④数据驱动下的露天矿无人驾驶多车协同智能调度技术；⑤云服务下的金属露天矿无人开采一体化管控平台。

（二）煤矿开采

1. 采动应力与岩体力学理论

研究了深井超长工作面采动应力分布及其旋转规律，提出了分数阶模型描述采动应力衰减模式，揭示了旋转性采动应力驱动采动裂隙扩展机理。基于地应力和采动应力旋转轨迹，提出了工作面推进方向优化原则。阐述了巷道周围主应力旋转与巷道非对称破坏的联系，发展了蝶形破坏理论。开发了煤层上覆厚硬顶板区域水力压裂技术，提出了“人造解放层”卸压防冲方法。

2. 厚煤层开采基础理论与关键技术装备

基于顶煤裂隙分布、应力加卸载和应力旋转复合效应，构建了顶煤破碎块度预测模型，建立了大采高综放采场覆岩“悬臂梁 - 铰接岩梁”结构模型，得到了煤壁稳定性与顶板压力、煤壁强度、煤壁高度、支架阻力等参数的关系，揭示了煤 - 矸 - 岩放落流动的时序规律。开发了特厚煤层大采高综放开采技术、急倾斜厚煤层综放开采技术等，在智能化综采与智能化放煤技术和装备等方面取得了突破性进展。

3. 岩层运动与围岩控制关键理论与技术

建立了采场“多参量感知 - 分析模式判别 - 设备自主决策 - 工序快速执行 - 效果动态评价”的智能控制技术构架，创新了开采系统智能化、装备自适应巡航、岩层运动原位感知等智能控制技术；开发了井下智能化分选及就地充填技术，实现了“采选充 +X”一体化安全绿色高效开采；提出了地面钻孔压裂与井下顶板预裂相结合的坚硬顶板远、近场协同预控方法。

4. 煤矿智能开采装备与技术

智能化工作面“采、支、运”三大装备都有了长足的发展与进步，采煤机借助控制器局域网总线及工业以太网技术，实现油位、变频器、电机、传感器等多种参数检测、控制与保护。一次采全高液压支架的最大支护高度已经达到 10.0 米，综采放顶煤液压支架的最大支护高度已经达到 7.0 米，单台液压支架重量达到 100 吨。提出了基于“数字孪生 +5G”的智能矿山建设新思路，构建了全域感知、边缘计算、数据驱动和辅助决策的智慧矿山平台。

5. 绿色开采理论与技术

研究了充填开采矿压显现规律、关键岩层运动及地表变形特征，构建了以控制关键岩层弯曲变形为目标的关键岩层充填控制理论。在地下水保护、地表低损伤、矸石和瓦斯近零排放的工程需求下，形成了“采选充 +X”的绿色化开采模式，开发了“采选充 + 控”“采选充 + 留”“采选充 + 抽”“采选充 + 防”“采选充 + 保”等关键技术，促进了井下分选与原位充填在煤矿的应用。

三、本学科国内外研究进展比较

随着碳达峰、碳中和战略的推进，世界各国在提供安全稳定的能源保障基础上，加快向生产智能化、管理信息化、煤炭利用洁净化转变。目前，国际研究前沿主要有自动化与智能化技术、低成本采矿技术、绿色采矿、机器学习和数据分析、深海采矿。结合我国本学科近年的最新研究进展，本学科国内外研究进展比较如下。

国内外煤炭行业高质量发展存在较大差异。经过 40 多年的发展，我国综放开采技术已经达到世界领先水平。但是我国煤矿矿井开采效率和技术水平不平衡、市场结构不平衡、煤炭利用清洁化程度与质量水平不平衡。同时，我国煤炭行业安全发展不充分、绿色发展不充分、低碳发展不充分、人力资源发展不充分、对外合作不充分、企业转型升级不充分。

国内外金属矿的开采深度存在较大差异。当前我国金属矿山的开采深度最大达到1990 米，尚未达到 2000 米。而南非、印度、加拿大、美国、智利等多个国家金属矿的开采深度已经达到或超过 3000 米，尤其南非已经有 10 余座金属矿山的开采深度达到或超过3000 米，最深可达 4800 米。由此可见，我国金属矿山的开采深度与国外矿业发达国家存在巨大的差距。

我国金属矿深部智能开采存在较大的差距。目前，我国矿山多数已建成了井下光纤主干通信网络，大部分矿山有环境感知设备、自动化控制系统、灾害监控系统和矿山管理软件等。但是，与国外矿山相比，智能化发展相对滞后，开采环境智能感知技术薄弱；缺乏深部高应力、高温条件下的高效采矿技术，采矿成本高，井巷工程推进速度慢；矿山机械化装备配套性差，井下大型采掘设备的制造水平低；采矿生产管控一体化综合信息平台开发相对滞后。

四、本学科发展趋势和展望

（一）煤矿负碳高效充填开采理论与技术

创新负碳高孔隙充填材料结构拓扑构型与强度理论以及充填体固碳理论、快速黏凝胶结材料反应动力学理论、矿区充填开采防治冲击地压等负碳高效充填理论构想，研发矸石快速高效胶结高孔隙充填材料制备技术、快速黏凝胶结材料绿色高效制备技术、充填体负碳高效充填开采技术、多面并采高效充填开采技术与工艺、全周期立体高效充填开采防冲技术等关键技术体系。

（二）矿产与地热资源共采理论与技术

加强矿 – 热共同赋存区勘探，尽快开展矿 – 热资源共采试点。发展深部高温坚硬岩层破岩与掘进技术。加强深部多场耦合环境岩石力学理论与试验研究。建立矿 – 热资源共采热能分级利用体系。

（三）深部固体资源原位流态化开采理论与技术

突破固体矿产资源临界开采深度的限制，构建在井下原位实现煤矿无人智能化的采选充、热电气等转化的流态化开采技术体系。研发深部金属矿原位溶浸开采和采选充一体化开采理论与技术。

（四）月球火星等深空资源探测与开采的探索研究

研究月球火星资源的遥感探测与获取技术，创新月球及火星大深度原位保真取芯技术，研究月球和火星地下恒温层空间利用构想及技术实施方案，研发月球和火星地下活动空间、地下热量存储、温差发电、矿物开采结构等配套技术，形成月球/火星地下空间开发利用的新理论、新技术、新方案。

（五）深海资源开发理论与方法

研究深海采矿系统水动力学，研究深海矿物岩石－采掘剥离设备的耦合响应特征，建立岩性探测汇报与数值在线分析联合系统，开发采掘机智能决策与智能采掘剥离方法，建立深海采矿系统对多变海洋环境响应机理与模型。

第十二节　种子学

一、引言

“国以农为本，农以种为先”，种子是农业的“芯片”，是农业生产中不可替代的、最基本、最重要的生产资料，也是人类生存和发展的基础，在整个农业与国家经济中有不可替代的作用。种子学是通过研究农作物种子的特征特性、生理功能和生命活动规律，为农业生产服务，解决种子生产中存在的各类科学技术问题。种子学是一门既古老又年轻的学科，作为一门科学被系统研究距今约一百多年的历史。早期的种子学主要包括种子生物学和种子生理学。20 世纪，随着人们对种子在农业生产中的重要地位的认识，推动了种子科学研究的发展。近年来，随着全世界种子产业发展和种子产业化进程的突飞猛进，种子科学在传统学科和新兴学科（如分子遗传学、分子生物学和基因工程等）的基础上，已扩展为种子科学与技术，研究已从群体拓展到个体，从细胞水平拓展到分子水平。种子学属于基础研究和应用紧密结合的学科，种子研究与生产需要紧密结合，在更广阔的范围内为农业生产服务。

本节对近五年来我国种子学科的发展进行评述和归纳，回顾、总结我国近几年种子学科的新理论、新技术、新方法和新成果等发展现状，结合 2022 年以来农业领域的重大专项，对涉及作物育种等方面的关键技术进展进行凝练，简要介绍种子学研究领域取得的进展，并分析比较国际上本学科最新研究热点、前沿趋势和发展动态。根据近五年种子学发展现状，对比国内与国际农业技术发展差距，分析我国种子学未来 5 年发展战略和重点发展方向，提出相关发展趋势和发展策略。

二、本学科近年的最新研究进展

（一）种子学科最新理论与技术研究进展

1. 种子与农作物产量品质

种子作为农业的“芯片”，是农业增产、粮食安全重要的保障。良种对农业的增产稳产起到关键性作用，据测算，良种的选育和推广对单产提升的贡献率在 50% 以上。例如，我国大豆、玉米受育种及栽培等因素影响，单产水平只有世界先进水平的 60% 左右，产量差距背后是品种的耐密性和抗逆性差异。近十年来，国家不断出台一系列促进粮食增产的政策，我国种业稳步发展和持续增强，农业高质量发展成效显著。我国种子相关专利数量复合增长率达到 28.8%，良种对粮食增产贡献率超过 45%。随着生活水平的提高，人们对于优质粮的需求增加，科研育种方向逐步从注重产量向产量品质并重转变。从吃得饱向吃得好的转变，再到吃出营养吃出健康，对于良种的选育提出了更多的需求。

2. 种子生产

种子生产是农业生产的第一步，对保障粮食安全和农业可持续发展作出了重要贡献。我国主要农产品的种子自给率处于世界领先地位，目前，我国已建立了超级稻、矮败小麦、杂交玉米等高效育种技术体系，农作物自主选育品种面积占比超过 95%。近年来，随着科技的不断进步和农业生产的不断发展，种子生产也在不断出现新的进展，包括基因编辑技术、单倍体育种、分子设计育种、智能化农业和植物工厂等。这些新技术和新模式的应用，为种子生产带来更加广阔的发展前景和更多的机遇。科技的进步为种业创新赋能，实现了育种模式和流程的变革，为保障我国粮食安全起到了重要的支撑作用。

3. 种子贮藏与加工

种子贮藏技术是通过将种子保存在适宜的环境条件下，以延长种子的保存期限，保证种子的质量和可用性。种子贮藏技术的关键是控制温度、湿度和氧气浓度。2021 年 9 月，新建成的国家农作物种质资源库正式投入运行，基本实现了种子的超低温保存，还可以保存试管苗和 DNA，保存全过程实现了智能化、信息化，种子贮藏寿命可以达到 50 年。种子贮藏技术的应用不仅可以提高农业生产的效益，也可以减少资源的浪费和环境的污染。

种子加工技术是指对种子进行处理，以提高其品质和适应性。近年来，我国对于种子加工设备的研究重视程度明显提升，我国农业已基本实现了全程机械化作业，烘干机、清选机、精选机、包衣机等设备基本实现了数字化控制，在智能控制下，实现传感器测量，提高了种子处理的精确性。同时，风力平衡式清选、自平衡重力精选、光学色选、精确丸粒化处理等新技术逐渐得到更广泛应用，加工处理后的种子品质得到了进一步提升。

4. 种质资源收集与保护

作物种质资源是作物种业创新的源头。种质资源是生物携带遗传信息的载体，具有实际或潜在利用价值，其形态包括种子、植株、茎尖、休眠芽、花粉甚至是 DNA 等。“十三五”时期我国加快种质资源保护与利用体系建设，建成完善了由 1 座长期库、1 座复份库、10 座中期库、43 个种质圃、205 个原生境保护点以及种质资源信息中心组成的国家作物种质资源保护体系。截至 2022 年年底，我国收集保存资源总量突破 54 万份，保护了一大批珍稀濒危资源；每年向科研、育种和生产提供有效利用 10 余万份，有力地支撑了我国的作物育种和农业科技创新。

（二）种子学科在种子产业中的应用

种子产业是一个系统工程，主要是包括：优良品种选育、大田生产用种的生产繁殖、优质种子的加工包衣处理、种子质量检验包装贮藏流通销售和种子产业管理五大系统。经过多年发展，我国在主要作物种子领域的研究取得了较大突破，极大地推动了种子产业的发展。在种质资源鉴定评价和创新，在种子发育、种子休眠和萌发、种子劣变和耐贮藏理论，新品种选育理论与技术、种子繁育技术、高活力种子生产关键技术、杂交制种技术与关键设备、种子加工取得了系列成果。近年来，我国种业科技创新取得显著成效：一是开展了农作物种质资源的规模化表型精准鉴定与全基因组基因型鉴定，为新品种选育及基础研究提供了重要物质基础；二是精细定位和克隆了一批重要性状的有利基因，为农作物分子定向设计育种提供了重要基因资源与路径；三是在农作物基因编辑、单倍体育种等核心关键育种技术上取得明显突破，加速了农作物高效育种体系的快速应用；四是综合利用植物分子设计、染色体细胞工程、诱变生物工程及杂种优势利用等技术，培育出一批精品优质水稻、抗旱节水小麦、高产机收玉米、早熟优质蔬菜等农作物新品种。我国种业企业依靠科技创新和产品创新已跻身全球种业前十强，为保障国家粮食安全和促进农业绿色发展提供了强有力的科技支撑。

三、本学科国内外研究进展比较

近几年，种子学无论是推动学科自身发展的基础与应用基础研究，还是与相关学科的交叉发展及新应用领域的拓展，均获得了显著成就，对于推动我国种业科技自立自强、种源自主可控产生了重要影响。我国农作物选育水平、良种水平和供应能力显著提升，自主选育的品种种植面积占到 95% 以上，做到了“中国粮主要用中国种”。猪牛羊等畜禽和部分特色水产核心种源自给率分别达到了 75% 和 85%，这些都为粮食和重要农副产品的稳产保供提供了关键的保障和支撑。当前我国种业市场规模已居全球第二，仅次于美国，但从种业自身来看，种质资源保护利用还远远不够，自主创新能力还不强，特别是在育种的理论和关键核心技术方面，与前沿国家相差还甚远。

在种质资源的发掘、鉴定方面，尽管我国保存的种质资源总量超过 52 万份，位居世界第二，但我国现有种质资源仍存在许多不足，现存种质资源结构较为单一，多以本国资源为主，国外资源占比较少，资源多样性较差，许多抗病、抗逆、优质育种材料还未收集到。在种子学基础理论研究方面，我国在生物育种科技创新和专利发明方面取得明显进步，在部分细分领域与以美国为代表的发达国家的差距明显缩小，在某些方面甚至已经超越发达国家，但具有重要使用功能基因的发现较少，重要经济性状形成的生物学基础及调控机制，品质、产量与抗性协调改良以及非生物逆境与作物发育的相互作用机理等方面研究薄弱。在育种关键技术方面，我国育种技术与国际先进水平存在明显的代际差距。国际育种技术基本已经进入以生命科学、信息科学与育种科学深度融合的智慧育种“4.0 时代”，建立了基于基因型－表型－环境数据采集与模拟分析的智慧育种研发体系，并已从温室表型技术拓展到大田表型技术，每年研发数据处理能力均在成倍增长。而中国育种大多处在以杂交选育和分子技术辅助选育为主的 2.0 时代至 3.0 时代之间，育种质量、效率以及品种地域适应能力等与国际先进水平还存在较大差距。在种子学研究领域所用到的科研工具与仪器设备研究方面，当前我国高度依赖进口，部分技术设备仍无法从国外获得。先进科研工具和设备的缺乏严重影响了我国种子学研究的发展。

四、本学科发展趋势和展望

当今世界正经历百年未有之大变局，新一轮科技革命和产业变革突飞猛进，科学研究范式正在发生深刻变革，学科交叉融合不断发展，新一轮科技革命和产业变革重塑全球经济格局，国际力量对比深刻调整，国际环境日趋复杂，保障粮食安全成为全球共同面临的挑战。“要开展种源‘卡脖子’技术攻关，立志打一场种业翻身仗”，我国农业生物育种技术研发及其产业化发展已进入自立自强、跨越发展的新阶段。但在生物育种研发方面还面临一系列包括原始创新薄弱、关键技术缺乏和创新链条脱节在内的制约因素。如此，种子学的发展显得尤为重要。只有大力发展种业科技创新，才能加快我国生物育种技术研发与产业化进程，增强我国现代农业核心竞争力，实现科技自立自强，保障国家粮食安全、生态安全与国民营养健康。

随着人工智能、大数据等技术的发展，种子学研究领域也将逐步实现智能化和精准化，以提高农业生产效率、降低资源浪费为主要研究目标，通过先进的生物信息学数据分析、分子生物学技术以及人工智能，快速改造种子的性能，提高种子应对气候变化和环境恶化所带来的挑战。在种子学人才培养方面，现代种业和数字化信息时代的到来，种业学科人才越来越倾向于应用型和交叉型，近年来各高校通过设立新型技术交叉学科专业，重塑种业人才知识培养体系，为现代种业人才培养提供了良好的沃土。在种子学研究基础条件建设方面，建设国际一流的种业科技创新平台，通过开放共享支持和引导种子学领域的基础研究，创造种业科技自立自强的沃土。

国家政策在种业发展中占有导向性的作用。因此在国家政策层面需要有更多的支持和倾斜，通过持续强化生物育种发展的战略意义，完善国家生物育种创新发展体系，通过种子学研究领域科技创新来推动我国由生物育种产业大国向强国的快速转变，确保关键共性技术自主可控，“中国碗装中国粮”。此外，种子学研究应紧紧围绕我国种业发展和创新中亟须解决的生产问题，在技术层面有更多的原始创新，并进一步全面发展和推广。针对我国农业农村现代化对粮食安全、绿色发展、健康生活、极端气候响应和战略新兴产业发展的重大需求，精准培育和创造增产提质、减投增效、减损促稳的新型农业资源，实现对现有品种的跨越升级，引领精准农业发展。

第十三节　材料学

一、引言

学科领域是知识创新的主战场，材料学科是研发物质产品的知识体系，产品创新是推动产业发展和社会进步的强大动力，也是建设创新型国家的关键所在。材料学科是兼具基础性和先导性的新兴大学科，有机综合了基础科学（物理、化学、数学、生物等）和工程技术（冶金、化工、机械、电子信息等），旨在研究材料性质、成分与结构、合成与加工、使役性能四大要素关系。在基础研究领域，既具有其他基础学科的共性、关键性，也为各大学科提供直接的技术支撑。无论事关国家安全的国防军工及武器装备，还是经济社会与生命健康的基础工业建设，材料科学发展水平已成为衡量一个国家经济发展、科技进步和国防实力的最重要的指标之一。

材料学科涉及领域广泛，主要包括金属材料、无机非金属材料、有机高分子材料、复合材料、生物医学材料、能源材料、环境材料、电子信息材料、纳米材料、材料基因工程、材料表面与界面、材料失效与保护、材料检测与分析技术、材料合成与加工工艺等。本节内容主要聚焦部分战略前沿方向，分别从二维半导体材料、能源转换与存储材料、超材料与超构工程、空间医药微纳材料、极端环境服役材料、材料基因组工程、材料合成制备与表征这 7 个方向概述近年来的新观点、新理论、新方法、新技术和新成果。

二、本学科近年的最新研究进展

（一）二维半导体材料

二维电子学多是以过渡金属硫化合物为代表的二维半导体材料。凭借原子级的超薄厚度、高载流子迁移率、层数依赖的可调带隙、自旋－谷锁定特性、超快响应速度以及易于后端异质集成等优点，突破主流硅基互补金属氧化物半导体 CMOS 芯片技术在进一步微缩时面临的短沟道效应等物理限制，是后摩尔时代替代硅的候选芯片材料之一。大面积晶圆级材料与多层复杂结构异质结的制备技术，为高密度低功耗存储、高效光伏、高灵敏度光电探测、超短沟道、超快运算等器件应用提供了发展的原动力，并有望应用在可穿戴电子器件、传感器、生物医疗方面，代表了新型电子学器件的发展方向。

我国学者取得了系列重要进展，主要包括：北京大学物理学院刘开辉团队发展了一套适用于二维材料的原子制造技术，实现了晶圆级过渡金属硫化合物的调控生长。南京大学王欣然团队通过改变蓝宝石表面构筑“原子梯田”，在国际上首次实现了 2 英寸晶圆级二硫化钼 MoS_2 单层单晶薄膜的外延生长。北京大学彭海琳团队在国际上率先开发了超高迁移率二维硒氧化铋 Bi_2O_2Se 半导体芯片材料，建立了一系列晶体可控制备及表界面调控方法。中国科学院物理研究所刘广同课题组开展了高质量外尔半金属和伊辛超导体的合成，并发展了一套普适性的熔融盐辅助化学气相沉积策略；在 $MoTe_2$ 和 $Mo_xW_{1-x}Te_2$ 体系中开展了一系列物性研究，发现了新型的伊辛超导、两带超导、维度依赖的外尔半金属态输运证据等重要现象，处在过渡金属硫族化合物低温量子输运领域的研究前沿。

（二）能源转换与存储材料

随着全球对可持续能源的需求逐渐攀升，能源材料领域正在高速发展，目前已成为科研和产业界的主要焦点领域之一。能源材料转换材料发展应着眼于优化电极材料的设计和合成，以提升性能、可持续性和环境友好性。

我国学者在能量存储方面取得重大进展，北京理工大学吴锋院士、南开大学陈军院士、中国科学院物理所陈立泉院士等的高比能电池、高活性燃料电池体系，浙江大学叶志镇院士、南京大学谭仁海教授等的高效率钙钛矿太阳能电池，均是适应不断多样化的能源需求，促进可再生能源广泛应用的关键技术。电催化氧化制氢、析氧反应电催化等技术则是能源转化和储存领域的重要研究方向。电化学能源，将重点关注轻元素多电子电池、本征安全水系锌电池等关键领域。清华大学何向明团队研究的全固态金属锂电池，是实现高能量密度与提高安全性的未来电池技术。重庆大学潘复生、上海交通大学丁文江团队的镁基固态储氢材料，具备最高储氢密度，而且金属镁在储氢领域也因其低成本、轻质量和无污染等优势而备受青睐，是最有发展潜力的固态储氢材料之一。

（三）超材料与超构工程

超材料是由人工微结构单元组成的宏观电介质。这一超构工程，可以通过调节人工单元的尺寸、结构及空间周期或非周期排布，可按需调节宏观电磁参数（包括介电系数、磁导率系数、折射率系数、吸收系数等），从而获得电磁波的不同传播性质（包括折射、反射、透射、吸收、波矢色散、各向异性等），以实现各种新颖奇异的电磁应用（包括负折射、完美透镜成像、隐身斗篷完美吸收、辐射制冷等）。超材料发轫于微波和可见光波段，目前研究热点集中在太赫兹、红外和极紫外等电磁波频率范围，同时正开始应用于深空深海深地探测、高定向电磁对抗、5G/6G 无线通信、绿色能源等国家重大需求的国防与经济建设领域。

超材料基础研究依然强劲，每年有百篇顶刊及千篇高水平论文发表。研究热点主要集中在动态可调及可重构超材料（基于相变材料的辐射制冷、红外微波太赫兹吸波、人工单元耦合的 Fano 谐振）、量子超材料（石墨烯拓扑超导量子模拟、基于二维材料的光学调制，全光芯片计算、超材料模拟轴子暗物质）、片上结构设计超材料（电子学超材料，极紫外硅基直超表面，倾斜扰动结构的超表面）、多维复用技术超透镜（多维光场多功能，局域共振微腔涡旋光，双曲超材料合成复频波、超表面偏振复用）。在开放系统的量子实际应用过程中，同时产生了非厄米超材料和时间晶体等前沿研究方向。

（四）空间医药微纳材料

医药微纳材料是纳米科学与技术的重要战略前沿方向之一，由于具有独特的物理、化学性质及生物效应，在疾病预防与诊断、治疗及预后监测等方面发挥着重要作用，是面向人民生命健康、关乎国计民生的重大战略性新材料。近年来，医药微纳材料已被广泛用于疾病诊断、药物靶向递送与控释、组织修复与再生、智能型生物器件等前沿领域，在医药与健康、医疗器械等各方面都具有非常重要的应用价值，对未来空间生存的生命保障也具有十分重要的意义。

我国在医药微纳材料领域具有深厚的研究基础，是全球最活跃也是最有影响力的国家之一。然而，临床转化却严重不足，高端的纳米药物甚至一些传统的纳米药物都处于零的阶段。一方面，是对相关医药微纳材料合成机制的研究和理解不够深入，导致产品性能难以控制；另一方面，是对于纳米药物的开发缺乏创新。中国空间站的建成，将为空间医药微纳材料的研究发展带来新的契机。医药微纳材料的空间研究，为我国攻克微纳材料合成的机理及技术难题并开发创新型纳米药物创造了优越的条件。未来空间医药微纳材料的发展重点应集中在两个方面：医药微纳材料空间合成机制与性能的研究；医药微纳材料面向空间的应用研究。当务之急是研制配套的空间载荷装置，以用于材料合成、材料性能及相关形成机制等的空间研究。

（五）极端环境服役材料

极端环境服役材料涉及航空工业（超轻、超强、超高温的材料性能）、载人航天和深空探测功能智能化材料、海洋工程的材料腐蚀与防护性能、原子能技术应用材料的抗辐照和极端服役性能，主要应用于深空、深海、深地的极端高压、极端温度、极端辐射、极端冲击和应变速率、极端能量和燃烧反应、极端腐蚀和氢环境等诸多战略国防领域，也是新材料研发的必争之地。航空发动机、航天推进器、核聚变反应堆以及海洋极寒高原沙漠用电子设备，急需高性能、长寿命、高可靠、多用途、经济性、绿色化的新材料。我国学者在高温耐蚀合金、高熵合金、先进陶瓷涂层及碳碳复合材料等领域取得重要进展，包括中国科学院金属所韩恩厚院士耐蚀镁合金及涂层、航空材料研究院、北京航空航天大学及钢研院的高温合金，北京科技大学高熵合金，西北工业大学李贺军院士的碳碳复合材料等。

（六）材料基因组工程

随着生成式 AI 大模型 ChatGPT 已风靡全球，高通量动态实验融合数据驱动技术（如机器学习模型）和人工智能，正迅速成为快速筛选数千种微观结构和 / 或化学物质的高效和经济方法。2023 年度，材料基因工程在高效材料计算与设计、革命性实验技术、材料大数据技术、先进材料研发、工业应用等领域已经取得的重大成果。

材料高效计算与设计方面包括材料性质高通量计算工作流（如 Pymatgen、AFLOW、AiiDA、ALKEMIE 等），以及 Materials Project、OQMD、JARVIS、AFLOWlib 等系列计算材料数据库，集成计算材料工程运用第一性原理计算、材料热力学相图计算、分子动力学模拟、相场模拟、有限元模拟等多尺度计算模拟技术。更为聚焦多维度、多尺度、全流程材料设计，更为关注自主计算、集成计算、跨尺度建模设计的核心算法与软件开发。高通量计算技术、实验表征技术、传感器技术、图像识别技术、文本挖掘技术的快速发展，显著提高了数据资源积累效率。我国学者应用材料组合制备技术，结合 X 射线衍射高通量结构表征和电阻测量技术，获得具有优异力学性质的高温块体金属玻璃。中国重燃集团与北京科技大学联合打造了重型燃气轮机专用材料数据库，开发仿真 – 设计实时数据交互技术、材料服役损伤大数据分析技术，推动重型燃气轮机的数字化、智能化设计和制造。

（七）材料可控制备与表征

高通量可控制备和表征技术的快速发展，变革了新材料发现技术，大幅提高了实验研究效率。增材制造、微反应器阵列、连续掩模、定向凝固和梯度热处理等高通量制备技术，实现了系列样品平行制备，加快复杂多元新材料体系成分空间探索。基于微区集成、连续扫描、多功能叠加等技术原理的材料理化性质高通量实验技术，显著提升了材料表征效率。人工智能有机反应搜索引擎、自主实验机器人、可重构全自动实验等智能实验技术应用于化学合成的实验设计、设备操作、数据采集与分析，显著提升新材料发现和验证的效率。我国在大科学装置和实验仪器建设上持续发力，目前我国在北京、上海、合肥、粤港澳等地等同步辐射光源的大科学装置，已经拓展应用到磁学和自旋电子学材料；X 射线衍射现已广泛应用于无机薄膜和有机光电材料等领域研究；X 射线小角散射应用在高分子材料、高性能碳纤维、纳米材料、纳米薄膜等的原位研究。

三、本学科国内外研究进展比较

随着国家国力增强及相关产业布局持续加码，材料发现与智能制造发展迅速。在国际材料基础研究领域，我国科研队伍人数居全球第一。在学术成果产出方面，我国所发表的论文数近十年间稳居论文量的世界首位，学术影响力具有领先优势。在化学与材料科学优势突出，物理学等物质科学与工程研究前沿热度仅次于美国，但尚未在生物医学材料与临床科学方面取得突破性贡献，原创性基础研究仍处于跟跑位置。

在材料产业方面，金属材料、纺织材料、功能陶瓷、化工材料等传统领域产业规模稳居世界第一，产业规模约占全球产业 30%。不过我国材料检测仪器行业起步较晚，高精度科学仪器依赖进口，尚未形成产业化规模。近年来在语言模型软件与自动智能实验技术以及国产仪器创新发展，需要突破物理系统和信息系统的虚实映射和实时交互技术，构建数字孪生系统，实现材料制造过程的在线快速优化和精准控制，以支撑材料数字化和智能化发展。

我国新材料科学和技术及产业发展，正处于历史转型的关键时期。材料发展已经从解决供需短缺问题为主的发展阶段转向满足国家战略需求和提升国际竞争力的高质量发展阶段。实现我国新材料科学和技术快速、可持续发展，必须建设和完善材料知识创新体系，推动学科系统机制和研发创新，加强人才培养和国际交流与合作。

四、本学科发展趋势和展望

（一）以问题导向引领材料科学基础研究

未来我国材料科学基础研究，需要以原创性思想、变革性实践、突破性进展、标志性成果为导向，从国家重大战略和国民经济发展需求中凝练基础科学问题及关键技术卡点。重点关注的方向包括二维半导体及量子材料、能源转换与存储材料、超材料与超构工程、空间医药微纳材料、极端环境服役材料等。

（二）以数据基础构建新材料自主创新体系

未来材料科学的新研发范式以融合计算 – 实验 – 数据为基础，发展变革性实验技术，从而加速新材料的研究与应用。这需要大量积累材料基础数据，制备加工多维参数、多场环境、多过程数据，服役行为数据，融合集成计算材料工程和机器学习建模，构建可动态描述材料成分－组织－工艺－性能的构效关系、交互作用和演变行为的数字模型库和知识库，从而整合数据为基础的研发平台，构建数据驱动的新材料创新体系。

（三）以战略高度打造关键原材料供应链韧性

面向我国关键原材料的战略需求，统筹加强各领域计划和政策体系，持续打造产业链供应链韧性，强化全球矿产资源供应链的主导力。重点关注稀土及功能材料的高效利用，高性能碳纤维生产线的自主可控，金属材料的关键研制技术等。

（四）以零碳理念提升材料产业高质量发展

利用人工智能等新一代信息技术，以零碳理念赋能绿色冶金矿山，智能制造，提升能源、资源和环境化工等产业的数字化转型升级，全力推进材料产业高质量发展。

（五）以学科交叉加大生物医药材料转化力度

融合化学、材料学、工程学、生物学、医学等多学科发现，未来重点提升医药微纳材料的临床转化，力争高效高端纳米药物零突破，加大高性能有机/无机医药微纳材料研发力度，以增进民生福祉，保障生命健康。

第十四节　粮油科学技术

一、引言

近 5 年，我国粮油科学技术学科坚持“四个面向”，立足自主创新，攻克多项科技难题，在粮食储藏、粮食加工、油脂加工、质量安全等分支领域科技创新成果丰硕，赋能产业发展成效显著，学科建设整体水平大幅提升，为保障国家粮食安全作出了重要贡献。

今后 5 年，本学科将主动适应新一轮科技革命和产业变革与建设农业强国的战略部署，聚焦粮食科技创新和产业发展存在的主要问题和需求，强化基础和重点领域关键技术研究，不断提高自主创新能力，以学科建设支撑引领产业高质量发展，助力国家粮食安全。

二、本学科近年的最新研究进展

（一）学科研究水平稳步提升

1. 粮油储藏科技创新能力显著提高

建立了“中国储粮害虫 DNA 条形码鉴定”“常见储粮害虫 3D 数字标本”以及“储粮害虫线粒体基因鉴定分析”等储粮应用系统，研究了储粮害虫猖獗的生殖和低氧适应性的分子调控机理、主要储藏害虫赤拟谷盗信息素生物合成的分子调控机制、储粮害虫磷化氢抗性产生的分子机理、阐明了红外和微波干燥阻控稻谷产后损失的分子机制、开发了降温通风智能监测系统。

2. 粮食加工技术与装备水平大幅提升

建立了稻谷和小麦适度加工技术体系，实现了稻谷柔性化碾米和刷米 / 抛光，降低了碾米工序的增碎和电耗；小麦制粉智能粉师系统使小麦粉产品出率及营养物质存留率明显提高。玉米加工科技提升，有效解决了我国玉米深加工的高能耗、高水耗、高排放和低效能等突出问题。面制品、高活性酵母以及杂粮精加工装备等方面取得了新突破。

3. 油脂加工技术与装备全面推进

精准适度加工取得重大进展，引领科技创新。食品专用油脂加工技术实现对部分氢化油的全替代；功能性油脂产品走向市场，新油源得到大力开发，植物油料蛋白产品趋向系列化；油料油脂综合利用研究取得新突破；智能化、数字化技术在油脂加工和油脂装备制造企业获得应用。

4. 粮油质量安全标准及评价技术全面提升

粮油标准体系、标准化工作体制机制、标准导向性以及国际标准化工作进展显著。粮油物化特性评价技术、粮油品质特性评价技术以及粮油食品安全评价技术体系快速发展。

5. 数字化、智能化塑造粮油物流新业态

开展了数字化赋能粮食供应链创新、粮食物流新业态和新模式、粮食跨省调运网络布局优化、疫情背景下的应急物流和“一带一路”跨国粮食物流通道研究。NFC、RFID、二维码、AR 和 AGV/AMR 新技术在智能包装和无人叉车、机械臂等方面得到应用。

6. 饲料加工技术与装备持续创新，推动行业高水平发展

标准化建设稳步推进，饲料加工基础研究不断深入。饲料加工装备设备制造质量明显提升。饲料原料加工等多种饲料加工工艺技术注重全产业链和可持续发展。饲料、饲料添加剂资源开发与高效利用技术水平不断提高。快速、在线检测技术为饲料质量安全提供了有力保障。

7. 粮油营养基础研究及食品开发取得新成就

完善了粮油营养和功能成分数据库。建立了以营养保留和风险控制为核心的健康粮油适度加工新模式。全谷物和高杂粮含量主食及营养方便食品实现产业化，个性化粮油产品的开发趋向多样化。

8. 粮油信息与自动化学科的基础和应用研究达到较高水平

粮油收储、加工信息和自动化显著提升。粮油物流信息和自动化技术逐渐成熟，为粮食流通提供了新的模式和路径。建立了“早知道”“北粮南运”散粮集装箱物流信息追溯平台、“全程不落地”收储信息平台。

（二）学科发展硕果累累

1. 科学研究成果优良

（1）科技创新赋予产业发展新动能

①获得国家科学技术进步二等奖 4 项、国家技术发明二等奖 3 项；省部级奖励多项；中国粮油学会科学技术奖特等奖 4 项、一等奖 25 项；②共申请专利 17159 项，获得授权 1946 项；③学科刊发的论文数量总数 25910 篇；出版专著多部；④立项或发布 190 多项国家标准，发布 164 项团体标准；⑤开发了系列粮油新产品、新装备及粮食物流设备，成果丰硕。

（2）承担国家科技专项提升创新能力

“十三五”期间多项国家重点研发项目通过绩效评价；“十四五”期间承担了多项国家重点研发项目；实施了多项省市自治区重大项目。

（3）科研基地与平台建设继续深入

建设了 20 个国家及部委级科研基地与平台，4 个原国家工程实验室纳入国家工程研究中心新序列管理。研发能力与世界先进水平的差距明显缩小，部分领域达到世界领先水平。

2. 学科建设固本强基、行稳致远

5 年以来，粮油学科形成了较为完善的基础、应用基础和工程技术应用协调发展的学科发展体系；形成了完整的职业教育、本、硕、博人才培养体系；多种形式的学术交流日益活跃；团体标准建设引领行业发展、科技奖励促进成果转化；团队建设持续加强（拥有 30 余个重要科研团队）；学术出版方面形成多部系列专业教材、专著和学术期刊，如《小麦工业手册（四卷本）》《中国粮油学报》《粮油科技（英文版）》等；通过公众号、科普书籍、建设科普基地、讲座等多种形式开展科普教育和宣传。

（三）学科在产业发展中的重大成果及应用

粮油科学技术创新的一批重大成果得到推广应用，产生了显著的经济和社会效益。具有代表性的成果包括：粮食气膜钢筋混凝土圆顶仓建成试点成功；粮食库存数量网络实时监测关键技术及系统研发与推广，实现了“人防”向“技防”的突破；稻米低温仓储成套技术装备集成与示范和大米适度加工产业化技术装备的应用示范；小麦制粉智能粉师系统得到推广应用；油脂加工一体化、数据化管理体系建设；粮食和食用油的精准适度加工技术广泛推广；构建了全球最大的乳脂数据库；创制粮食收购智能扦样系统和散粮堆表面三维点云数据采集技术；畜禽饲料非法添加物与有毒有害物质检测、加工过程质量安全控制和追溯体系等关键技术实现突破；各类预制粮油食品逐渐兴起，并得到迅速的发展。

三、本学科国内外研究进展比较

（一）国外研究现状

1）粮食储藏基础与技术研究并重，形成了较为完善的粮食收储运技术研究体系。

2）粮食、油料油脂和饲料加工领域装备制造更为精细，新产品和引领性技术不断涌现，质量检测设备与检测技术处于领先地位，资源开发和加工副产品高值化利用充分。

3）粮油质量安全标准和评价体系比较完善，先进、高效和无损化的检测仪器设备保障检测能力不断提升。

4）粮食物流高效智能化创新装备不断涌现，区块链物联网等物流系统数字化技术应用步伐加快，粮食物流系统效率明显提升。

5）深入开展粮油食品健康作用、膳食模式和功能因子的研究。对功能因子的研究更加深入，其多元化影响被挖掘。

6）粮油收储信息和自动化方面注重区块链、大数据与云计算的结合应用，并随着粮食领域监管信息化不断深化。

（二）国内研究存在的差距和原因

与发达国家相比，我国粮油产前和产后的结合不紧密，粮油生产未充分考虑终端消费的需求，全产业链中薄弱环节明显；跟踪国际前沿的时候，忽略了中西方饮食文化和结构的差别，只是模仿或照搬，没有把重点放在中国传统粮油食品的研究上；基础理论研究不够深入，加工副产品高值化利用和技术创新动能不足，高附加值产品种类少；交叉学科融合深度不够，装备的原创性开发和创新意识不强，许多装备核心部件仍需进口；“专精特新”企业少，产品同质化严重；粮油标准体系不够完善，标准体系建设滞后，覆盖面较窄。

产生差距的原因有以下几方面：①高层次科技创新人才培养机制不完善，职业技能型人才和学科带头人等人才缺乏，团队建设相对滞后；②基础相对薄弱，理论研究和实际应用缺乏有效对接；③科研经费投入有限，持续性差，科研与实际成果转化脱节等现象依然存在；④产业链数据共享交换标准化水平低，粮食信息聚合程度不高。

四、本学科发展趋势和展望

（一）战略需求

学科建设和发展围绕"全方位夯实粮食安全根基"和"乡村振兴"等国家战略需求，以保障国家粮食安全为宗旨，践行"大食物观"，加快转变经济发展方式，支撑"碳达峰、碳中和"重大战略目标，积极推动粮油产业的高质量发展。

（二）研究方向及研发重点

1）粮食储藏学科。强化基础研究和应用基础研究；聚焦新型绿色储粮技术与装备的研发；开展节能降耗技术和装备研发；储粮技术集成配套，开展典型区域应用示范。

2）粮食加工学科。加强碾米装备的数字化和智能化开发，拓展米糠等副产物的功能性利用；推进玉米的梯次化、精准化加工；开展专用小麦粉等产品的加工和保鲜技术研究；建设自动化和智能化生产线以及酵母菌种资源库；开展杂粮和薯类食品的基础和高值化加工关键技术研究。

3）油脂加工学科。聚焦油料资源开发、油脂加工技术等传统科学研究内容，结合合成生物学、组学分析等新兴学科，推进油脂加工学科交叉融合发展。

4）粮油质量安全学科。完善粮油标准体系；发展完善快速、智能化的检测技术；完善粮油质量安全数据库和预警模型，提升粮油质量安全检验监测和预测技术水平。

5）粮食物流学科。开展适应新发展格局的粮食物流布局、供应链理论应用研究、"一带一路"背景下发展战略、粮食安全应急物流保障体系构建、粮食物流技术标准体系完善和粮食物流新技术、新装备等项目研究。

6）饲料加工学科。加强饲料应用基础研究，对影响饲料加工技术水平的瓶颈问题进行攻关；多途径加强饲料资源、新型饲料添加剂的研发与产业化应用研究；提升饲料加工装备与工艺技术智能化水平。

7）粮油营养学科。注重应用基础研究，搭建粮油营养健康大数据共建共享平台；攻克关键技术，努力提升健康食品开发能力；完善具有中国特色的营养健康粮油食品标准体系，引领粮油制品产业健康发展。

8）粮油信息与自动化学科。建立粮油收储全流程自动化体系；构建粮油加工智慧工厂和智能仓储管控系统；安全生产 AI 监管预警；AI 智能预警分析和研判决策；构建粮食全过程数据中心，促进粮食质量安全的全链条追溯。

（三）发展策略

一是持续深化粮食体制改革，优化收储制度，强化定价能力；二是积极参与全球粮食治理；三是搭建粮食科技创新大平台，充分发挥科技创新的引领支撑作用；四是持续开展粮食产业“五优联动”“三链协同”，深入推进“六大提升行动”和“优质粮食工程”。

第十五节 基础农学

一、引言

本节梳理了作物栽培与耕作学、植物保护学、农业信息学、农业资源环境学、农业生物技术、农产品贮运与加工学近年来的最新进展与代表性成果，对国内外研究进展进行分析比较，并展望了未来的发展趋势。

二、本学科近年的最新研究进展

（一）作物栽培与耕作学

围绕作物生产丰产、提质、增效、绿色发展新趋势和技术创新需求，在挖掘高产潜力、提高水肥资源利用效率、作物产量－品质协同提升、精确智慧化栽培耕作技术构建、作物布局配置和农作制度优化等方面取得显著进展。一是与基因组学、蛋白组学互相渗透交融，揭示作物高产生理及其分子机制，开辟了栽培机理认识与调控新领域；二是不断深化作物－环境－栽培措施的互作机制解析，创建适合不同区域特点作物栽培技术模式；三是作物生产农机农艺深度融合和智慧农作技术水平不断提升；四是兼顾产能提升和生态效益的新型多熟种植得到发展；五是探索气候韧性与低碳农作技术，构建抗逆丰产和绿色栽培技术体系；六是完成基于大数据平台的耕作制度新区划，重新界定了我国耕作制度的熟制界限和区域划分。

（二）植物保护学

植物保护科学发展已进入复杂性研究的新领域，通过多学科联合攻关，近年来取得了大量高水平原创性研究成果。一是构建草地贪夜蛾监测预警与可持续控制体系，以新型生物农药、种衣剂和植保无人机撒施微型颗粒剂施用技术为主，结合高效低毒化学农药应急防控为辅，精准及时有效控制草地贪夜蛾危害；二是发掘及利用植物抗病及感病基因，鉴定发掘出多个水稻、小麦、玉米等重要农作物相关的抗病和感病基因；三是杂草危害机制和防控技术持续深入，建立了稻田杂草群落消减控草技术，基本实现了麦田杂草防治策略精准、防治药剂精准和施药时间精准；四是原创性农药分子靶标发现取得新突破，几丁质合成机制、植物免疫作用机制等相关研究和 RNA 生物农药研发等方面有新进展；五是空地一体化精准施药技术迅速发展，植保无人机已成为我国病虫害快速、高效防治的主要作业模式之一。

（三）农业信息学

农业信息科学在基础理论、技术研发和装备应用等方面取得了一系列重要进展，信息获取手段不断创新，信息分析模型取得新的进展，应用场景不断拓展。一是农业信息获取手段不断创新，智能搜索引擎、天 – 空 – 地一体化遥感信息获取技术、新型农用传感器等得到研发应用；二是农业信息分析模型取得重要进展，农业生产模型技术、监测预警技术、人工智能技术得以研发；三是农业信息技术得到广泛应用，农业机器人、植物工场、智能养殖、智慧物流等相关技术和产品得以研发应用。

（四）农业资源环境学

在土壤培肥与退化耕地修复、雨水高效利用和智慧灌溉、丰产增效与绿色低碳协同发展等方面取得显著进展。一是土壤改良培肥、节水节肥节药、废弃物循环利用等农业绿色生产技术稳步发展，推进耕地土壤培肥改良与退化耕地治理修复；二是突破了农田精量高效灌溉、灌区高效输配水、旱地雨水高效利用等关键技术；三是深化农业生产应对气候变化的适应机制解析，阐明了气候 – 作物 – 管理交互作用机理，揭示了气候变化对粮食作物生产的影响程度和过程，构建了碳排放监测网络和减排固碳核算评价体系；四是农业废弃物资源化与高值化利用技术得到研发，创新了农林废弃物资源化利用及环境污染治理新途径；五是创新集成全过程全链条面源污染防控技术，构建种植业全过程全链条、生态种养结合和流域农业面源污染分区协同防控的面源污染防控技术体系。

（五）农业生物技术

农业生物技术已成为重构全球生物育种创新版图和重塑世界种业新格局的革命性技术。初步形成了“自主基因、自主技术、自主品种”的发展格局，已进入世界第一方阵。一是理论研究取得系列突破，种质资源收集和鉴评更加全面，育种性状形成基础研究更加系统，重要农作物生物品种或性状演化的遗传基础研究越发深入；二是方法原始创新取得系列突破，表型组、智能设计等新型育种技术不断涌现，基因工程育种技术、基因编辑技术、全基因组选择技术、合成生物技术等突破推动生物种业进入新一轮技术变革；三是学科交叉融合促进生物技术潜能的快速突破和扩大，基因组技术、表型组技术、生物技术、人工智能技术、机器学习技术、物联网技术等跨学科技术深度交叉，正实现作物新品种的智能、高效、定向培育；四是新品种培育实现突破，培育出一批玉米、大豆、水稻、小麦、油菜等作物新材料和新品种，研发出一批新型动物疫苗。

（六）农产品贮运与加工学

在强化粮食产后减损、生鲜农产品仓储物流、粮食油料加工提质增效、特色农产品品质评价与高值化利用以及预制菜肴与传统食品加工等取得积极进展。一是在玉米、花生真菌霉素防控技术，淀粉、油脂、蛋白质精细化提取和加工技术，粮油加工副产品高值化利用技术等方面取得一系列进展；二是在果蔬产地商品化和保鲜贮运，果蔬非热杀菌、绿色提取、节能干燥等新型加工技术以及番茄红素、蓝莓花色苷等活性组分高效分离和评价等方面取得一系列进展；三是在畜产品品质评价与智能仓储物流保鲜、风味与健康双导向的传统肉制品绿色制造、畜产品及副产物高值化利用、细胞培养肉等方面均取得突破性进展；四是水产品生产、加工、流通全过程的品质和安全控制技术取得突破。

三、本学科国内外研究进展比较

（一）基于论文专利的国际比较

我国基础农学学科科技论文“量质”双优，发文总量和高被引数量均排第一位。2018—2022 年，6 个基础农学学科全球共发表科技论文成果 656649 篇，我国共发表论文 218972 篇，占全球发文总量的 33%，保持稳步增长。高被引论文量累积 3200 篇，占全球高被引论文数量的 50%。我国发文总量和高被引论文量均排第一位。

我国农业专利成果产出水平不断提升。2013—2022 年，我国农业领域发明专利申请数达 142.93 万件，位居全球第一。2018—2022 年，我国农业领域授权发明专利数量为 12.33 万件，约占全球农业授权发明专利的 48%，授权量居全球首位。在植物保护技术、农业信息技术、农业生物技术、农产品贮运与加工等技术领域的专利产出保持领先优势。

（二）作物栽培与耕作学

我国作物栽培与耕作学对于以高产为目标的高效栽培、耕作理论与技术研究已处于世界前沿。水稻、小麦、玉米三大粮食作物高产创建处于国际领先水平。但在现代农业新技术、新装备应用及综合推广上与国际先进水平存在一定差距。我国水稻、小麦、玉米、大豆大面积生产的产量水平提升不明显，与国际最高单产水平差距显著。优质专用型作物生产和绿色标准化技术方面相对落后，标准体系不健全，技术产品覆盖率不高。作物精准化智慧化现代生产技术刚刚起步。

（三）植物保护学

我国植物保护学整体处于世界先进水平，其中监测预警、生物防治处于国际领先水平，转基因安全评价与控制、基因编辑等处于国际先进水平。农作物病虫灾变规律研究越发深入，在小麦赤霉病、稻瘟病、作物疫病、棉铃虫和稻飞虱等部分重大病虫领域的病虫害 - 作物 - 环境互作机制研究与国际先进水平差距逐步缩小。农作物病虫灾变监测预警技术快速发展。农作物病虫害防控核心技术及产品更新换代，但自主创制农药大面积推广不足我国农药使用量的 10%。农作物病虫害防控理论体系日渐完善，综合防控国际领先。外来入侵、新农药创制和智慧植保等方面还存在一定差距。

（四）农业信息学

我国农业信息学科的发展已取得重要成就，与发达国家相比有一定的特色优势，在高端农业传感器、农业人工智能多模态模型、农业高端智能装备等方面还存在差距。农业环境信息传感器的国内市场占有量超过进口产品，但在精度、稳定性、可靠性等方面与国外产品差距较大，核心感知元器件、高端产品严重依赖进口。在农产品产量预测、消费量分析、价格信息建模方面具有特色，在农产品全产业链监测预警集群建模方面具有先进性，但在大数据深度学习、数据挖掘以及模型算法等方面原始创新不足。基本构建了适应我国农业生产需求的智能农机装备技术及产品体系，但农业智能控制与农业机器人关键技术及核心零部件落后于美国、德国、日本等发达国家。

（五）农业资源环境学

我国在耕地质量提升、旱地适水种植、农业废弃物高值化利用等领域达到国际先进水平。在土壤有机质提升、旱作农业与节水农业、农业农村减排固碳和全链条面源污染防控研究方面取得一系列理论和技术突破。突破了水田有机质动态无法准确预测的难题，在作物水分生理生态、生物性节水理论与技术、旱作农业与节水农业模式，农业资源与环境和农业绿色发展等研究均取得国际领先的科技成果。但与国际相比，仍需在农业资源环境信息化技术融合和集成创新、场景驱动应用等方面加强基础理论和技术研发。

（六）农业生物技术

我国农业生物技术在基础理论、技术创新和产品创制方面都取得长足进步，整体上已经迈入国际前列，但原始创新及发展深度还不够，成果应用能力需进一步加强。前沿农业生物技术的拓展开发和迭代升级不断突破，但核心算法和模型缺乏，对外依存度高。农业生物技术产品的储备充足、成熟度不断提升，但产品产业化及重大产品迭代升级滞后。

（七）农产品贮运与加工学

我国农产品贮运与加工学科建设取得重要突破，与发达国家相比有一定的特色优势，尤其农产品加工技术装备不断创新，生鲜农产品保鲜和贮运技术迭代升级，农产品精深加工向着全组分梯次利用和绿色低碳适度加工方式转变，农产品加工技术与装备向着智能化、精准化方向发展，精准营养的个性化未来食品不断涌现。但在原始创新和核心技术装备上还存在一定的差距，亟须在多维品质评价、低碳化和智能化加工、全组分利用、精准营养的基础理论和前沿技术探索等重点方向发力。

四、本学科发展趋势和展望

（一）作物栽培与耕作学

破解作物丰产增效协同和降低资源环境代价一直是国际农业科技的研究热点与前沿。积极探索作物生产精确化、智慧化技术，不断提升农作物生产技术水平。将现代作物生产理论、信息技术、农业智能装备等综合应用于作物生产管理过程，实现作物生产管理从粗放式到精确化、从经验性到智慧化的方式转变。

（二）植物保护学

植物保护学的未来研究方向应着眼于适应农业生产新形势的农作物病虫害新规律、新对策，适应现代科技新发展的植物保护新理论、新方法，研制满足大区域、长时效要求的农作物病虫害检测、监测和预警技术，研发满足农产品安全需求的农作物病虫害防控新技术、新产品及满足自动化、智能化要求的智慧植保新装备与施用技术。

（三）农业信息学

农业信息学的未来研究集中于农业信息获取、处理、利用与服务等重大科学问题和关键技术难题。在农业数据获取技术方面，朝着智能化、自动化、网络化和大数据化的方向发展。在农业信息分析技术方面，朝着精细化、模型化、集成化的方向发展。农业信息技术更深度融入农业生产、经营、管理、市场、服务等各个环节的应用场景。

（四）农业资源环境学

生物与信息技术在农业资源环境学中的应用，将为世界性农业生产中的资源高效利用和环境难题提供革命性解决方案。强化生物技术、信息技术和农业资源环境学的交叉融合，尤其是数据密集型科学研究范式在农业资源环境学中的应用，攻克农业资源环境基础性和技术性难题，突破农业资源、环境要素、废弃物资源理论创新和关键技术。系统探索农业资源环境场景驱动创新的内涵特征、理论逻辑、实践路径，深入研究典型场域资源环境要素时空演变规律及环境响应机制，强化场景驱动式的集成创新。

（五）农业生物技术

农业生物技术相关的基础理论研究将更加系统和深入，原始创新能力的提升，突破重大育种价值基因克隆，阐明复杂性状形成的遗传基础，解析分子调控网络。推动基因编辑技术原始创新，开发新型基因编辑核心工具。推动构建多维数据收集挖掘为基础、数据建模预测为指导的智能化育种技术体系。实现农业生物新产品逐步向规模化和工程化转变。突破性种源的创制以及转基因、基因编辑等产品产业化应用推进，是未来的突破重点。

（六）农产品贮运与加工学

生物、信息、工程科学等多学科、多技术的交叉融合将开辟新领域、提出新理论、发展新方法、催生新业态。加快创建未来加工理论和技术，拓展食物来源渠道。建设农产品冷链物流、仓储和加工技术体系。发展高效低耗、绿色低碳加工技术。促进农产品加工的营养化、功能化转型，创新个性化营养健康产品。

第十六节　植物保护学

一、引言

植物保护学科是控制农作物生物灾害、保护农业生态系统、控制环境污染和外来生物入侵、遏制生物多样性不断丧失的一门一级学科，为保护国家农业生产安全、保障农产品质量安全、减少环境污染、维护人民群众健康、促进农业可持续发展提供重要的科学支撑和技术保障。近年来，植物保护学科在基础科学和应用科学领域双向发力，坚持“四个面向”，不断创新和丰富农业重大生物灾变的微观解析方法和理论，创新发展植保绿色防控技术和产品，不断发展和完善农业重大生物灾变的防控技术体系，为促进农业农村高质量发展和乡村振兴作出了切实的贡献。

二、本学科近年的最新研究进展

（一）提出一批新理论、新方法，引领学科创新发展

1. 植物病理学

（1）病原物致病性

发现了稻瘟菌效应蛋白 MoCDIP4 在水稻中靶标线粒体分裂相关的 OsDjA9-OsDRP1E 蛋白复合体，通过影响 OsDRP1E 的蛋白丰度调控水稻的线粒体分裂和免疫反应。发现条锈菌效应蛋白 Pst_A23 直接与可变剪接位点特异 RNA 基序结合调控寄主抗病相关基因的可变剪接，抑制寄主免疫反应。

（2）植物抗病性

发现水稻广谱抗病 NLR 受体蛋白通过保护免疫代谢通路免受病原菌攻击，协同整合植物 PTI 和 ETI，进而赋予水稻广谱抗病性的新机制。成功克隆了来源于长穗偃麦草的抗赤霉病主效基因 *Fhb7*，揭示了其抗病分子机理和遗传机理。鉴定了多个小麦感病基因，通过基因编辑技术突变，显著提高了小麦对条锈菌的抗性，实现了小麦对条锈病的广谱抗性。

2. 农业昆虫学学科

（1）害虫变态发育与滞育调控机制

发现了表皮生长因子受体（EGFR）对 JH 合成的促进作用及 JH 对昆虫卵形成和产出的机制。解析了多巴胺受体、GPCR 和转录因子 *KLF5*（Krüppel-like factor 15）介导的 20E 信号通路调控害虫变态发育的机制。

（2）害虫迁飞机制与规律

通过连续 18 年（2003—2020 年）对夜间迁飞过境昆虫进行持续监测发现了迁飞植食性昆虫和天敌昆虫的丰富度的下降趋势。筛选出欧洲小红蛱蝶（*Vanessa cardui*）迁入量的 2 个关键环境因子，提出了褐飞虱（*Nilaparvata lugens*）迁飞的控制模式。

（3）害虫抗性与化学通讯机制

发现了细胞色素 P450 基因 *CYP6CM1* 和 *CYP4C64* 的过量表达导致烟粉虱对烟碱类杀虫剂吡虫啉和噻虫嗪的抗性增强。对棉铃虫的气味受体基因家族功能进行了系统研究，阐明了昆虫编码寄主植物挥发物的基本原理。

3. 杂草学科

发现差异耐寒性与 CBF 冷反应通路基因的相对表达水平以及该通路的调控因子 *OsICE1* 启动子区甲基化水平密切相关，阐明了杂草稻和栽培稻对低温的适应机制。发现除草剂安全剂解草啶能够加速水稻体内丙草胺的降解，降低丙草胺引起的水稻植株脂质过氧化和氧化损伤，并鉴定出保护水稻免受丙草胺药害的关键基因，揭示了通过选择性诱导水稻 GSTs 基因上调表达从而缓解除草剂药害的机理。

4. 鼠害学科

开展鼠害治理生态阈值研究，客观评价鼠类在不同生态系统中的功能。在农区重大鼠害褐家鼠的入侵暴发机制、洞庭湖区东方田鼠种群数量暴发机制及该地区鼠类群落演替规律等方面取得了重要进展。

5. 绿色农药创制与应用学科

（1）农作物病虫草害的分子靶标及作用机制

在免疫诱抗剂调控植物抗病导向的原创分子靶标发现、农药活性探针分子导向的原创分子靶标发现、农药分子靶标的结构生物学等方面形成了自身的特色，发现了一些潜在原创分子靶标和绿色农药品种。

（2）农药抗性与治理的研究

发现 m^6A 能够调节烟粉虱中的细胞色素 P450 基因表达，从而导致对噻虫嗪产生抗性。棉蚜对氟啶虫胺腈的抗性机制和摄食行为、生活史等潜在的适应性成本，抗性种群对新烟碱类、拟除虫菊酯和氨基甲酸酯类杀虫剂产生交叉抗性，更积极寻找食物，繁殖力显著高于敏感种群。

6. 生物防治学科

发现寄生蜂的 miRNA 能够跨界调节寄主蜕皮素受体的表达来抑制寄主的生长发育，通过水平转移从细菌中获得毒液蛋白基因，调控寄主的免疫反应。

7. 入侵生物学学科

以苹果蠹蛾、番茄潜叶蛾、薇甘菊、紫茎泽兰、福寿螺、非洲大蜗牛等农业入侵生物为对象，构建了外来入侵物种组学数据库，揭示了苹果蠹蛾和番茄潜叶蛾等入侵害虫的内在优势和竞争力增强的入侵机制，薇甘菊在全球入侵过程中的内在优势和化感作用的入侵分子机制等。创建了潜在和新发入侵物种数据库及预判预警平台，完成了 100 余种潜在和新发重大入侵物种的全程风险驱动综合定量评估预判，初步实现了信息的即时可视化显示。

8. 农作物病虫害监测预警学科

通过不断完善旋转极化垂直昆虫雷达软硬件技术，将昆虫雷达的距离测量精度由原来的 50 米提高至 1.25 米，雷达盲区由 200 米降低至 80 米左右。研发了兼具扫描模式和波束垂直对天观测模式的 Ku 波段相参高分辨全极化昆虫雷达，将距离分辨率进一步提高到约 0.2 米。

（二）研发一批关键技术和产品，推动产业发展升级

1. 关键防控技术研发取得重大进展

首次研制成功将全相参、高分辨、全极化等新技术用于昆虫雷达，研发了兼具扫描模式和波束垂直对天观测模式的 Ku 波段相参高分辨全极化昆虫雷达，将距离分辨率进一步提高到约 0.2 米，并能获得目标昆虫高精度的体轴朝向、体重 / 体长、振翅频率、速度

和上升下降率等参数，已在草地贪夜蛾、黄脊竹蝗等境外重大害虫迁飞监测中发挥了重要作用。针对主要蔬菜卵菌病害研发了品种抗灾和检测预警 2 项核心防控技术及高效栽培防病、生态控害、生物防治和精准用药减灾 4 项关键防控技术，创建综合治理技术体系。建立了三唑磺草酮、双环磺草酮等一批除草剂新药剂在水稻、小麦、玉米等作物上的田间应用技术。

2. 实用新型防控产品取得重要突破

历经 7 年，选育出我国首个携带抗赤霉病基因 *Fhb7* 的小麦新品种山农 48。通过转基因技术将 *RXEG1* 分别导入 3 个赤霉病易感小麦品种：济麦 22、矮抗 58 和绵阳 8545，显著性提高了小麦对赤霉病的抗性。开发出了防治茎腐病的 9 种木霉菌生防制剂，10 种低毒化学种衣剂。研发了我国首例基因工程微生物农药——苏云金杆菌 G033A。创制了甲磺酰菌唑、香草硫缩病醚、环氧虫啶等一批新型高效低毒杀菌、杀虫剂。根据我国草地贪夜蛾性信息素的地域特异性，开发了适用于我国草地贪夜蛾监测和防控的性诱剂产品。

（三）集成一批绿色防控新模式，服务现代农业主战场

面向现代农业主战场，以服务农业农村为目标，着力解决农作物重大病虫害危害，集成监测预警、生物防治、理化诱控、化学防控等主要防控措施，形成了草地贪夜蛾分区治理技术体系、盲蝽区域防控技术体系、地下害虫韭蛆绿色防控技术体系、麦蚜精准化防控技术体系、重要农业入侵生物的持续治理技术体系等，并加强试验示范与推广，近年来取得了良好成效。

三、本学科国内外研究进展比较

（一）基础理论研究取得了系列突破，但与国外仍有差距

在病虫生物学、生态学等基础理论研究取得了系列突破，一些基础研究领域重要研究成果，在《自然》（*Nature*）、《科学》（*Science*）、《细胞》（*Cell*）等顶尖综合期刊上发表，具有一定的国际影响力。现代生命科学、信息科学等基础学科的新理论与新技术不断融入农业有害生物的检测、监测、预警与控制各个阶段，我国对此研究相对滞后。欧美发达国家聚焦利用国家地理信息系统、大数据、云计算等现代信息技术显著提升了农业病虫害监测预警能力，我国相关研究刚刚起步，尚有大量空白。缺乏转基因技术、基因编辑技术、纳米材料和药物分子设计等现代科技前沿技术的深度研究，利用上述技术研发新的植保产品差距明显。

（二）绿色防控水平整体提升，但诸多技术瓶颈有待突破

当前，我国农业病虫害绿色防控产品具有了一定规模，天敌昆虫、微生物农药、理化诱控产品等种类不少，也取得了一定的防治效果。然而，相关绿色防控产品的核心技术仍未突破，例如天敌昆虫产品缺乏高效繁育技术装备、长时间贮存技术，尚未解决工厂化生

产和货架期等技术问题；微生物农药需要解决效价提升、防治效率不高的技术问题；高效低风险绿色农药需要解决新结构研发、合成技术问题；害虫诱杀新型光源与应用技术、害虫化学通信调控物质利用技术、害虫辐照不育技术等，都需要突破技术瓶颈。

（三）资源碎片化问题突出，生物资源平台有待加强

我国在开发利用生物资源方面主要集中于粮食和重要经济作物以及重要动物品种，对植物有害生物和天敌等植物生物安全资源的评价和挖掘利用重视不够。对于具有潜在生防价值的物种和基因的研究尚在起步阶段，未形成特种资源的专库，资源碎片化问题突出。欧美日等发达国家先后制订并实施了植物生物安全重大科技基础设施发展规划，并投入巨资新建或扩建植物生物安全资源保藏基地与基础设施条件，以抢占植物生物安全科技创新的制高点。植物生物安全资源的深度开发，建立稳健的技术制高点，形成生物安全科技能力的“杀手锏”，能够有力推进我国植物生物安全领域科技自立自强。

四、本学科发展趋势及展望

在全球气候变化和种植业结构不断调整的大背景下，加重和突出了农业生物灾害问题。同时，多学科交叉给植物保护学科带来了新的发展机遇。未来 5 年至更长时间内，夯实重大植物保护理论创新体系，优先发展前瞻性绿色植保技术，健全绿色植保技术的推广体系，补齐平台建设短板，进一步提升重大植物疫情防控能力和水平，为我国粮食安全、生物安全和生态安全作出更大贡献。

（一）夯实重大植物保护理论创新体系

未来植物保护基础理论研究将更深入、更新颖，突出从 0 到 1 的原始科技创新工作，应从多个方面推动我国植物保护理论创新体系的建立。一是要加强交叉学科的发展，近年来，交叉学科研究的优势逐渐显现，引起各个研究领域的持续跟进；二是要重视新兴领域的拓展，新兴领域的不断涌现不仅开拓了病虫害发生规律与灾变机制的研究，同时也为防控理念和技术创新提供了新的增长点；三是要强化产业需求的导向，植物保护学的基础研究需要紧密结合农业生产实际，通过解析重大害虫发生新规律与新机制，促进更加科学、更加精准的害虫防控技术创新。

（二）优先发展前瞻性绿色植保技术

应充分利用监测预警、植物对病虫抗性、生物多样性、生物农药等，研发经济、有效和环保的控害技术。围绕前瞻性绿色植保技术研究，优先支持植物抗病性利用与诱导技术、农作物有害生物生态防控技术和生物防治技术、RNA 干扰与遗传精准控害技术等新型绿色防控技术的基础与应用研究。支持利用人工智能发展现代病虫害控制技术，集成构

建设施作物病虫害绿色防控技术体，克服连作障碍。从多物种的生态食物网出发，加大在宏观调控、宏观网络研究上的投入力度，突破宏观生态学瓶颈。

（三）健全绿色植保技术的推广体系

建立公共植保服务体系，围绕监测预警体系和专业防治队伍两个支撑，形成联防联控机制。倡导做有组织的科研，各单位加强交流，企业与农户也要多沟通交流，将问题反馈至科研系统。在景观格局水平，把农业与非农业间的生态学服务功能发掘出来，将自然生态资源优势转化为经济效益。夯实政府、农民农户、企业、科教工作者共同贯彻绿色发展新理念，加强绿色植保宣传引导、责任履行和社会监督。不断完善“公共植保、绿色植保、科学植保、智慧植保”的中国特色农业病虫害防控新理论，同时研究农作物病虫害绿色可持续控制新模式、新体系，进而保障生物、粮食、食品、生态等的安全。

（四）加强植物保护资源等条件平台建设

未来 5 年，资源优势将是病原物核心致病因子及作物重要抗病基因鉴定的关键，美欧等发达国家建立并运行了数十年生物防治种质资源库，我国至今尚无国家级天敌昆虫及生物防治微生物资源库，亟须依托这些资源优势，鉴定和挖掘大量有害生物核心致害因子，寻找防控新靶点。农作物有害生物监测预警已成为草地贪夜蛾等迁飞性害虫防控的重要手段，但目前监测系统较为分散，缺乏全国布局的监测预警平台，尚需整合现有资源，合力开展长期监测和精准预测预报。

第十七节 作物学

一、引言

粮食安全事关国计民生，是国家安全的重要基础。2023 年中央一号文件首次提出“农业强国”，把抓紧抓好粮食和重要农产品稳产保供、坚决守牢粮食安全底线作为建设农业强国的首要任务，再度凸显了粮食安全作为“国之大者”的重要性。作物学是农业科学的核心学科之一，其根本任务是研究作物生长发育与重要农艺性状遗传变异的基本规律，揭示作物生长发育和产量、品质形成及其与环境的关系，培育优良品种，采取农艺措施将优良品种的遗传潜能转化为现实生产力，实现高产、优质、高效、生态、安全的生产目标，为保障我国粮食安全和农产品有效供给、现代农业绿色高效可持续发展提供可靠的技术支撑。本节主要总结分析了近年来作物学领域的新见解、新观点、新技术、新理论、新成果与发展前沿，提出本学科在我国未来的发展趋势、研究方向和重点任务，为社会各界准确了解作物学学科发展态势提供重要窗口，为优化布局中国特色学科专业体系、合理配置创新资源、实现农业产业链自主可控提供科学决策依据。

二、本学科近年我国取得的最新研究进展

（一）作物种质资源收集保存与创新利用取得新进展

资源总量突破 54 万份，位居世界第二。抢救性收集到疣粒野生稻、多年生野生大豆等种质资源 1000 余份。国家作物种质库新库于 2021 年 9 月开始试运行，可长期保存农作物种子等资源 150 万份。

（二）作物重要性状遗传解析取得新突破

发现同时提高玉米和水稻产量的趋同选择基因 *KRN2*/*OsKRN2*，在水稻中发现高产基因 *OsDREB1C*，克隆了小麦主效抗赤霉病基因 *Fhb7*，从野生玉米中克隆到了首个控制玉米高蛋白含量的主效基因 *THP9*，克隆了水稻中控制纤维素和氮利用效率的转录因子 *MYB61*，克隆了籼粳杂交不育性基因 *RHS12*，发现了水稻广谱抗病 NLR 受体蛋白赋予水稻广谱抗病性的新机制。

（三）作物育种关键技术创新与应用成效显著

在水稻中同时创制 4 个基因的突变，基本解决了“一系法”的生产结实与繁种相统一的技术问题。玉米单倍体育种技术实现工程化应用，利用细胞染色体工程与诱变技术创制了小麦新品系。转基因抗虫玉米、耐除草剂玉米和耐除草剂大豆获批生产应用安全证书。基因编辑底盘工具 Cas12i 和 Cas12j 获得中国内地、中国香港地区及日本专利授权，标志着我国基因编辑技术迈入国际第一梯队。

（四）新种质创制与新品种培育取得重大进展

创制了聚合 *Yr30*、*Lr27*、*Sr2* 等 10 多个抗病基因的优异小麦新种质。扬麦 33 等抗病小麦新品种实现了黄淮麦区抗赤霉病品种零的突破。裕丰 303、登海 605、东单 1331、MC121、中玉 303、川单 99 等玉米品种被遴选为主导品种。创制培育高抗南方锈病骨干自交系京 2416K，并组配育成京农科 767 为代表的 40 多个高抗锈病杂交品种，获得大规模

推广应用。大豆新品种黑农 84 兼抗灰斑病、病毒病、胞囊线虫病三大主要病害。培育出耐密、高产油菜新品种中油杂 501。棉花品种中棉所 113 扩大了北疆植棉北界，品质超越“澳棉”。

（五）作物高产潜力突破与优质高产协调栽培理论和技术取得明显进展

水稻、小麦、玉米等作物高产潜力不断突破，分别实现了百亩方亩产 1000 千克、850 千克、1600 千克以上的超高产。宽幅精播是北方小麦生产主推技术之一，与常规条播相比，宽幅播种实现了提高产量、氮素利用效率条件下强筋小麦品质的稳定。阐明了玉米产量潜力突破的主要途径，创新了密植栽培、水肥一体化精准调控、机械粒收等关键技术，2020 年创造亩产 1663.25 千克的全国高产纪录。

（六）作物肥水资源高效利用与“双减”绿色栽培技术得到新发展

因地制宜发展了水稻轻简氮肥管理技术、小麦养分优化管理方案、稻田节水灌溉措施、小麦全膜微垄沟播技术、玉米覆膜滴灌技术等。集成玉米密植滴灌精准调控高产技术、冬小麦节水省肥高产栽培技术，集成旱耕旱整、控水增氧、增密调氮等技术，创建了高产、低碳排放稻作新模式。基于自然解决方案的禾 – 豆等间套作模式获得大面积应用，实现了稳粮增收和固碳减排协同。

（七）作物生产全程机械化、信息化、“无人化”智慧栽培技术逐步推广应用

“玉米密植高产全程机械化生产技术”入选 2020—2022 年全国农业主推技术，基于无人机数码影像等监测平台对作物叶面积指数动态等提供了技术支持，“稻麦绿色丰产‘无人化’栽培技术”入选 2021 年农业农村部重大引领性技术，无人农场实现了耕、种、管、收全程无人化农机作业。

三、本学科国内外研究进展比较

（一）作物种质资源研究

我国国家种质库（圃）保存种质资源数量位居世界第二，种质资源精准鉴定水平和规模接近国际领先水平。水稻、马铃薯、棉花和食用豆等作物的基因组解析、作物驯化改良和种质形成规律等方面研究处于先进水平。作物野生资源利用取得了重大进展，特别是利用长雄野生稻成功创制出多年生水稻。与国际水平相比存在以下差距：一是我国种质资源数量质量尚需同步提升，库存种质资源中物种多样性和国外资源占比较低；二是优异等位变异的挖掘和育种应用滞后，需要加快从野生种质资源和地方品种中鉴定优异等位变异；三是种质资源的创新利用平台还不完善，需要打造种质资源大数据系统。

（二）作物育种基础研究

我国作物育种基础研究方面，水稻和小麦基因组研究位居国际前沿。但与发达国家相比，生物育种重大原创性基础理论研究不够深入，玉米、大豆等作物基础研究尚在跟跑阶段，重要性状形成的遗传基础与调控网络研究不系统，具有重大利用价值的关键基因在数量上尚不能满足生物育种的需求。

（三）作物育种方法与技术创新

我国建立了自主全链条转基因技术体系，抗虫耐除草剂玉米和耐除草剂大豆具备产业化应用条件，基因编辑育种技术研究进入国际第一方阵。与发达国家相比，我国水稻、小麦、大豆、油菜等杂种优势利用居国际领先，分子育种、细胞工程、倍性育种等育种技术获得突破，但作物遗传育种理论和重要技术创新能力及对种业的支撑能力还有待提升，育种技术原始创新能力仍然薄弱，部分关键技术受制于人。

（四）作物新品种培育

发达国家的作物育种目标呈现以产量为主向优质专用、绿色环保、抗病抗逆、轻简栽培等方向多元化发展。我国的育种目标中产量仍是核心指标，并开始向绿色、优质、高效的方向转变。作物单产水平与发达国家相比仍存在差距，玉米和大豆单产不足美国的60%，大豆进口规模仍很大，重大品种研发滞后，品种亟须升级换代。受益于知识产权保护，发达国家种企已成为投资和创新主体，我国尚未形成以企业为主体的商业化育种体系。

（五）作物栽培与耕作基础研究

与发达国家相比，我国作物栽培与耕作正加快由传统向现代化的转型中。发达国家作物栽培与耕作更注重优质、绿色、低碳、生态、安全，并加强了应对气候变化带来的高低温、旱涝灾害、病虫草等非生物或生物灾害频发的作物抗逆抗性栽培生理机制研究，生产全过程资源高效利用机制、碳足迹量化评估方法的研究与应用。我国亟须加强这方面的研究，并在作物高产优质高效协同的栽培耕作基础研究上形成中国特色。

（六）作物栽培与耕作关键技术

与发达国家相比，我国因区域多样性、种植制度的复杂性，在保护性栽培耕作，资源高效利用、规模化生产等方面存在明显的差距，严重制约着耕地综合生产能力、效率和效益的提升，亟须创新集人、机、艺、智一体的规模化机械化作物高效生产技术与装备，以满足我国新时期产业结构调整、劳动力锐减、新型经营主体规模化生产对现代作物栽培耕作的要求。

（七）作物栽培新技术创新

欧美等发达国家已实现了机械化、智能化、标准化生产，日本小型智能化农机装备种类齐全、作物生产实现了智能控制作业，我国迫切需要研究数字化感知、智能化决策、精准化作业和智能化管理的农艺、农机、信息融合的关键技术及其整合应用。此外，我国高通量作物表型组学技术研发及装备研制相对滞后，缺少精准的作物表型及生长调控模型等，需要建立作物表型高通量精准鉴定的现代技术体系，实现我国作物栽培学研究的快速迭代突破。

四、本学科发展趋势及展望

（一）未来 5 年的重点发展方向

作物遗传育种基础理论与关键技术创新方面，将注重作物育种基础研究与核心技术创新、作物种质资源保护和利用、作物基因资源深度挖掘以及作物重大新品种培育。作物栽培耕作和生理学基础研究与关键技术创新方面，将注重发展优质丰产高效协同规律与关键栽培技术创新、旱作节水高产高效作物栽培耕作的新模式与技术创新、大田作物固碳节能减排绿色栽培关键技术创新以及作物智慧化精准栽培关键技术创新。

（二）发展趋势与发展策略

1. 生物育种基础理论的原始创新

启动实施农业生物育种创新行动，重点解决好生物种业科技基础研究和前沿技术的源头创新，支撑突破性重大品种的培育。利用多组学协同技术持续挖掘重要性状的主效基因并解析遗传调控机制，加强复杂性状基因组选择技术和基因编辑技术创新，提高分子育种自主研发与创新能力。

2. 农作物优异种质资源收集和利用的持续创新

通过实施优异种质资源创制与应用行动，完善创新技术体系，规模化创制突破性新种质，推进良种重大科研联合攻关，创制目标性状突出、综合性状优良的新种质，培育具有自主知识产权的高产、优质、适应性强的新品种（系）。

3. 作物育种重大关键技术和体制机制的突破

为大力推进种业创新攻关，启动种源关键核心技术攻关，实施生物育种重大项目，有序推进产业化应用，推动要素聚合、技术集成、机制创新，促进种质资源、数据信息、人才技术交流共享，加快突破一批重大新品种。

4. 作物抗逆减灾栽培技术研究

全面深入地阐明作物感受、应答和适应非生物逆境胁迫的机制和调控网络，明确能够提高作物抗逆减灾能力的关键栽培和耕作技术，揭示植物抗逆应答和生长发育的协同关系，研究灾害性天气发生特点及对作物的伤害以及绿色调控机制。

5. 丰产高效智能化栽培新技术与绿色发展新模式构建

驱动我国作物生产由资源消耗型向绿色高效型转变，研发精准高效施肥施药以及减控污染的新理论与新技术，加快推进作物资源高效、减肥减药、节水固碳绿色生产技术研发，构建绿色生态系统和优质丰产绿色发展新模式。

6. 作物一年多熟种植模式创新与高效配套栽培技术研究

将一年多熟种植与现代农业新技术充分结合，逐步形成类型丰富的粮、经、饲复合高产高效种植技术模式，农牧结合、农林复合等高效种养技术模式。通过轮作轮耕、间混套作、种养结合等模式有效解决种植结构单一、地力消耗过大、化学投入品过多、生产成本过高问题，优化作物结构、布局及模式调整。

第十八节　恶性肿瘤

一、引言

肿瘤学是研究肿瘤发生发展规律，提出预防、诊断和治疗手段的学科。近一个世纪，特别是 20 世纪 40 年代以来，恶性肿瘤发病和死亡率有所增高，提高了人们对肿瘤危害性的认识和重视，对肿瘤研究加大了力度。随着科学技术的发展，肿瘤的基础理论和临床研究都有了迅速的发展。它不仅成为一门独立的学科，并已形成许多分支学科。研究的范围涉及与肿瘤相关的基础医学、临床医学、预防医学、生理学、生物化学、心理学、社会学、经济学等多个学科领域。近 5 年来，中国抗癌协会积极践行“健康中国”战略，秉承“肿瘤防治，赢在整合”的核心理念，完成“建大军，办大会，写大书，立大规，创大刊，开大讲”六件大事，进一步促进了我国恶性肿瘤研究和防治工作的快速发展。在樊代明院士提出的“扩大队伍”“提升学术”两大发展目标引领下，为推动我国本土指南的建设，中国抗癌协会组织 13000 多位专家，编写完成我国首部肿瘤整合诊治指南，有效推动了恶性肿瘤规范化治疗进展。本节系统性总结了我国 2018 年 1 月至 2023 年 6 月恶性肿瘤学科在技术方法、学术理论、研究进展、人才培养、学术建制、研究团队等多方面的重要学科发展概况。同时，对国内外研究进展进行了比较，提出学科未来发展趋势及展望。

二、本学科近年的最新研究进展

（一）肺癌

近年来，随着靶向、免疫治疗的发展及基因检测、放疗技术的创新，肺癌 5 年生存率有所提升，为 17%~32%。肺癌靶向治疗方面的进展主要集中于适应证扩展、常见突变靶点耐药后治疗选择及少见突变靶点新药研发，免疫治疗方面的进展主要集中于适应证扩展、与放疗有机结合的多学科诊疗模式等。

（二）乳腺癌

乳腺癌是中国女性发病率最高的恶性肿瘤。乳腺癌治疗进展主要集中在 DAWNA-1 研究、PEONY 研究等国产靶向药物治疗方面，临床试验得到广泛开展，并不断改写国际指南。基础研究方面，三阴性乳腺癌复旦分型为三阴性乳腺癌精准治疗提供了新的研究方向。

（三）胃癌

近年来，对胃癌发病机制的基础研究不再局限于某个基因的突变或表达改变，而是已经转变到多基因 / 多位点，甚至全基因组水平。胃癌微创外科治疗经过 30 多年的探索，尤其是近十年中国腹腔镜胃肠外科研究组（CLASS 研究组）先后启动了系列高水平腹腔镜胃癌外科临床研究，引领了该领域的范式革新。联合治疗模式及新的治疗靶点的发现进一步提高了胃癌靶向治疗的疗效。ADC 药物已成为 HER2 阳性晚期胃癌后线治疗标准方案。免疫检查点抑制剂在晚期胃癌一线和后线治疗中均展现了较传统治疗更优的疗效。

（四）结直肠癌

目前，微创手术在国内各级医院所占的比重不断增加，已成为结直肠癌手术的主流。优化直肠癌新辅助化放疗模式，强化同步放化疗方案、全程新辅助治疗、短程放疗联合化疗，可以进一步提高疗效，使更多患者获得器官保留的机会。靶向治疗依然是转移性结直肠癌重要的治疗手段。作为结直肠癌发生发展的主要信号通路，以 RAS-RAF-MEK 通路为靶点的转移性结直肠癌分子靶向治疗研究持续深入。

（五）食管癌

近年来，微生物在癌症的发生和发展中的作用成为研究热点。ERAS 在减少术后并发症、加速患者康复、缩短住院时间、降低医疗费用及增加患者满意度等方面具有明显优势；全胸腹腔镜食管癌根治术成为食管癌根治术中目前最能体现微创理念的一种术式。针对 EC 基因组驱动因素的靶向治疗成为研究热点。作为继手术、放化疗之外的新的治疗选择，免疫治疗已改写食管癌的治疗格局。免疫联合治疗模式既是当前临床中主要的应用模式，也是未来亟待深入探索的方向。

（六）原发性肝癌

近年来，人工智能辅助病理学诊断在肝癌中得到了迅速发展，包括自动提取病变区域、判断疾病类型，对疾病的分析更加准确，处理一些肉眼难以分辨的细节和特征纹理。目前，肝癌的外科治疗仍是肝癌患者获得长期生存的重要手段，主要包括肝切除术和肝移植术。分子靶向药物是治疗中晚期肝癌的主要手段，近年来，肝癌分子靶向药物研究取得众多新进展。随着临床试验的进一步探索，免疫治疗也成为新辅助治疗及转化治疗的重要治疗手段。国内现已有多个免疫检查点抑制剂单药获批用于肝癌二线治疗。

（七）胆道恶性肿瘤

目前，如何针对个体化患者实施恰当的区域淋巴结清扫及确保相应切缘阴性是精准胆道外科重要的工作组成部分。同时随着 3D 腹腔镜、机器人等新型医疗器械的发展，BTC 的微创化治疗不断实现该领域的术式革新。近年来，临界可切除或局部晚期的 BTC 可从新辅助治疗和多模式治疗中获益。随着对 BTC 基因突变谱及不同亚型特征、免疫逃逸等的认识逐渐深入，化疗联合靶向免疫等其他治疗对提高疗效值得期待。

（八）泌尿系恶性肿瘤

前列腺癌方面，随着外科技术的发展及机器人腹腔镜技术的普及，机器人辅助腹腔镜根治性前列腺切除术可以缩短手术时间，减少术中失血。以 PSCA 为靶点的 CAR-T 治疗研究正在开展。KLK2 CAR-T 的 Ⅰ 期临床研究也在进行。RC48-ADC 在 HER2+ 局部晚期 /mUC 患者中显示出良好的疗效和安全性。免疫检查点抑制剂已用于不能切除和转移的 MIBC 患者二线治疗，以及无法耐受铂类且 PD-L1 阳性患者的一线治疗。肾癌方面，3D 重建联合机器人辅助下的肾部分切除术是过去几年保留肾单位手术的热点。靶向治疗仍是晚期肾癌的一线治疗策略。

（九）宫颈癌

继 LACC 试验以来，多项研究均证实开腹根治性子宫切除术在治疗早期宫颈癌中较微创手术有较高的安全性。宫颈癌的放射治疗包含外照射及近距离放疗，近年来兴起的三维组织间插植近距离后装放疗，能够充分保证高危临床靶区的剂量覆盖，从而提高局控率、改善生存。免疫疗法代表了宫颈癌一种新的治疗选择，在复发患者中具有生存获益。

（十）卵巢癌

为延缓上皮性卵巢癌的复发，目前多项药物获批用于上皮性卵巢癌的一线维持治疗。目前，各种临床证据及国内外指南显示，推荐用于上皮性卵巢癌一线维持治疗的药物主要包括 PARP 抑制剂及抗血管生成药物。免疫检查点抑制剂在晚期卵巢癌一线和后线治疗中均开展了临床研究。遗憾的是在一线治疗中尚未取得突破性疗效。铂耐药复发卵巢癌，叶酸受体 α 高表达的患者可以考虑使用靶向 FRα 的抗体偶联药物 Mirvetuximab Soravtansine。

（十一）骨与软组织肉瘤

揭开骨肉瘤免疫微环境特征景观对于提高骨肉瘤免疫治疗效果至关重要，基础研究不断取得进展。现有医学 3D 打印技术的发展为骨盆肿瘤精准切除和个性化重建提供了思路。对进展期骨肉瘤，抗血管生成的靶向药的治疗基本已经达成共识。GALLANT 研究提示节拍化疗联合免疫检查点抑制剂用于晚期肉瘤二 / 三线治疗。细胞治疗有望开启晚期肉瘤治疗新征程。

（十二）血液系统肿瘤

经过数十年的停滞，几种针对新诊断和复发 DLBCL 的有前景的疗法已获得 FDA 批准或处于研发的最后阶段，包括增强型单克隆抗体、抗体药物偶联物（ADC）和双特异性抗体等等。目前，CAR-T 细胞疗法作为 R/R DLBCL 的二线治疗受到了广泛的关注。免疫检查点抑制剂近年来在淋巴瘤领域取得显著进展。霍奇金淋巴瘤近年来主要进展集中在晚期或难治性患者。近年来，靶向治疗、免疫治疗和细胞疗法为急性白血病提供了新的治疗策略。

（十三）头颈恶性肿瘤

目前，已有一些西妥昔单抗联合免疫治疗在复发 / 转移头颈鳞癌中获得较好疗效的研究，并写入指南推荐。免疫治疗在 HNSCC 围手术期治疗中的初步成效。新辅助靶向治疗作为一种新兴的治疗模式，有望提高局部晚期甲状腺癌的 R0/1 切除率，改善患者预后。近年来，分子检测技术的进步、靶向药物和免疫检查点抑制剂的临床突破，使甲状腺未分化癌的管理已然步入精准化诊疗时代。

三、本学科国内外研究进展比较

通过近些年医学的发展，恶性肿瘤的治疗水平得到显著提高。恶性肿瘤整体治疗水平的提高，得益于我们对其生物学行为认识的不断深入，以及包括外科手术、化疗、放疗、内分泌治疗、靶向治疗、免疫治疗在内的综合治疗的进展。在基础研究方面，尽管和国外还有一定差距，但是近年来我国学者在恶性肿瘤发病、分子分型方面取得显著进展。以乳腺癌为例，复旦大学附属肿瘤医院精准肿瘤中心自主研发了多基因测序平台，并绘制了中国首个千人乳腺癌基因突变图谱，全面分析了中国乳腺癌的临床特征和基因组特征，发现了中国乳腺癌特有的精准治疗靶点。邵志敏教授团队绘制出了全球最大的三阴性乳腺癌多维组学图谱并提出“复旦分型”，为三阴性乳腺癌精准化诊治指明了新的方向。在新药临床研究方面，长期以来由于欧美等国家在新药治疗研究领域的领先地位，导致相关药物在海外的临床试验质量及新药审评速度均占有一定优势。近年来，国内在药政、临床试验、新药审批等多方面的改革，尤其是 2017 年以来，中国正式加入人用药品注册技术要求国际协调会议（ICH），为中国研究者提供了更好的环境。在一大批中国研究者的共同努力下，伴随中国临床经验和数据的不断积累，中国患者的临床试验数据无论从质量还是影响力均处于快速提升之中。伴随国内创新药物研发实力的增强与本土研究者国际影响力的提升，越来越多“中国数据”亮相世界舞台，并不断改写国际指南。

四、本学科发展趋势和展望

中国恶性肿瘤的每年新发病例将近 400 万，癌症死亡病例 234 万。整体来看，恶性肿瘤发病率呈上升趋势。特别是与环境生活方式相关的肿瘤，死亡率整体上呈上升趋势。《“健康中国 2030”规划纲要》提出，2030 年实现总体恶性肿瘤 5 年生存率提高 15% 的重要目标。为了实现“健康中国 2030”的目标，恶性肿瘤学科的高质量发展十分必要。恶性肿瘤的发展趋势和展望主要体现在以下方面：精准预防深化发展，个体化、高效筛查策略的优化，早期可手术恶性肿瘤辅助 / 新辅助治疗的精准化探索，分子靶点探索及基因检测助力实现恶性肿瘤精准治疗，恶性肿瘤手术治疗向精准、规范、微创、重外形、保功能发展，免疫治疗方兴未艾，国产原研药物不断问世、临床研究蓬勃发展，基于不同治疗策略的联合治疗有望成为优选治疗方案，人工智能在恶性肿瘤学科会有更大发展空间，肿瘤科普在恶性肿瘤防治中继续发挥重要作用。

第十九节 产科学

一、引言

产科学是保障孕产妇健康，协助新生命诞生的传统医学学科。现代产科学实现了临床与预防医学理念的深度融合，并随着分子生物学、遗传学和免疫学等医学基础学科以及临床诊疗技术的进步，其内涵与外延不断发展。目前，我国生育水平持续走低，总人口增速明显放缓，人口老龄化趋势显著。规划好与发展好产科学对于保障我国人口安全和经济社会健康持续发展具有重大意义。本节从妊娠生理机制、产科重大疾病防治、产科医疗服务模式演变与创新三方面，梳理总结近五年（2018—2023 年）我国产科学领域的主要进展。

二、本学科近年的最新研究进展

（一）妊娠生理机制研究

在早期胚胎发育和胚胎植入研究方面，我国学者从遗传学角度发现了导致卵母细胞成熟障碍和精子发生障碍的致病基因；从转录—翻译—蛋白多维度探索早期胚胎发育和胚胎植入的分子机制，绘制了早期胚胎发育的时空图谱；利用多组学技术揭示了胚胎着床失败的潜在原因；同时，建立了非人灵长类动物胚胎体外培养系统，为理解胚胎植入和发育提供了技术支持。在胎盘发育方面，深入阐释了胎盘细胞谱系发育，发现胎盘滋养层细胞分化的多种精细调控机制；成功利用人诱导多能干细胞构建了胎盘 / 子宫类器官，为深入探究胎盘发育关键事件提供重要手段。全方面描述了内分泌时钟、胎膜时钟、蜕膜时钟和子宫肌层时钟等妊娠生理时钟，绘制了妊娠维持与分娩启动的时空景观。

（二）产科重大疾病的机制、预防和诊治研究

1. 产前筛查、诊断和出生缺陷防治

我国政府及医疗机构采取了多种措施促进和规范产前筛查。通过开发孕妇外周血胎

儿游离 DNA 的检测方法，实现常 / 性染色体非整倍体及染色体微缺失微重复综合征检测。产前诊断手段如 CMA、CNV-Seq、WES 甚至 WGS 的应用日益成熟和普及。新的产前诊断技术将胚胎诊断从单基因病迈向多基因病，为与遗传高度相关的慢性疾病源头防控提供了支持。

2. 母体疑难危重疾病防治

产后出血方面，我国学者发现孕晚期异常的免疫炎症凝血状态可能与宫缩乏力性产后出血发生密切相关；开展随机对照研究，为产后出血预防性和治疗性药物遴选、止血球囊研发、止血操作技术的临床应用提供高证据强度；在管理方面，国家产科专业医疗质量控制中心综合国内外进展、运用新的策略及工具建立全孕期多维有序、防治结合的产后出血综合管理策略并进行全国各基层医院的推广。对于胎盘植入性疾病，通过建立胎盘植入评分系统，进行妊娠期胎盘植入风险动态评估、制定围产期尤其围手术期精准个体化管理；九步手术法等手术技术的创新显著降低了胎盘植入患者输血率和子宫切除率。对于妊娠期高血压疾病，我国学者率先在单细胞水平解析胎盘形成的细胞亚型及滋养细胞分化在子痫前期发病中扮演的角色；开创性揭示肠道 - 胎盘轴在子痫前期发病中的作用；初步开发了基于临床高危因素及生物学指标的子痫前期妊娠早中期预测模型和围产期管理方案。妊娠期糖尿病领域，建立妊娠合并糖尿病规范化诊疗管理模式，制订我国妊娠期糖尿病临床诊治规范和行业诊断标准，并参与国际 GDM 筛查和诊断标准和临床指南的制定。产科危急重症方面，多学科联合救治模式发展日益成熟和普及，一些新的诊疗技术应用如体外膜肺氧合（ECMO）等的应用极大提高了抢救成功率。母体疑难危重疾病代表性指南或共识发布为其规范化诊疗提供了依据，其中《妊娠期急性脂肪肝临床管理指南（2022）》被指南 STAR 评级为 2022 年产科领域最佳指南。

3. 胎儿及新生儿疾病防治

早产领域，随着我国早产率逐年上升，国内学者通过多组学技术发现与分娩启动调控相关的蛋白、细胞分子信号等，为自发早产防治提供干预靶点；国内多团队通过构建高危妊娠风险预测最大程度降低医源性早产的发生；极早产儿救治能力大幅提高，26~28 周胎龄超早产儿存活率已接近发达国家水平。死胎死产领域进展集中在流行病学、妊娠管理、胎盘病理学等研究方面；然而，死胎尸体解剖的低送检率以及临床对于胎盘、脐带等病理学结果重视度不足限制了对死胎病因的深入探索。在胎儿生长受限研究方面，北京大学第三医院团队建立了全球规模最大的双胎胎儿生长受限出生队列，首次揭示了遗传因素及宫内不良环境对生命早期菌群的塑造，并发现关键肠道微生物及其代谢产物与生长受限新生儿生理及神经行为发育的关系；在胎儿生长受限宫内干预方面，动物实验发现血小板源性生长因子纳米颗粒可以通过加强胎盘血液供应及促进胎儿骨骼肌中的营养摄取能力，从而实现宫内时期的胎儿发育干预。双胎妊娠方面，我国已建立双胎妊娠的规范诊治及转诊流程；利用胎盘灌注、胎盘血管吻合和多普勒血流特征进行复杂性双胎妊娠的病因探索，聚焦胎盘多组学、双胎新生儿肠道菌群宏基因组及代谢组学等多组学分析对复杂性双胎的发生机制进行多层面研究。

4. 产科药理研究

我国制定了多项规范和标准，如《妊娠期药物致畸风险咨询技术规范》和《中国妊娠用药登记专家共识》，为开展妊娠用药咨询和登记建立了统一标准。首次在全国范围开展癫痫女性妊娠期用药登记，为建立专科疾病用药导向的孕产妇用药数据登记体系奠定了基础。基于 PBPK 模型体系进行孕妇用药的精准用药研究，建立药物通过胎盘屏障的体外模型用于评估药物对胎儿的安全性。胎盘屏障微电流仿生芯片、胎盘类器官模型等新技术研发有望用于评估药物妊娠期安全性。目前，妊娠合并免疫系统疾病用药缺乏高水平循证医学证据作为指导，需要多学科团队共同开展妊娠期自身免疫疾病多中心研究，制定标准化药物治疗方案。

（三）产科医疗服务内容和模式的创新与演变

近 5 年，我国产科医疗服务内容和模式发生了巨大改变。产科服务内涵也在不断延伸和拓展，从单纯的产科医疗处理转变为生理、心理、社会适应的全面支持，从重视个体医疗转变为重视健康教育、重视群体保健，提高孕产妇的自我保健能力；服务对象也从孕产妇及胎婴儿，扩大到所有家庭成员。互联网医疗、人工智能等近代科学新技术的开发和应用使产科医疗服务迈向新的发展阶段。在国家政策层面，我国一方面推动国家医学中心建设，让高水平医院发挥国家医学中心前沿引领作用；另一方面建设国家区域医疗中心，促进优质医疗资源纵向和横向流动，缩小区域间、省域间医疗技术和服务水平差距。同时，实施母婴安全五项制度，构建以危重孕产妇和新生儿转诊救治中心为核心的区域母婴安全保障网络。产科医疗服务体系日臻完善，产科质量管理理念、制度和体系不断健全，以证据驱动和目标导向的产科质量改进工作正在全国推进。

三、本学科国内外研究进展比较

（一）研究投入及数量和质量比较

从我国国家自然科学基金和国家重点研发计划资助情况来看，产科学领域科研项目的资金投入呈现逐年增加的趋势。科研产出方面，国内文章数量在国际排名较为领先，但是高质量文章数量尚有待提升，特别是高水平临床研究如随机对照研究极少。与欧美国家相比，我国科研产出中与国际合作文章占比不足四分之一，国际合作与交流有待加强。

（二）研究方法和技术进展比较

近年来，国内多个基础与产科临床团队采用多种新兴前沿科学技术包括单细胞多组学技术、类器官技术等对产科相关疾病进行深入的机制探索，处于国际领先地位，但现有科研成果的临床转化率不足。国外研究更注重从产科临床问题为导向，通过多中心的大型队列或多学科交叉融合对于母胎相关疾病机制及诊疗策略进行高质量纵向深入研究。

（三）研究热点领域和竞争态势分析

研究热点聚焦于高危妊娠的机制和防治方面。我国在人类和非人灵长类早期胚胎发育和细胞分化研究方面居国际前列，在母胎界面的代谢与营养物质传输的机制、母胎界面免疫豁免调控等方面与国际并跑，产前诊断技术、胎盘植入手术、胎儿手术等部分技术已达到国际先进水平。但妊娠并发症和合并症的发病机制研究不够深入，全国多中心大样本队列研究、随机对照设计的临床试验数量有限，科学研究成果转化和应用推广不足。高水平研究团队和高质量的研究成果数量落后于美国、英国等发达国家，在全球范围内影响力有限。

四、本学科发展趋势和展望

面临当前人口发展态势下的新挑战与新机遇，产科学需从人才培养、学科规划、科研转化、服务模式、质量管理等多层面创新，以实现学科发展新突破。

我国产科学发展面临妊娠人群特征复杂变化带来的突出挑战。未来应以产科重大问题和需求为牵引开展基础研究，实现应用牵引基础研究、基础研究促应用的双向驱动发展；同时开展多中心随机对照研究，积极探索诊疗新技术和多学科合作救治等管理模式，以高质量循证医学证据改善我国产科临床诊疗策略。

一些重大问题，如高龄妊娠的健康影响、产科重大疾病管理、产科大数据信息资源整合、胎儿医学等亚专科发展等，迫切需要对学科发展和产业布局进行整体规划和顶层设计，搭建全国性或区域性的跨学科联合研究网络，为深入探索和源头创新提供强大支撑平台。以国家重点项目或计划为抓手推进协同创新攻关，支持多学科、多中心凝聚科研力量和资源攻坚克难，促进临床实践和卫生决策。

在产科质量管理及医疗服务方面，需进一步规范指南 / 共识的制定及实施，加强医疗服务质量监测，优化质量评估体系和绩效管理机制，促进医疗质量同质化发展，探索质量提升创新模式。通过优化医疗资源配置，整合服务内容，促进妇幼健康服务效率和质量提升，打造生育全程优质服务链，促进母婴健康。

第二十节　城市科学

一、引言

城市科学研究作为整个科技工作的一部分，紧紧围绕建设中国特色社会主义这一目标，力求在学科发展和政府决策中发挥作用，同时围绕城镇化发展中的热点、重点、难点问题，开展多视角的学术研究活动。随着城市环境的发展，党的十八大以来，我国城镇化已进入以提升质量为主的转型发展新阶段。

在当今这一信息时代，数据成为关键生产要素，城市不再被简单地理解为空间中的场所，而应当理解为城市系统各要素之间联系的网络和表达联系之间强度的流共同组成的系统。城市科学在这一新阶段，成为与复杂适应系统理论、信息技术、社会物理学、城市经济学、交通理论、区域科学及城市地理学等相关研究相结合研究城市运行深层结构的理论。

二、本学科最新研究进展

城市作为人类与众多其他有机系统共生的复杂自适应系统（CAS），具有该系统的一般特征，运用复杂自适应系统理论研究城市科学成为这一新历史阶段的主流。在信息化的支持下，城市科学系统论的思维是落实习近平总书记提出的城市化战略在理论与方法上的探索，为研究当前城市发展问题提供了思路。城市科学的研究充分注重传承与创新结合，以城市为对象的学科研究既体现了城市科学发展的延续性，又满足了当今时代的发展需求。

在当今城镇化的发展过程中，城市正不断地吸引着越来越多的人口与资源积聚，城市作为当今生产和消费的主要场所，包含城市的各类子系统，城市研究在可持续发展、应对气候变化等方面呈现大量研究进展，但研究分散在各个领域，为聚焦城市科学，《自然》（*Nature*）旗下形成子刊《自然（城市）》（*Nature City*）这一综合性期刊，标志着城市科学以城市为对象，综合性开展各类学术研究活动成为必然。

（一）理论基础

城市作为复杂系统，运用复杂性科学研究城市系统成为城市研究的基本思路。近年来，基于复杂适应系统理论的城市研究成为城市科学学科研究的热点；城镇化是我国经济社会发展的必然趋势，也是工业化、现代化的重要标志，是城市发展的宏观背景与趋势，城镇化政策和理论成为研究城市学科发展的基础；探索人与居住环境之间的科学构成关系，建设符合人类理想的聚居环境，是人居环境科学研究的目标，也是城市科学发展的理论基础及未来的演变趋势。

1. 复杂适应系统理论用于城市科学

“适应性造就复杂性”，城市是为了人及其群体间相互作用而成长进化的。盖尔曼（Murray Gell-Mann）在谈到复杂适应系统的共同特征时说：CAS 系统的适应过程是系统获取环境及自身与环境之间相互作用的信息，总结所获信息的规律性，并把这些规律提炼成一种“图式”或模型，最后以此为基础在实际行动中采取相应行动的过程。对城市系统而言，则是通过其发展战略、城市规划、文化资本、产业结构、历史文化习俗的传承创新与政治体制和管理制度等“城市结构基因”的优化来实现这一过程。

城市发展战略、规划、政策等实践活动的反馈，通过改进和深化决策者和市民对外部世界及自身发展的规律性认识，从而改善规划决策和行为方式。城市的规划实施本身也是规划编制、实施、修改、再实施的动态反馈过程。城市具有能动性，城市发展轨迹就能够主动地适应环境，成为人类和自然界共同创造的最具能动性的系统。正因为城市是由这些学习型和可适应性的市民组成的，城市本质上具有“学习”与“适应”的能力。

2. 城镇化的研究进展

全球已有一半以上的人类居住在城市，城市成为人类生产和消费的主要载体，对城市的认识和研究对于人类前景至关重要。城镇化是指非农产业在城镇集聚、农村人口向城镇集中的自然历史过程，是人类社会发展的客观趋势，是国家现代化的重要标志。城镇化的研究以产业经济、人口和土地 3 个方面为主。在人口城镇化方面，2024 年美国学者在研究成果中预测至 2100 年，将有一半的城市面临人口下降，这些城市资源的利用、城市经济的增长将面临巨大挑战，将迫使城市寻求创新的发展方式。中国学者近两年以新型城镇化为研究主题开展研究，主要聚焦在人口土地的城镇化、产业结构、城镇化空间形态等方面。至 2022 年年末，我国城镇常住人口总数为 9.2 亿人，年末常住人口城镇化率达 65.22%，至 2023 年年末，我国人口总量减少 208 万，但城镇常住人口达 93267 万人，比 2022 年增加 1196 万人，我国常住人口城镇化率为 66.16%。

3. 应对气候变化的人居环境研究

人居环境问题是城镇化进程中面临的主要障碍之一。随着对气候变化的持续关注，人类居住环境出现了各种问题，全球气温上升、城市热岛效应、各种极端气候，改善人居环境成为全球关注的重点，气候变化建模评估城市热效应成为研究的热点，提升人居环境、应对气候变化可发挥城市适应与缓解政策的协同作用，采取热缓解与能源适应技术的综合实施方案等。

（二）重点领域

在复杂适应系统理论支撑下，新型城镇化与人居环境研究背景下，城市科学以城市为研究对象，针对我国城市科学的相关重点、热点领域：人文城市、智慧城市、低碳城市、健康城市、韧性城市、海绵城市、生态城市、宜居城市、城市更新、新城市科学十大领域，开展专题研究。

1. 新技术、新方法用于历史文化遗产保护

人文城市的建设包括传统文化的传承、人文城市的建设及历史文化遗产的保护。城市科学研究的人文城市研究聚焦历史文化遗产保护，近年，众多学者开展了大量研究和保护实践工作，研究方法不断创新，研究对象和内容不断延伸。

新时代背景下，历史文化遗产保护的研究和实践注重与新方法、新理念、新技术相结合。计算机、信息学、互联网技术、数字孪生、城市信息系统等多种技术方法运用于历史文化遗产保护。历史文化资源数字化、信息化工作大量开展，在部分城市已形成融合城市信息模型的历史文化三维信息平台，为保护、规划、管理形成技术底板，并与数字孪生等技术结合，为历史文化遗产的活化利用、城市的文旅发展提供有效支撑。历史文化遗产保护的研究对象从传统历史文化名城（例如北京、西安）、历史街区、历史建筑的保护延伸到近、现代的工业遗产保护，从历史城镇、历史城区、历史街区、历史建筑物质实体的保护拓展到历史文化、风土习俗、传统工艺等非物质文化遗产的保护，从狭义的名城、名镇、历史文化街区和历史建筑扩展到广义的遗产保护，如非法定的历史城区、街区和既有建筑的改造和公共空间的营造。历史文化遗产保护从传统的“自上而下”的管控式保护方式转变为“自上而下”结合“自下而上”的公众参与的方式展开。

2. 智慧城市发展进入新阶段

智慧城市是城市全面数字化基础之上建立的可视化和可量测的智能化城市管理和运营，包括城市的信息、数据基础设施以及在此基础上建立网络化的城市信息管理平台与综合决策支撑平台。其本质是通过综合运用现代科学技术整合信息资源、统筹业务应用系统、优化城市建设和管理的新模式，是一种新的城市管理生态系统。随着人工智能、数字孪生、元宇宙等技术的发展，智慧城市的建设已经从前期构建基于城市信息系统的智慧城市技术系统，转向具备一定的复杂适应系统特性，能够自主更新、迭代的智能城市系统。基于复杂适应系统理论，智慧城市的设计与建设将自上而下的“构成”与自下而上的“生成”进行有机组合。政府通过自上而下提供“智慧城市公共品”，构建智慧城市基础设施成为智慧城市主要载体，智慧城市终端获取各类信息、智慧城市分析民众的需求，引导城市主体运用人工智能机制，自下而上生成自主演化的智慧城市，从而提升城市智慧水平。

随着城市感知、各类视觉分析等智慧城市技术的发展，智慧城市建设不可避免地也面临技术滥用、个人数据泄露、信息安全等数据监管方面的问题。近年来，各国的城市政府和各类研究机构对技术的私用、数据安全等方面开展了大量工作。例如纽约市为各类非政府机构制定了一系列规则，例如生物识别隐私法、自动驾驶车辆许可制度等；俄勒冈州波特兰市则禁止面部识别的某些商业用途；洛杉矶则要求拼车公司共享实时数据。城市通过司法系统进行创新，提出针对这一新形势的法律理论体系。城市作为行业监管的先行者，同样主导和肩负着科技监管重任，在约束下发展的智慧城市能有效提升人居环境，实现城市的高质量发展。

3. 低碳城市、健康城市、韧性城市的研究进展

低碳城市理论的发展经历了从概念提出、内涵界定、评价指标到实践探索等阶段，目前仍在不断完善和创新中。低碳城市理论的核心问题是如何构建低碳城市发展模式，即如何在保障城市功能和民生福祉的前提下，优化城市产业结构、空间结构、交通结构和能源结构，实现城市碳排放与经济增长的脱钩。碳减排成为近几年城市发展的关键问题，相关的碳减排技术、新能源技术等迅速发展，我国新能源占比逐年提高，至 2022 年，新能源占比已达 50%。

进入 21 世纪，随着城镇化的快速推进，人类社会面临着人口老龄化以及疾病谱、生态环境、生活方式变化等带来的新挑战，诸多国家从提升全民健康福祉的角度，在国家战略层面提出健康城市的建设目标和实施策略，认为健康不仅仅是一种发展状态，更是一种重要的治理能力，进而将城市健康问题纳入政府的社会、经济和政治议程中。健康城市建设重点包括全面促进健康公平、创造支持性环境、健康影响评估等。国际上学者关于城市环境与健康关系的研究较多，国内健康城市的研究从健康社区、健康标准、城市交通等多方面展开。

最新研究认为韧性城市是应对黑天鹅式风险的必然选择，其韧性体现在结构韧性、过程韧性和系统韧性 3 个层面。城市的新不确定性来源可概括为极端气候、科技革命、冠状病毒和突发袭击 4 个方面。在新形势下，有学者在 2023 年发表的文章中提出建设韧性城市十大步骤：一是转变思想观念；二是创新设计研究机构；三是制定治理方案，列入五年计划；四是编制生命线工程分组团化改造方案；五是在每一个社区、每一个组团补充原来没有的微循环；六是利用信息技术来协调各个组团和组团内的微循环设施；七是改造老旧小区，补足社区的单元短板；八是新建和改造公共建筑均考虑“平疫结合”“平灾结合”；九是在超大规模城市的周边布局建设“反磁力”的“微中心”；十是深化网格式管理智慧城市建设。

4. 海绵城市、生态城市和宜居城市进展

海绵城市是指城市能够像海绵一样，在适应环境变化和应对自然灾害等方面具有良好的“弹性”，下雨时吸水、蓄水、渗水、净水，需要时将蓄存的水“释放”并加以利用，提升城市生态系统功能和减少城市洪涝灾害的发生。海绵城市基于多技术、跨尺度的水生

态基础设施建设与规划，是对城市排水思路由传统的快排模式向雨水资源化利用的转变，是城市建设在价值观上对雨水资源、生态环境、旱涝灾害等问题的弹性适应与灵活应对，通过对原有生态系统的保护、修复、低影响开发等途径，以及机制建设、规划调控、设计落实、建设运行管理等过程，实现径流总量控制、径流峰值控制、径流污染控制和雨水资源化利用。海绵城市理论逐步完善，近几年，理论转向实践，全国各城市开展了大量的海绵城市建设工作，有效缓解了城市内涝问题。

全球持续推进生态城市发展。作为广为认可的城市发展模式之一生态城市，在众多示范城市建设后，近年来，开始从示范建设走向理念共识，融入各类城市的规划建设中。为应对气候变化，作为生态城市的重要内容之一，生物多样性不断受到国际和国内的关注，2022 年《昆明 – 蒙特利尔全球生物多样性框架》（简称“昆蒙框架”）通过，确立了全球生物多样性新一阶段的目标。

中国快速的城镇化发展带来了环境污染、交通拥堵、服务设施缺失、城市历史和文化特色消逝等一系列的城市病，制约了城市宜居性的发展。建设宜居城市已成为现阶段我国城市发展的重要目标，对提升城市居民生活质量、完善城市功能和提高城市运行效率具有重要意义。宜居城市是指经济、社会、文化、环境协调发展，人居环境良好，能够满足居民物质和精神生活需求，适宜人类工作、生活和居住的城市，即人文环境与自然环境协调，经济持续繁荣，社会和谐稳定，文化氛围浓郁，设施舒适齐备，适于人类工作、生活和居住的城市。

5. 城市更新与新城市科学研究进展

近年来，我国城市建设从增量转存量，城市更新是这一时期城市建设的主要任务，包括城市老旧街区、老旧厂区、老旧小区和城中村的更新改造。城市更新的实践从局部的老商业地带的更新、城市工业厂房的更新和城市老旧小区和城中村的局部更新开始走向城市片区的整体更新。城市更新的研究聚焦城市转型过程中的融资问题、城市更新的技术体系、城市更新的制度设计以及涉及历史文化遗产保护区域的更新方式等方面，城市更新实践案例不断增多、类型丰富、更新主体多样，包括政府主导、企业主导和居民自发开展的城市更新。

21 世纪以来，随着人工智能、传感器、物联网、云计算等技术的迅速发展，图灵奖得主吉姆·格雷认为，当下科学研究在技术的推动下已经进入第四次范式，即数据驱动型。这些颠覆性技术将进一步在不同层级作用于城市空间。新城市科学，即依托深入量化分析与数据计算途径来研究城市的学科模式，在多种新技术和新数据的支持下，以城市计算、虚拟现实、人机交互等方向为代表的多学科交叉的新城市科学正在为城市设计带来革新的可能性。2019 年，英国学者迈克尔·巴蒂出版了《新城市科学》，强调新城市科学在促进人们更好地理解城市系统和结构方面的作用。新城市科学提供关于城市面临的限制和挑战的新见解，丰富当前的城市规划方法，并用有利于所有城市居民的现实城市规划取代传统的自上而下规划。新城市科学从城市规模、内在秩序、特征性交通路线、构造网络的区位等多方面，介绍了从简单随机模型到自下而上的进化模型，再到交通与土地使用整合模型的模拟方法。借助这些理论和方法，新城市科学提出了设计和决策模型，以预测城市未来的互动和流。国内大量学者就新城市科学开展研究，主要聚焦在量化分析城市空间形态、城市模拟仿真、大数据和开源数据在城市规划中的运用等方面。

三、本学科国内外研究进展比较

国际上，可持续发展已成为共识，信息与数字技术不断崛起，促使城市走向智慧；国内以新型城镇化为引领，以“双碳”为目标，推动智慧城市的高质量发展。互联网、大数据、人工智能、信息系统、数字孪生等技术不断融入城市发展，复杂适应系统理论与城市研究紧密联系，成为城市科学发展新趋势。

（一）以复杂适应系统理论研究城市成为共识

以美国圣菲研究所为代表的复杂科学学派作为最早开始复杂科学研究的机构，近年相继出版了《复杂经济学》《规模》等复杂科学丛书。其中，圣菲研究所的外聘复杂系统教授路易斯·贝当古，作为芝加哥大学城市创新研究所主任，于 2021 年出版了《城市科学导论——城市作为复杂系统的理论与实证》（*Introduction to Urban Science: Evidence and Theory of Cities as Complex Systems*）一书，以复杂适应系统理论为基础，构建了城市和城镇化研究框架，并阐述了什么是城市、什么是城市科学。

美国伊利诺伊大学城市及区域规划专家路易斯·霍普金斯为代表的伊利诺伊规划学派，从城市规划的角度出发，分析城市规划理论发现，相关学科理论丰富，而城市规划学科本身的核心理论空心化现象明显，认为制定城市发展计划的逻辑是面对城市复杂的行动指引，收集规划行为中的主要变量，认为城市发展的复杂性特征可归结为相关性、不可预见性、不可分割和不可逆 4 个特征，城市发展无法实现均衡。

以英国学者迈尔克·巴蒂为代表的新城市科学学派，近年来出版了《新城市科学》一书，巴蒂运用现代复杂科学理论中的复杂网络分析城市的网络结构，并基于复杂科学、城市经济学、交通理论、区域科学以及城市地理学等相关研究，提出了分析城市结构的理论方法和预测城市未来发展方向的设计决策模型。

国内，钱学森最早提出城市作为复杂系统，应采用复杂科学研究城市。之后，随着密切的国际国内学术交流，国内相关研究机构持续关注复杂科学的最新研究，其中北京师范大学的系统科学院、中国科学院数学与系统科学研究院始终追踪复杂科学学术前沿，研究成果逐步与国际接轨。复杂系统理论在城市科学的应用在浙江大学、清华大学、东南大学、中国城市科学研究会等多所高校和学术机构开展。研究内容关注复杂适应系统理论与城市规划的关系，以复杂科学理论为基础的新技术、新方法，以及复杂适应系统理论在智慧城市、韧性城市建设等方面的创新与应用。

（二）在城市科学新技术方面国际相对领先

国际前沿科学技术已全面渗透植入城市研究之中，其中遥感、人工智能、地理信息系统、智慧城市、数字孪生、大数据等技术领先于国内水平，并应用于城市的信息收集、模型构建和决策指导。国内的城市研究逐步向国际接轨，逐步开始将大数据、地理信息系统、智慧城市理论运用到现代的城市研究与规划建设中，但因研究基础和起点的不同，与国际领先水平仍存在差距。国际上，伦敦大学学院巴特莱特高级空间分析中心成立于1995 年，麻省理工学院感知城市实验室成立于 2004 年，纽约大学城市科学与发展中心、芝加哥大学城市计算与数据中心均建立于 2012 年，国际上各高校均迅速建立与计算机、数据科学、信息科学相结合的中心和机构。而国内高校相应的研究机构较少，城市科学与计算机、信息系统等科学的交叉融合不足。武汉大学城市设计学院的城市规划与地理信息系统相结合城市研究，起步较早；中国城市科学研究会自 2012 年起，相继成立了数字城市工程研究中心、智慧城市联合实验室、城市数据安全管理中心、大数据专委会、海绵城市专委会、健康城市专委会、老旧小区改造专委会等分支机构，成为国内为数不多在城市科学新技术、新领域的研究组织。此外，北京城市实验室、北京城市象限、广州国地科技及华为的智慧城市、阿里的智慧交通均在城市科学前沿领域持续发力。

四、本学科发展趋势与展望

（一）城市科学学科发展的主要问题

城市科学研究需要突出综合性和战略性。吴良镛院士明确指出，开展城市科学研究要多学科协同，需要进行城市发展综合性重大问题的研究。城市科学研究应当坚持从实际出发，密切联系我国实际情况，“城市科学”这一学术概念研究城市比较合适、比较恰当。

城市科学学科发展存在的主要问题有以下几个方面。首先，城市科学理论体系不足，一直以来城市科学相关性较高的城乡规划这一学科，从城市规划到城乡规划，再到今天的国土空间规划，更多是在应用层面开展工作，基于我国过去几十年的快速城镇化过程中城市建设用地的增量需求，这一学科多与工程相关学科联系，以应用为主，学科基础科学理论研究不足，城市科学的学科理论支撑体系方面需要完善。其次，城市科学相关各专业分散在工程学、理学、社会学、管理学等多学科，这种不同类别下的学科对于统一的研究对象城市而言，难以形成合力。最后，学科体系缺少系统性，正如《自然（城市）》（*Nature City*）的主编在创刊开篇语所讲，尽管城市作为人、环境与技术紧密联系的关键所在，但大多数城市的研究各自独立，例如关于城市健康的研究，需要到公共卫生领域寻找相关研究，城市历史、城市生态和城市社会也是如此，这就为城市科学综合研究的推进带来了阻碍与不便。

（二）城市科学学科发展的趋势及展望

1. 城市科学逐步形成学科理论体系

城市科学依托传统的经济学、地理学、社会学及新兴的复杂科学、信息学、人工智能等研究基础，结合复杂适应系统理论、城镇化及人居环境理论，逐步综合形成城市科学以城市为研究对象，以复杂、多样、流动性为主要特征，多学科理论研究为基础的理论体系。

2. 城市科学科学性不断突出

研究城市的科学不断兴起，将城市地理学、城市社会学、城市经济学、城市管理学、城市历史学、城市生态学、城市地理信息系统、复杂科学等理论运用到城市研究中，引导城市从规划建设为主转向兼顾学术研究的建设实践。这些学科在自然科学、社会科学、技术科学、应用科学等多种分类中交叉、融合，它们都是以现代城市为研究对象，从多个侧面对城市进行研究。近年来，我国城市科学的发展不断从以城市的物质空间规划为主的城乡规划，转向区域规划和国土空间规划，与国际城市规划趋势相近；城市规划越来越关注经济、社会、环境等内容，工程性内容的比重下降，科学性内容不断突出。

这种国际、国内的学科转变，多学科共同研究城市成为发展趋势，丰富了城市科学的内容，推动了城市科学的发展。

3. 城市科学学科的横向复合与纵向交叉

随着全球化的进程，以实践为主的城乡规划学科面临理论创新的挑战，这也就赋予了偏重理论研究的“城市科学”新时代下的学科发展使命。学科纵深方向呈现明显的交叉融合趋势，尤其在智慧城市、生态城市、低碳城市、海绵城市、韧性城市等多种研究领域，互联网、大数据与城市规划结合，促使了智慧城市的发展；生态技术研究与城市研究结合，形成了生态城市；市政水务、海绵理念与城市雨洪管理结合，催生了海绵城市；研究城市应对自然灾害的规划设计，衍生了韧性城市的研究。

城市科学在多种学科的交叉复合下，形成了新时代的学科群特征，城市科学以人居环境改善为目标，复杂适应系统为理论，以城镇化为基础，聚焦在人文城市、智慧城市、低碳城市、健康城市、韧性城市、海绵城市、生态城市、宜居城市、城市更新和新城市科学十大方向，形成包涵人文城市、生态城市、海绵城市、智慧城市等多个二级学科的交叉学科群。

第三章

相关学科进展与趋势（英文）

1 Mathematics

1.1 Introduction

Mathematics is the basic subject which mainly studies on numbers and forms. It is vital to the foundation of natural sciences. Mathematics has provided a precise language, strict method and new research paradigm for scientific research. Looking back on the history of sciences development, it is not difficult to find that almost all major discoveries were related to the development and progress of mathematics. Mathematics is also an important theoretical basis for major technological innovation. Many of our country's "bottleneck" technologies are due to unsolved mathematical problems.

To implement the policy of accelerating the realization of high-level scientific and technological self-reliance proposed by the 20th National Congress of the Communist Party of China, this study, under the support and promotion of the "Discipline Development Leading Project" given by the China Association for Science and Technology (CAST), the Chinese Mathematical Society has set up a special research group to carry out a comprehensive study on the current development of mathematics in China through expert discussions, mathematical journal biometric study, and expert interviews. This section discusses the current situation of discipline development, makes an international comparison, thinks about problems and challenges, and puts forward countermeasures and suggestions.

1.2 Latest Research Progress of Mathematics in Recent Years

1.2.1 The importance of mathematics as a knowledge and tool has become increasingly prominent

Today, mathematical science permeates almost every aspect of daily life and is an indispensable and important support for many research fields such as Internet search, medical imaging, computer animation, numerical weather forecasting and other computer simulations, various types of digital communications, business, military optimization, and financial risk analysis. For example, there are still unsolved mathematical problems in general relativity, the mathematical description of black holes and their rotation, and the Hilbert space operator provides a natural framework for quantum mechanics. In addition, statistical research as a data science methodology has accelerated the completion of human genome sequencing. The great challenges facing astrophysics depend partially on advances in scientific computation and engineering. The rich data resources provided by satellite clusters and large ground–based instruments require integrated analysis with statistical, scientific and engineering calculations. Mathematics is an important theoretical basis of natural sciences, and the interaction between mathematics and natural sciences provides tools and knowledge to promote the development of natural sciences.

Mathematics, as the most universal tool and method, changes the way of scientific research, and provides a new paradigm for scientific research through the intersection with computer science. For example, emerging interdisciplinary disciplines such as computational fluid mechanics, computational materials science, and econometrics have emerged. Scientific and engineering computing can simulate physical processes that cannot be solved analytically and are difficult to experiment with, such as tsunamis, climate, nuclear explosions, etc. Therefore, simulation is recognized as the new paradigm of scientific research after experimental and theoretical reasoning. The analysis and use of big data is becoming an important means of scientific discovery, known as the fourth research paradigm after experiment, theory and simulation. The development of mathematics has promoted the emergence of data science and has become the theoretical basis and methodological means that big data analysis must rely on.

1.2.2 High-quality development of mathematical scientific research in the new era

At present, China has formed a multi–level collaborative innovation system of "government–universities–scientific research institutions–societies–journals–awards" in mathematics. The 5 societies relevant to mathematics are *Chinese Mathematical Society, China Society for Industrial and Applied Mathematics, Operations Research Society of China, Systems Engineering Society of China, and Chinese Association for Applied Statistics.* There are 9 relatively high–level journals of mathematics in China include *Acta Mathematica Sinica, Science China Mathematics, Acta Mathematica Scientia, Chinese Annals of Mathematics, Peking Mathematical Journal, Communications In Mathematics And Statistics, CSIAM Transaction on Applied Mathematics, Mathematica Numerica Sinica, Journal of the Operations Research Society of China*, etc.

By examining the development of mathematics research and the scale of output in recent years, we can focus on 4 important mathematical journals. These include the *Annals of Mathematics*, *Journal of the American Mathematical Society*, *Inventiones Mathematicae* and *Acta Mathematica*. These have great influence in the mathematical community, although they can not cover all the important fields of mathematics, they are recognized internationally to reflect the mathematical research level of the research object to a large extent. From the output data of these four journals, from 2001 to 2005 and from 2006 to 2010, the number of Chinese publications was 13 and 21 respectively, accounting for 2% of the total number of publications of the four journals. From 2011 to 2015 and from 2016 to 2020, the number of Chinese publications of the four journals of mathematics increased to 44 and 39. The proportion of the total number of papers published in the four major journals increased to 5%, an increase of 3 percentage points compared with the previous two periods, indicating that the number of papers published in China's top mathematical journals in the past ten years has significantly improved than before.

1.2.3 High-quality development of mathematics science popularization in the new era

In recent years, we have promoted the popularization of mathematical science and the cultivation of mathematical culture in different ways. Taking the Chinese Mathematical Society as an example, the society has actively innovated ways to popularize science, spread mathematical culture, and improved scientific quality.

1.3 Comparison of Research Progress in Mathematics Domestically and Abroad

Since the 18th National Congress, the level of mathematics research has been steadily improved. According to the statistics from WoS (Web of Science), since 2019, the total number of mathematical papers issued in China is more than that of the United States, ranking first. However, in the four top mathematical journals recognized by the global mathematical community, the number of publications in the United States is still the first, followed by France, Germany and the United Kingdom, China ranks the fifth. In terms of paper output efficiency, China also ranks the fifth, with an output efficiency of 4.21% from 2011 to 2020. Here the output efficiency of important journals refers to the share of the number of papers in important journals published in a country to the total number of papers published in that country[①]. In addition, the citation impact index of China's mathematics discipline standardization is 1.03, ranking 10th in the world and exceeding the average.

The development of mathematics at home and abroad can also be viewed from the International Congress of Mathematicians (ICM). The International Congress of Mathematicians is the largest and most important conference in the international Mathematical community sponsored by the International Mathematical Union (IMU). The conference is an international conference for mathematicians to exchange mathematics, demonstrate and discuss the development of mathematics. It is the grand meeting of the international mathematical community. The conference is held every four years. Presentations are divided into one-hour presentations and 45-minute presentations. It is a high honor to be invited by the General Assembly to give a one-hour or 45-minute presentation From 2010 to 2022, 195, 211, 238, and 226 people gave presentations respectively, including 20, 20, 22 (21 sessions), and 175, 191, 216, and 205 people who gave 45-minute presentations. According to the statistics of the nationalities of the scientists who gave presentations at the four mathematical congresses, the top ten are: 301 people from the United States,124 people from France, 58 people from the United Kingdom, 50 people from Germany, 34 people from Israel, 33 people from China, 29 people from Canada, 26 people from Russia, 25 people from Japan and 25 people from Switzerland. Among them, the United States ranks first with a global proportion of about 34.6%. China ranked sixth, with about 11 percent of the number of speakers in the United States and about 3.8 percent of the world's total. In the four conferences, the United States ranked first in one-hour speeches, and the number of speakers fluctuated but remained flat; China had only one one-hour speaker in 2010 and one in 2022.

In short, the development of Chinese mathematics has made many breakthroughs, but at this stage, there is still a gap of mathematical strength between China and the world's first-class countries, facing the transformation from a big country to a strong one.

1.4 Development Trend

Looking forward to the future of mathematics, we aim to build a strong country in mathematics, train international mathematics masters, and put forward mathematics problems with high international attention. In the field of mathematics, can China open up new topics, new directions and new branches that can lead the future of international mathematics? Can a Chinese School be

① The definition of important journal data in this book: Refer to the international journals in the Classification of Mathematical Science and Technology Journals of the Chinese Mathematical Society (2020 edition) T1, and the comprehensive, high-level journals recognized by the academic community.

formed? Can we cultivate and build mathematics awards and competitions with a high international reputation? These are worth thinking about and looking forward to.

First, the development of global mathematical science shows an obvious trend of inter-crossing and interpenetration among various branches of disciplines. On the one hand, there are many new knowledge growth points between the branches of mathematics. On the other hand, the extensive cross-integration of mathematics with natural science and engineering technology provides important support for aerospace, national security, bio-medicine and other fields. It is hoped that the quantity and quality of high-level mathematical papers in China can be further improved in the future, and the internationally influential Chinese mathematical journals and mathematical science popularization brand activities (such as IDM) can be continuously developed and consolidated, so as to improve the overall level of mathematical research and science popularization in China.

Second, China's mathematical research will maintain a rapid development trend during the "14th Five-Year Plan" period. In the field of basic theory, it is hoped that a certain number of Chinese mathematicians will become leaders in mathematical research, make China one of the countries leading the development of international mathematics, continue to train and bring up some young mathematicians with global competitiveness, and hope that local mathematicians will make breakthroughs in the Fields Medal, Wolf Prize and other internationally famous mathematics awards as early as possible; in the field of practical application, mathematicians will continue to be encouraged to care about practical problems, undertake and solve the country's major urgent problems, promote the rapid transformation of China from a mathematical big country to a mathematical great power. They will promote the research of key mathematical problems in science and technology which are related to national defense and people's better lives.

The third is to look forward to more science and technology policies to support the development of mathematics. Until today, China has formed a top-down policy environment from the three levels of laws and regulations, policy measures and industry norms, and constantly improve the system mechanism in mathematics evaluation, discipline layout, ethics and other aspects, creating and consolidating the cultural atmosphere of the whole society to respect mathematicians and attach importance to mathematical research. The future development of mathematics must be a strategic basic discipline that matches the overall development of science and technology in the country and is coordinated with the development of national science and technology policies, laws and regulations. It is necessary to put an end to short-sighted policies that rush for quick success and short-term benefits, and to have concentrated efforts to do major things. It is expected that more policies conducive to the development of mathematics and the cross-application of mathematics will be introduced and implemented.

2 Advanced Scientific Instruments and Equipments for Integrated Circuits

Instrumentation and equipment serve as the "tools" for people to understand and transform the world within a technological society. They are not only the outcomes of technological innovation but also the foundation that supports the progress of technology. The theoretical prototypes of advanced instrumentation typically originate from physics, and through integration with fields such as electronics, automation, and software, they form systems that can be applied in cutting-edge areas such as materials science, chemistry, biology, and integrated circuit (IC) manufacturing. The demand for advanced scientific instruments is growing, but the market is largely monopolized by foreign enterprises, leading to a "bottleneck" issue in domestic technology. Particularly in recent years, the IC industry has become one of the key areas of economic and trade disputes between China and the United States. Advanced instruments and equipment are closely related to the IC industry. To develop an independent and controllable IC industry, China has increased its investment and determination in the research, development, and industrialization of domestic instruments and equipment. Meanwhile, the United States has implemented strategic restrictions on the export of advanced manufacturing and testing equipment. This has intensified competition between the two countries in the field of IC equipment, creating a "Cold War" of science, technology, and trade.

The development of disciplines related to advanced scientific instruments and advanced IC equipment, and the progress of their industries, are mutually reinforcing and interdependent. In this era of globalization, advanced IC products rely on advanced manufacturing equipment, and mastering the technology of advanced instruments and equipment has become a significant reflection of a country's comprehensive strength and strategic competitiveness. The development of advanced instrument and equipment technology cannot be separated from a favorable scientific research environment and professional technical talent. Ultimately, in this competition, the cultivation of talent and the capacity for scientific research and technology are closely related to the development of disciplines related to advanced scientific instruments and advanced IC equipment. Addressing the current challenges faced by the discipline, such as talent shortages, weak basic research, and the disconnection between industry, academia, and research, will promote the high-quality development of the discipline and greatly enhance the overall strength and international competitiveness of China's IC industry. This concerns not only national security and scientific and technological progress but also promotes global scientific cooperation and benefits human society.

Based on the classification of IC manufacturing equipment and the technological progress of advanced instruments, we provide a concise description of the core technologies and corresponding research progress in front end of line (FEOL) and back end of line (BEOL) equipment, respectively. We analyze the technical characteristics, development trends, and core technical challenges, and point out the domestic and international development status of related technologies and the technical bottlenecks faced by China. FEOL equipment primarily includes photolithography machines, etching machines, coating equipment, ion implantation equipment, cleaning equipment, heat treatment equipment, CMP equipment, and measurement equipment. BEOL equipment mainly includes thinning equipment, wire bonding equipment, flip-chip bonding equipment, electroplating equipment, wafer bonding equipment, sorting equipment, dicing equipment, test machines, and probe stations.

In terms of the global development and competitive landscape, the IC equipment field generally maintains a high growth momentum, especially for equipment that is automated, intelligent, and efficient. This provides foundational support for the rapid growth of the IC industry under the drive of Moore's Law. However, there is an inversion in the consumption and production of IC equipment, with the development of China's IC industry facing a significant "shortcoming" in the equipment segment, which has become an important hidden danger affecting industrial safety and the stability of the industrial chain. Looking at the leading companies in the global IC equipment industry, the top giants have absolute control capabilities within the industry and have become the dominant forces in determining the technological direction of industry development. However, due to the multi-process, high-precision, and high-reliability requirements of IC manufacturing, driven by specialized division of labor, small and medium-sized enterprises have gained certain competitiveness in some niche areas. This also provides an "opportunity window" for the development of China's IC industry, which is in the stage of catching up from behind.

Scientific research cannot proceed without the support of disciplines related to Advanced instruments and equipment. In the context of IC research, we have primarily analyzed the instruments and equipment used for manufacturing nanoscale devices and exploring new materials and phenomena. This involves instruments designed based on various physical principles such as photon, electricity, magnetism, and force, as well as trends in related technologies. The equipment includes electron beam lithography systems, magnetic measurement devices, atomic force microscopes, high-frequency electronic instruments, X-ray devices, pulse laser deposition (PLD), and molecular beam epitaxy (MBE). We discuss the potential application of scientific research instruments in the IC industry and analyze the future trends of scientific instrument technologies in IC applications.

The application of scientific equipment in the IC field is a significant factor in promoting industrial upgrading and serves as the "maternal technology" driving the development of IC equipment. Therefore, it is a crucial element in analyzing the development and technological research of the equipment industry. There are clear differences between traditional IC equipment and scientific instruments in terms of functionality, design, performance, and application areas. In terms of functionality and application, traditional IC equipment is mainly used for large-scale manufacturing processes, such as lithography machines and vapor deposition equipment, while scientific instruments are used in scientific research and laboratory environments, such as atomic force microscopes and Raman spectrometers, for exploring new materials and phenomena. In design and manufacturing, traditional IC equipment focuses on engineering needs to meet large-scale production, while scientific instruments are more flexible to adapt to diverse research objectives. In terms of performance and precision, IC equipment needs to ensure consistent product quality, while scientific instruments must provide high-resolution measurement and analysis. As for automation, traditional IC equipment is highly automated in the production process, while scientific instruments usually emphasize the flexible operation by researchers.

From the perspective of the development process of IC equipment technology, the advancement of traditional IC equipment has been inseparable from the influence and inspiration of scientific equipment. With the rapid development of information technology, integrated circuits, as the core of modern electronic products, have become increasingly complex in manufacturing and research. The use of advanced scientific instruments has provided the IC industry with more precise measurement, analysis, and control methods. Especially in the research and development of the next-generation integrated circuits, scientific equipment has significantly propelled technological progress in the industry, helping to ensure the performance and quality of integrated circuits and promoting technological innovation. The potential applications of these scientific instruments in the IC manufacturing process encompass process research, quality control, and fault analysis, playing a vital role in improving production efficiency, reducing costs, and optimizing product performance. Although the scientific instrument industry is not large in scale, its role in the overall development of the IC industry is crucial. In the early 1990s, the U.S. Department of Commerce's National Standards Bureau issued a report stating that the total output value of the instruments and meters industry accounted for only 4% of the total industrial output value, but its impact on the national economy reached 66%. With the continuous advancement of technology, the application of scientific instruments in the IC industry will continue to evolve, bringing new opportunities and challenges to the industry's development.

Instruments empower human beings to observe and transform the world, giving rise to the birth of new fields and technologies. The innovative development of instruments and equipment is essential for the sustainable progress of science and technology. It not only caters to the needs of various industries but also allows scientists to explore the unknown, significantly promoting the advancement of human civilization. Advanced instruments and equipment represent the crystallization of interdisciplinary and multidisciplinary basic research, as well as cutting-edge engineering technology. However, China's development in the field of IC equipment is relatively insufficient, with gaps existing in basic research and key technology areas. Hence, it is imperative to support industrial technology accumulation and strengthen basic research while constantly embracing cutting-edge engineering technology to bridge these gaps.

China is grappling with a scarcity of talent in the field of advanced instruments and equipment, particularly in the cultivation of top-tier talent, which requires substantial time and investment. There is a discernible discrepancy between the quality and quantity of talent trained domestically and that trained abroad. A significant challenge exists in the siloing of academic disciplines, with a particular shortage of foundational disciplines. The advanced instrument industry is characterized by long production cycles and a limited ability to attract talent, both of which impact the size and composition of the R&D talent pool. Consequently, there is an urgent need to bolster the development of relevant disciplines and expedite the training of high-demand, high-level talent to underpin China's aspirations to become a manufacturing and scientific powerhouse. To this end, it is imperative to encourage collaboration between universities, enterprises, and investment institutions to foster industrial incubation and enhance the cultivation of scarce talent.

Advanced instruments and equipment are instrumental in driving scientific and technological innovation and national development. While China faces challenges in this domain, it also encounters tremendous opportunities. By observing the development of the IC industry, it is evident that government-led technology alliances have been pivotal in swaying the geographical competition of the IC industry at various stages. Examples include Japan's VLSI Program, the United States' SEMATECH, and other technology development alliances. In light of this, China should enhance top-level planning and establish its own technology alliance to navigate and capitalize on the opportunities in this field.

Furthermore, China should maintain strategic focus and persist in promoting the development of independent and controllable advanced instruments and equipment. The IC industry chain requires significant investment and has a long payback period, necessitating continuous dedication and financial support. It is essential to reinforce the industry chain, prioritizing autonomy and control over immediate commercial gains. The advancement of IC equipment should outpace IC manufacturing by three to five years, with due attention given to forward-looking layout strategies. The development of EUV stepper, for instance, has spanned 20 years from the initial stages to another 20 years for mass production, underscoring the need for sustained investment support. We should leverage cutting-edge applications as a driving force, integrate the upstream, midstream, and downstream industries, enhance the stability and competitiveness of the industrial chain, and prioritize the establishment of an independent intellectual property system.

Assessing the recent development and technological trends of China's advanced scientific and IC equipment, it is evident that China's strength in scientific and technological innovation is on the rise. However, the market for advanced scientific and IC equipment remains predominantly controlled by foreign firms, posing a 'stranglehold' risk to domestic scientific endeavors and impeding the development of disciplines related to China's advanced scientific and IC equipment. It is imperative to acknowledge the strategic role of high-end scientific instruments in driving scientific and technological innovation and national economic growth. We must prioritize talent cultivation, refine scientific research system guidance, enhance policy support, and expedite the development of disciplines related to scientific instruments in order to achieve self-reliance and reduce dependency on foreign technologies.

3 Chemistry

The field of chemistry, as a fundamental discipline that investigates the properties, composition, structure, and dynamics of matter, serves as a primary tool for humanity to comprehend and manipulate the material world. It plays a pivotal role in addressing sustainable development challenges encompassing resources, information, environment, life sciences, and health. This section summarizes the achievements of Chinese chemists in scientific research and education in the past four years, covering six major branches of chemistry: physical chemistry, analytical chemistry, inorganic chemistry, organic chemistry, polymer chemistry, and chemistry education, as well as the latest developments in interdisciplinary fields such as nuclear chemistry and radiochemistry, chemical biology, public safety chemistry, environmental chemistry, organic solid chemistry, molecular medicine, energy chemistry, and combustion chemistry.

3.1 Physical Chemistry

Physical chemistry is a discipline that investigates the behavior, properties, and laws of chemical systems by employing physical principles and experimental techniques. The frontier of physical chemistry research internationally focuses on developing advanced in-situ characterization techniques, deepening understanding of the structure, processes, and mechanisms of chemical systems, establishing new theories and insights, and realizing rational design of materials at the atomic and molecular levels. Since 2020, China's research in physical chemistry has entered an era characterized by continuous breakthroughs through deepening comprehension of chemical systems, with deepening understanding of chemical systems and the emergence of new theories, concepts, and technologies.

In the field of chemical thermodynamics, Chinese scientists have proposed new concepts such as "spatiotemporal phase-change materials" "acidity heterogeneity in solutions" and "generalized excess spectrum", established quantitative measurement methods of ionic liquid ionization rate, new methods of characterizing thermal conductivity and interface thermal resistance of microscale materials with Flash DSC, and technologies for measuring thermal effects of chemical processes under extreme conditions.

China has achieved international leadership in chemical dynamics and kinetics. In the field of gas phase reaction dynamics, the quantum interference mechanism and geometric phase effect of important series of elementary reactions related to H_2 have been revealed, providing a new perspective for understanding the quantum effects in chemical reactions. In the field of cluster structure research, a new method using VUV free-electron laser light source-high resolution tandem mass spectrometry has been developed, providing a new technology for solving protein structure. In research of ultrafast dynamics of carrier in energy materials, a number of new femtosecond spectroscopy technologies have been developed, revealing the microscopic mechanism of energy transfer and charge transfer process of quantum dots, perovskite solar cells and other systems, realizing the rational design and performance optimization of energy materials. In the field of quantum computation and reaction dynamics, the construction of high-precision potential energy surface using neural network algorithm is in the international leading position.

In the field of catalytic research, Chinese scientists have made significant contributions to the optimization of catalyst selectivity and catalytic activity by quantitatively describing, redefining, and expanding important catalytic theories and concepts. Through deeper understanding of the essence of catalytic action, they have successfully resolved the challenge of achieving high selectivity and high catalytic activity simultaneously, while also developing various methods for high-throughput screening of advanced catalytic systems. Furthermore, by integrating chemical processes with biological engineering, they have achieved the catalytic preparation of high value-added substances such as starch from CO_2.

In the field of colloid and interface chemistry, Chinese scientists have enhanced their comprehension of the influence on the properties such as chirality, catalysis, and assembly of of micro- and nano-scale materials, and discovered the importance of nano-scale curvature matching for chiral transfer. They proposed the concept of "emulsion droplet fixed-bed catalysis", which solved the difficulties of immobilization and continuous use of enzymes, coenzymes, and homogeneous catalysts. They established molecular self-assembly in the solid-phase, and realizing the leap of molecular self-assembly from solution dispersion system to macroscopic continuous material.

In the field of electrochemistry, Chinese chemists have integrated advanced electrochemical research methods with theoretical calculations to demonstrate that changes in chemically adsorbed water coverage with potential contribute to a negative capacitance of the electrochemical interface. This finding expands our understanding of the classical double-electric-layer model for the electrochemical interface. Additionally, they have elucidated that the hydrogen bond network within the double-electric-layer may serve as a source for pH effects in the electrochemical hydrogen evolution reaction, providing a novel explanation for previously controversial issues in this field.

In the field of green chemistry, Chinese scientists have developed a series of advanced chemical processes aimed at achieving non-toxicity, harmlessness, waste reduction and maximum atomic efficiency. They have promoted the utilization of waste plastics as resources, designed a new green pathway for vitamin K3 production and achieved China's transition from import dependence to becoming a leading producer of caprolactam.

In the field of biophysical chemistry, Chinese scientists have elucidated the physical and chemical mechanisms underlying various biomolecules in vital processes, while also pioneering novel technologies and methodologies for advancing biophysical chemistry research. These contributions have significantly bolstered China's global standing and influence in this domain. Furthermore, they have independently developed SPONGE, a high-performance molecular dynamics simulation software that serves as an invaluable theoretical research tool and platform for investigating biophysical chemistry.

In the field of theoretical and computational chemistry, Chinese scientists have successfully developed and integrated mobile robots, chemical workstations, intelligent operating systems, and scientific databases, and have developed a data-intelligence-driven full-process robot chemist, demonstrating the enormous advantages of a new intelligent paradigm guided by the "strongest chemical brain".

3.2 Analytical Chemistry

Analytical chemistry is the scientific discipline that investigates the chemical composition, content, structure, and form of substances, serving as a crucial pillar in science and technology advancement. From 2020 to 2023, China has consistently demonstrated robust progress in bioanalysis and sensing research. Continuous advancements have been made in areas such as single molecule and single cell analysis, in vivo bioanalysis, functional nucleic acid-based bioanalysis, nano bioanalysis, and there has been a steady increase in the publication of research papers in top-tier international journals. China's chromatography research has achieved remarkable breakthroughs encompassing sample pretreatment techniques, chromatography stationary phase and column technologies, multidimensional and integration strategies, as well as complex sample separation and analysis (including proteome analysis, metabolome analysis, traditional Chinese medicine group analysis), making important contributions to the development of life sciences, precision medicine, new drug creation and public safety.

3.3 Inorganic Chemistry

In the field of inorganic chemistry, China has made outstanding achievements in the construction, regulation and application of framework materials such as COFs/MOFs/HOFs; engineering and performance of cluster/functional complex crystal; assembly and properties of supramolecular crystalline compounds; structural design, growth and performance of functional crystalline materials; preparation, structural regulation and performance of low-dimensional crystalline materials; and synthesis, structure and performance of other crystalline materials. Chinese scholars continue to take the lead in the research of COF/MOF/HOF materials. The research focus has shifted from precise analysis of crystal structures and discovery of novel structures to exploring the structure-activity relationship and industrial applications of crystalline compounds. Many high-quality achievements have emerged in the fields of high efficiency/high selectivity adsorption/separation, high efficiency and high selectivity catalysis, high efficiency luminescence and high sensitive photoelectric response, fuel cells/hydrogen energy/energy storage and so on. In the high purity separation of chemical materials and efficient storage of hydrogen/ammonia, MOF/COF materials have made further breakthroughs, which is expected to promote the development of China's fine chemical industry and new energy industry. In addition, significant advancements have been made in the exploration of crystalline compounds such as coordination polymers and inorganic non-metallic compounds, the synthesis of low-dimensional crystalline materials, and the functional group arrangement within crystalline materials. These achievements demonstrate promising applications in areas including new energy, quantum information, life sciences and healthcare, as well as national defense and security. In the field of molecular sieve research, Chinese scholars have achieved a series of important breakthroughs in synthesizing novel molecular sieve topological structures, developing advanced analysis and characterization techniques for molecular sieves, as well as applying molecular sieve catalysis and adsorption separation methods. Based on major national demands and the forefront of rare earth science and technology research, we have conducted investigations in four directions: green separation chemistry and purification processes for rare earth elements; advanced functional materials incorporating rare earth elements; structural materials involving rare earth elements with sub-microstructure regulation capabilities; chemical biology studies related to rare earth elements along with nano biomaterials development. Substantial progress has been made in these areas.

3.4 Organic Chemistry

In recent years, China's organic chemistry has made great progress, has approached and reached the international advanced ranks, and some areas have made leading research achievements. Numerous original ligands have been successfully developed, enabling efficient and highly selective transformations of a wide range of organic chemical reactions. These ligands are widely utilized by scholars and enterprises worldwide for academic research and industrial production. Notable examples include Zhou's catalyst, Feng's ligand, and the Ullmann-Ma reaction. Chinese chemists emphasize the originality and systematic approach in their work, proposing innovative concepts and strategies. A number of distinctive and influential chiral ligands and catalysts have been developed, reaching the international leading level in some research directions. Furthermore, remarkable achievements have been made in China's organic chemistry regarding the synthesis of important bioactive natural products, creation of novel structures, and discovery of new functions. For instance, natural product chemistry has gradually merged with chemical biology to explore small molecular biological functions within new structures as well as discover new targets, drugs, materials, and environmentally friendly bulk chemical manufacturing.

3.5 Chemistry of Polymers

The year 2020 marks the centennial anniversary of polymer science's birth. Over the past four years, China has witnessed continuous advancements in polymer science research, with a growing emphasis on cross-disciplinary integration. Notably, significant progress has been made in biomedical polymers and optoelectronic functional polymers, leading to breakthroughs in treating major diseases using polymer materials and organic photovoltaic materials. Furthermore, remarkable original research accomplishments have emerged in polymer synthetic chemistry, particularly within olefin coordination (co)polymerization, ring opening (co)polymerization, sulfur-containing polymer synthesis, and closed recyclable polymers. Additionally, several new catalysts (systems) with independent intellectual property rights have been developed; some are currently undergoing industrial transformation. These advancements lay a solid foundation for the production of high-end and high-performance polymer materials in China.

3.6 Chemistry Education

Chemistry education plays a pivotal role in fostering scientific literacy and critical thinking, promoting environmental conservation and sustainable development, enhancing health and safety awareness, as well as instilling scientific and social ethics. From 2020 to 2023, Chinese chemistry educators have made significant strides in advancing the curriculum reform of fundamental chemistry education, cultivating highly skilled secondary school chemistry teachers, and elevating the teaching quality of tertiary-level chemistry education.

China has revised and released the chemistry curriculum standards for general high school, chemistry curriculum standards for compulsory education and four versions of high school textbooks. In terms of curriculum materials research, in terms of textbook content analysis, Chinese scholars have deeply explored the knowledge structure, teaching objectives and learning content of textbooks. In terms of textbook design and evaluation, Chinese scholars have systematically analyzed the principles and methods of textbook design, evaluated teaching strategies, learning activities and organizational structure. The relevant research results provide guidance for textbook design and improvement, and promote textbook innovation. In terms of teaching design research, they have further deepened the structure and connotation of chemical subject ability literacy, and systematically developed assessment tools such as scientific ability test, chemical subject ability evaluation, and scientific attitude evaluation. In terms of student learning research, researchers have carried out brain science and brain cognition research, subject understanding research, scientific demonstration, scientific thinking and scientific identity research, modeling ability evaluation, experimental innovation, and gamification teaching research. By studying the process and characteristics of students' understanding of science and chemistry concepts, they have revealed the chemical concept structure and learning mechanism in students' minds, and provided ideas and approaches for optimizing teaching. In terms of interdisciplinary practice research, researchers have carried out in-depth research and practice of project-based teaching and deep learning, and have deeply explored the teaching practice to promote the development of core literacy and chemistry core literacy. In terms of information technology research, researchers have carried out middle school chemistry experiment research by means of implementing modern technical means such as handheld technology digital experiment, sensor, information technology, photography technology, and visualization technology. The design and development of chemistry educational games can provide support and guidance for teachers, and promote the innovation of chemistry education and the improvement of teaching effectiveness.

3.7 Interdisciplinary Subjects

3.7.1 Nuclear Chemistry and Radiochemistry

In the past three years, the overall level of basic studies on the terrigenous elements has been significantly improved by Chinese scientists, and major breakthroughs have been achieved in the theoretical chemistry of super plutonium elements, the synthesis of terrigenous elements compounds, and the cluster chemistry of terrigenous elements. The Cf–C bonding properties have been revealed and the UV–vis–NIR spectra of terrigenous compounds have been accurately simulated. The largest terrigenous–silver cluster [Th9Ag12] has been prepared by the induced decomposition–recombination strategy. A number of advanced separation technologies and new methods for the back–end of the nuclear fuel cycle have been developed. The protective–group that can accurately match the coordination configuration of hexavalent terbium has been innovatively designed, and the unconventional valence stabilization strategy of terrigenous elements has been developed. The long–term stability of hexavalent terbium in aqueous solution and solid has been realized, and a new strategy of safe, efficient, and green lanthanum–terbium separation has been derived, achieving the best separation effect between hexavalent terbium and trivalent lanthanide reported so far in the world.

3.7.2 Chemical Biology

In the past four years, Chinese chemical biology research has witnessed a significant advancement, with groundbreaking work conducted in the following pivotal and cutting-edge areas, leading to remarkable accomplishments: ①efficient synthesis and assembly of biomacromolecules and biological machinery encompassing protein ubiquitination, sugar biosynthesis, glycosaminoglycan chemistry, DNA chromosomal features, and metabolic pathway construction; ②in vivo regulation of biomacromolecules and life processes through utilization of small molecule probes, bio-orthogonal chemical reactions, and other innovative tools; ③dynamic chemical modification of biomacromolecules along with their labeling and detection within biological processes; ④mechanistic elucidation of bio-macromolecular machinery as well as life processes; ⑤target validation and development of targeted compounds based on strategies derived from chemical biology. These endeavors have exerted a profound international influence while attaining global parity across most research domains alongside international leadership in certain directions.

3.7.3 Public Security Chemistry

The major of public security chemistry has made obvious progress in new theories, new technologies, new materials and other aspects, and has made innovative contributions to national security, military security, anti-chemical terrorism, chemical accident prevention, drug prohibition, emergency response and other aspects. Notably progress has been achieved in the study of drug metabolism mechanisms. The development of rapid, accurate and intelligent detection technologies for hazardous substances has been emphasized, and novel technological breakthroughs such as single hair micro-segment drug detection, sewage-based drug monitoring systems and explosive safety detection methods have been preliminarily implemented.

3.7.4 Environmental Chemistry

In the past three years, Chinese scholars have carried out a large number of internationally advanced researches in the field of environmental chemistry. The researches in the following aspects have attracted extensive attention both domestically and internationally: identification of novel pollutants through non-target analysis technology, pollutant traceability using stable isotopes, elucidation of air pollution mechanisms, development and synthesis of new materials for water pollution control and soil remediation, assessment of quality and safety in agricultural products, as well as investigation into environmental toxicology and pollutant-related health effects. Several research domains have assumed a pioneering role in the world.

3.7.5 Organic Solid

Since 2020, there has been a rapid development trend worldwide in the field of organic solids. Currently, the overall performance of organic light–emitting diodes continues to optimize and has been widely used in the field of display. The key performance indicators of organic photovoltaic and thermoelectric materials are steadily improving, surpassing conventional limits on energy conversion efficiency for molecular materials. High mobility organic semiconductors have gradually achieved coupling regulation between mechanical and electrical functions, resulting in stretchable and repairable intelligent organic field effect transistors. Breakthroughs in organic bioelectronics and single molecule technologies are revolutionizing human–machine interfaces and molecular circuits, fostering new interdisciplinary growth points. Notably, China has made remarkable progress in the field of organic solids by achieving pioneering research accomplishments in high–performance molecular design, multi–scale molecular assembly techniques, and device function regulation.

3.7.6 Molecular Medicine

Molecular medicine research encompasses the molecular–level investigation, diagnosis, treatment, prevention, and mechanism exploration of diseases. The crux lies in the identification, design, and application of molecules capable of specifically targeting diseased cells and disease markers to significantly enhance the accuracy of disease detection and treatment efficacy. Major pharmaceutical companies in Europe and the United States have acquired a substantial number of pivotal core patents in monoclonal antibody research field, which has caused Chinese pharmaceutical companies to be at a competitive disadvantage in monoclonal antibody drug research and development, facing huge barriers and difficult to break away from the status of followers. In view of these factors, Chinese scientists have exerted tremendous efforts and achieved a series of original research breakthroughs. From 2020 to 2023, numerous cutting–edge innovations have emerged in China's molecular medicine research domain encompassing single–cell protein mapping based on nucleic acid aptamer sequencing technology, molecular diagnostics and therapeutics techniques, nano biomedicine advancements as well as in vivo bioanalysis.

3.7.7 Energy Chemistry

Energy chemistry uses chemical principles and methods to study the basic laws of energy acquisition, storage and conversion, laying a foundation for the development of new energy technologies and playing an important supporting role in the development of energy science. In the past three years, Chinese scholars have made a series of breakthroughs in the basic science and engineering application of energy chemistry. Fiber lithium–ion batteries, synthetic synthesis of carbon dioxide to starch and in situ direct electrolysis of seawater for hydrogen production have been selected as the top ten scientific progresses in China. They have attracted great attention from academia and industry, and greatly promoted the dissemination of basic knowledge and the development of energy chemistry.

3.7.8 Combustion Chemistry

Combustion chemistry is a multidisciplinary subject that faces the needs of major national strategies. The development of combustion chemistry for advanced power equipment is of great significance for breaking through the barriers between basic research and engineering applications. Chinese scientists have made significant progress in the fields of hydrocarbon fuels, combustion dynamics simulation, combustion mechanisms, and engineering computing software and databases in the field of aviation power.

4 Water and Environment

4.1 Introduction

The water environment is a general term for the environment in which water bodies in various layers of the Earth's surface system are located. The discipline of water environment is a discipline that studies the characteristics, evolution laws, mechanisms, and regulation theories, methods, and technologies of water environment systems. According to different research objects, the discipline of water environment can be divided into lake environmental science, river environmental science, groundwater environmental science, and urban water environmental science; according to different research purposes, it can be divided into water environment monitoring, water environment evaluation, water pollution control, water environment management, and water ecological protection. In recent years, the quality of water ecological environment in China has been continuously improving, and significant achievements have been made in water ecological environment protection.

Water environmental science has shown new development trends in theory, methods, technology, and other aspects. In theory, sustainable development theory has become a guiding ideology for studying the water environment. Traditional disciplines such as hydrology, hydrodynamics, hydrochemistry, environmental hydraulics, and water pollution treatment intersect deeply with ecology, economics, and other disciplines. Methodologically, water environment science integrates multiple water environment elements and conducts research on river basins and regional systems. In terms of technology, interdisciplinary and organic integration of existing professional technologies, high-tech is continuously applied in the field of water environment, and emerging technologies such as artificial intelligence and big data are constantly being applied in the field of water environment, greatly improving the depth and breadth of research in water environment science.

4.2 Recent Research Progress in the Discipline of Water Environment

In recent years, major progress has been made in the water environment discipline in the areas of water-quality and water-ecology evaluation and environmental benchmark and criteria establishment, water treatment theory and technology, and lake governance theory and technology, as briefly described below.

(1) The control object pays more attention to the identification and control of new high-risk pollutants and the synergistic control of compound pollutants. The types of pollutants in the water environment continue to increase, new pollutants continue to emerge, and the characteristics of compound pollution are increasingly prominent. Focus on conventional single pollutant in the water environment pollution formation mechanism and control principles, has been unable to meet the increasingly complex water environment pollution management needs. Therefore, the identification of new high-risk pollutants, microscopic transformation mechanisms and control principles, water quality standards set the basic theory, the theory and technology of synergistic control of complex pollutants has become the main direction of development of the discipline of water environment.

(2) Control objectives pay more attention to water ecological protection and restoration and water ecological health protection. In China, water environment management goals have been cut from conventional pollutant emissions to improve the quality of the water environment, water ecological protection and restoration and water ecological health security development. Therefore, water ecological restoration and safety and security theory and technology system, water ecological health evaluation theory, methods and technology has become the new development needs.

(3) The control means pay more attention to the new theory and technology of pollution reduction

and carbon reduction and resource recycling. Low–efficiency and high consumption of water environment pollutants at the end of the governance model, is not in line with the new needs of China's socio–economic high–quality development. The whole process of water pollutant prevention and control, pollution reduction and carbon synergistic theory and technology, water ecological recycling as well as efficient recycling of resources and energy theory and technology breakthroughs are receiving more and more attention to achieve the recycling of resources, the whole process of control and fine management is the inevitable requirements of the future development of water environment discipline.

(4) Research methods to the development of microscopic analysis and macroscopic simulation. Water pollutant migration and transformation research methods to electronic transfer tracking, ultrastructural analysis, micro and nano–interface observation, water pollutant ecological effects of research methods to the molecular, cellular, microbial community direction, the water environment system simulation methods to the regional simulation, watershed simulation and global scale simulation development, the application of information technology means of big data will be more extensive and in–depth.

(5) Theoretical innovation is more concerned about the in–depth intersection and deep integration with emerging disciplines. The development of basic theories and cutting–edge technologies of water and environmental disciplines is becoming more and more in–depth and closely integrated with modern biotechnology, information technology, biotechnology, new energy technology, new materials and advanced manufacturing technology. Diversified disciplinary crossover and the introduction of big data, artificial intelligence and other emerging technologies provide a strong impetus for the original innovation of the basic theories of water environment disciplines, breakthroughs in disruptive technologies and the development of comprehensive solutions to multi–scale, cross–basin and cross–regional water ecological and environmental problems.

4.3 Comparison of Recent Research Progress at Home and Abroad

From the perspective of the overall research progress of the water environment discipline, the development directions, including wastewater resources and energy recovery, low–carbon optimal design and operation of water systems, and the application of artificial intelligence, are both research frontiers and hotspots at home and abroad.

At the same time, for the "14th Five–Year Plan" period, according to the new needs of water and ecological environmental protection in China, the research focus has been from wastewater pollution control to synergistic governance of water resources, water ecology, and water environment.

In the field of water quality, water ecology evaluation and environmental benchmarking standards, comprehensive indicators for complex environments, regionally differentiated benchmarking standards and systematic evaluation studies are the main development trends. At present, the traditional water quality indicators at home and abroad focus on the evaluation of physical and chemical properties of water resources, and the evaluation results are difficult to provide a detailed and holistic analysis of the relevant environment. Chinese scholars face the non-linear changes in the complex system of the water environment and the qualitative characteristics of the "three water integration" as the core of the innovative concept of "water feature", to understand the water quality of the water ecological situation to provide strong support. Developed countries have established relatively perfect technical methods and systems for water quality benchmark standards, and Chinese research on water quality benchmark standards started relatively late, but developed rapidly. Since the "Eleventh Five-Year Plan", China's ecological environment benchmark work has made breakthrough progress, based on China's regional differentiation characteristics have been issued a series of water quality benchmarks and their development of technical guidelines; after many years of development and revision and improvement of China's water quality standards have been gradually formed in line with China's national conditions of the complete set of systems. The development level of China's water quality benchmark standards has basically reached or even exceeded that of developed countries or regions such as the United States and the European Union. The study of ecosystem health assessment began in the 1980s, but there is no uniform view of "ecosystem health" in the international community. In China, after continuous research, the research on ecosystem health assessment has been extended from rivers to lakes, reservoirs and other water environment types, and gradually formed its own system, which can provide data support for the development of ecological restoration objectives, assessment of ecological restoration effects, and environmental legislation and law enforcement.

In the field of water treatment theory and technology, the synergistic effect of pollution reduction and carbon reduction has been the hotpot in the international water treatment industry all over the world. With economic and social development, although the pollution intensity is increasing, and the types of pollutants are becoming more and more complex, the public environmental awareness has been enhanced, and the requirements for the quality of the water environment are constantly improving.

Therefore, wastewater treatment plants in some developed countries are developing from biological nitrogen and phosphorus removal to enhanced nitrogen and phosphorus removal. Meanwhile, the application of advanced treatment technologies such as advanced oxidation, nanofiltration and reverse osmosis is becoming more widespread to achieve the removal of emerging pollutants such as environmental endocrine disruptors, pharmaceuticals and personal care products, and to meet the demand for a healthier and safer water environment quality. Low-carbon treatment and energy development, climate change issues and energy crises require urban wastewater treatment to achieve low-carbon, treatment process to achieve energy conservation and improve energy self-sufficiency. Developed countries in Europe and the United States have carried out research focusing on wastewater reclamation and reuse, wastewater biomass recycling, nitrogen and phosphorus recycling, etc., in accordance with their respective national conditions. Domestic also carried out a series of research and development work represented by the concept of municipal wastewater resources plant, with a view to promoting the transformation and upgrading of the water treatment industry.

In the field of lake governance theory and technology, following the concept of ecological civilization and the harmonious development of man and nature, the health of lake ecosystems and the comprehensive restoration of ecological functions are becoming new goals and requirements. As a result, lake governance is changing from pure water quality management to the synergistic improvement of water quality and water ecology. How to achieve the new goal of higher-quality protection and governance of lakes along with rapid socio-economic development of the watershed is a key issue to be addressed at home and abroad. In recent years, the domestic lake governance research and practice, combined with traditional water quality management theory and technology system, drawing on the path of lake governance in developed countries, but also constantly exploring new mechanisms and methods of ecological restoration of lakes to meet the current and future governance needs, lake protection and restoration of research and practice level basically reached or even exceeded the level of developed countries or regions. On the one hand, the technical level of lake basin environmental management has been effectively improved, through the optimization of the traditional technology to obtain the best management efficiency while continuously integrating the latest theories and technologies, looking for the technical increment of lake management, and improving the scientific and technological support capacity of lake management. For example, new type of passivator, new type of wetland, bio-habitat restoration and constructive technology and new lake environment monitoring technology based on Internet of Things, satellite remote sensing, artificial intelligence, eDNA, etc., have achieved better application effect in the process of lake management. On the other hand, the concept of coordinated governance of lake basins has been strengthened, the industrial structure of the basin has been optimally adjusted, the scientific scheduling of water resources has been strengthened and lake ecological restoration projects have been carried out, thus promoting the organic combination of lake governance-related legislation, policies, planning, standards and other management measures and governance technologies, and gradually forming a systematic governance system.

4.4 Trends and Perspectives in the Discipline

In the future, the discipline of water environment will focus on the major strategic needs of national ecological civilization construction and international academic frontiers. In response to the water environment problems that have emerged in China's urbanization process and social development, we will implement the concepts of green, low-carbon, and sustainable development, promote forward-looking, original, and disruptive theoretical innovation and technological breakthroughs, continuously enrich and improve the basic theory, method system, and technical system of water environment discipline, and form innovative ideas. A new situation in the development of disciplines is characterized by the gathering of innovative talents, the emergence of original innovative achievements, and the improvement of the ability to solve complex water environment problems. During the 14th Five Year Plan period and beyond, the development of water environment discipline should focus on long-term layout, rational planning, and key breakthroughs.

To promote the future development of water environment discipline in China, the following suggestions are proposed.

4.4.1 Basic research

The research methods, water environment standard formulation methods, water environment measurement standard materials, basic concepts, terminology definitions, and basic data of the water environment discipline have a fundamental position in the field of water environment, and play an important foundational and driving role in the development of the discipline. However, China's research and accumulation in this area are still insufficient, and its fundamental contribution to the development of the discipline needs to be improved.

4.4.2 Systematic research

As of 2023, most of China's water environment discipline research has focused on the principles of water pollution control technology. The phenomenon of "technological isolation" is prominent, including the integration theory of water pollution control technology, combination process design and control theory, safety guarantee theory of drinking water throughout the entire process, construction theory of urban-rural integrated water supply and drainage system, and overall optimization theory of water environment treatment system in key river basins The research on the construction theory of regional recycled water recycling system is still insufficient, and the systematicity of water environment discipline research urgently needs to be improved.

4.4.3 Discovery based research

Discovering new water environment problems and pollution mechanisms has breakthrough or disruptive significance in ensuring water environment safety and promoting disciplinary development. China has obvious advantages in solving existing water environment problems, but the discovery of internationally recognized new problems, phenomena, and mechanisms is still very limited, catering to international academic frontiers and major needs in China. It is urgent to propose and implement a Chinese approach to solving water environment problems, leading future development.

5 Refrigeration and Cryogenics Engineering

5.1 Introduction

Refrigeration and cryogenics technologies use artificial methods to obtain temperatures below the ambient temperature. Its basic task is to study the principles, technologies and equipment to obtain and maintain a temperature and humidity different from the ambiance, and to use these technologies in different scenarios. Its scope includes the cooling output to maintain low temperatures, as well as dehumidification, environmental parameter adjustment, and heat pumps. Refrigeration and cryogenics technologies are closely related to almost all sectors of the national economy and people's daily life. In addition, cryogenics technology is an indispensable means for cutting-edge research in physics, while it is also used in medical and health care, industrial technology, space exploration, etc.

Under the background of environmental crisis and international technology competition, the implementations of many major global issues and national policies and the solution of scientific problems rely on the development of refrigeration and cryogenics technologies. Therefore, refrigeration and cryogenic technologies have attracted more and more attention. Currently, driven by policies and markets, traditional refrigeration technologies are facing mandatory upgrading; meanwhile, new refrigeration and cryogenics technologies are developing rapidly.

5.2 Recent Development

In recent years, the discipline of refrigeration and cryogenics engineering has focused on the forefront of international technology and major national needs. A number of significant achievements have emerged, supporting the development of urban and rural construction, clean heating, waste heat recovery, logistics, new energy vehicles, air separation, etc. The development of various fields such as natural gas liquefaction has also played a major supporting role, especially in assisting the rapid development of cutting-edge scientific and technological fields such as cutting-edge physics, quantum technology, and superconductivity in China.

In the field of refrigeration technology, the substitution of refrigerants is a research hotspot. Meanwhile, advancements such as the design and application of the cycle, the improvement of compressor and heat exchanger, and intelligent regulation have been made in the vapor compression refrigeration. In the field of heat-driven refrigeration and heat pump, absorption refrigeration and heat pump technology has mainly developed two types of absorption heat exchangers and promoted the engineering application of heat pumps and refrigeration systems. Moreover, sorption refrigeration and heat pump technology has evolved high-performance composite sorbents such as multi-halides and constructed multi-stage combined cycles and systems. In addition, thermoacoustic technology focuses on the development of dual traveling wave-type and gas-liquid/solid coupling thermoacoustic refrigeration systems. As for the solid-state refrigeration, magnesium-based thermoelectric materials have been developed; additive manufacturing has been demonstrated to produce elastocaloric materials with customized shape and functionality; durable, low field intensity, and giant caloric effect were discovered in a cutting-edge electrocaloric polymer material; fluid-free compact electrocaloric cooling devices have received much development and improvement; elastocaloric cooling is moving towards kW-range cooling capacity.

The application of refrigeration technology has rapidly developed. In air conditioning systems, controlling the indoor thermal environment is important, including independent control of temperature and humidity, system process innovation, and new desiccants. The research on key components of air conditioning systems is also significant, including the upgrade of chillers such as magnetic suspension and aerodynamic suspension, as well as the optimization of heat exchangers. Further, low-temperature air source heat pumps and multi-split air conditioner systems have been developed rapidly, while the development of online performance measurement and big data technology has reduced energy consumption and maintenance costs. In terms of cold chain equipment technology, the main advances include differential pressure precooling equipment based on flowing ice, mobile differential pressure precooling equipment and intelligent three-dimensional freezing tunnel. The latest research progress of high-temperature heat pumps mainly includes efficient high-temperature water vapor heat pumps using twin screw compressors and water spray cooling, large-temperature-lift cascading compression heat pump cycles or absorption-compression coupled heat pump cycles with both low-temperature heat source utilization and high temperature output, and direct steam supply technology of high-temperature heat pumps combining open and closed compression. As for the heat pump in vehicles, there have been new thermal management technologies such as vehicle thermal management frost-free operation control strategy. Some models have promoted battery direct-cooling or direct-heating technology. The use of highly integrated module products in thermal management loops becomes more popular.

The development of cryogenics technology has also enjoyed a fast development. In the field of air separation, recent research progresses have focused on high-precision prediction of the thermophysical properties of cryogenic mixed working substances, development of efficient components for air separation equipment, and implementation of flexible and reliable control for large-scale air separation systems. As for the LNG technology, the process selection of large natural gas liquefaction plants is highly concentrated; China has rich practices in the small-scale and unconventional natural gas liquefaction; floating LNG (FLNG) used in offshore gas fields has developed rapidly; large scale and offshore floating storage and regasification units (FSRU) are the focuses in the field of LNG storage and transportation. For the large-scale cryogenic devices, Chinese researchers have completed the development of hundreds of watts and kilowatts of liquid hydrogen and liquid helium temperature zone refrigerators. In the field of small-scale cryogenic devices, in terms of regenerative refrigerators, commercial products on miniaturization, low temperature and large cooling capacity refrigeration have been realized in China recently. In terms of ultra-low temperature refrigeration below 1K, adsorption refrigeration and magnetic refrigeration are still in the research stage, while the minimum cooling temperature of dilution refrigerator can reach below 10mK, and its commercialization has been started already. In the field of cryotherapy and cryopreservation, the domestic application of cryotherapy still lags behind that of America. We are the currently frontrunners in the development of biomimetic materials and the preservation of small-scale samples, but we still need to catch up with the world class level on the preservation of large-scale samples. Additionally, the construction of biobanks is rapid in China, but it is still necessary to learn from the advanced international institution.

5.3 Comparison Between Domestic and Foreign Research Progresses

In recent years, China's refrigeration and cryogenics technologies have continued to develop rapidly, narrowing the gap with foreign advanced levels, and even being in a dominant position in some fields. However, there is a lack of original innovation in basic mechanisms, and there are still "bottleneck" problems in key technical fields.

As for the refrigeration technologies, except for the development of low-GWP refrigerants and the recovery and destruction of waste refrigerants, China has taken the leading position in the field of the component development and integration innovation technologies for vapor compression refrigeration, and leads the world towards an advanced technology path of high efficiency and low carbon. As for the heat-driven refrigeration and heat pump, China is also leading in the engineering application of absorption heat pump systems coupled with compressed heat pumps and the miniaturization of absorption chillers. In the field of heat-driven refrigeration and heat pump, the development of sorption refrigeration technology in the progress of composite sorbents and multi-stage cycle and system construction is faster than in foreign countries. In thermoacoustic refrigeration and heat pump, China is dominant in the conventional thermoacoustic refrigeration systems, while the novel wet thermoacoustic refrigeration cycle is becoming an international research hotspot. Although all solid-state refrigeration technologies were born abroad, domestic activities in material and system research have been competitive with those from international leading groups in the field.

Refrigeration technologies serve many fields of the national economy. China is the world's largest country of manufacture of air conditioning systems, but many core technologies are still governed by overseas producers. As for the high-temperature heat pump, research has been conducted abroad on high-temperature heat pumps with natural and low GWP working fluids, as well as megawatt industrial residual heat source high-temperature heat pump. China started the research on high-temperature heat pump relatively late, but there has also been the emergence of novel technologies like air source high-temperature heat pumps. As for heat pumps for vehicles, with the introduction of the P-Fas Act in Europe, the popularity of HFO technology such as R1234yf has declined; domestic enterprises have dominated CO_2 refrigerant, while foreign enterprises also have CO_2 products, and have begun to carry out the research on R-290.

Cryogenics technologies provide fundamental support for the industrial field and cutting-edge sciences. As for the air separation, domestic companies have successfully achieved the indigenization of complete sets of large air separation equipment. In the field of LNG technology, there is a significant amount of engineering practice and exploration in small (including skid mounted) and unconventional natural gas liquefaction plants in China, but there is currently no practice of large-scale plant. A lot of attention are being paid to floating liquefied natural gas (FLNG) abroad, but there is no specific case in China. The construction of large LNG storage tanks and transport ships in China is approaching the world's advanced level. As for the large-scale cryogenic devices, countries in Europe and the United States have mature experience in the development of large-scale cryogenic systems. China started late, and now has the capacity of independent development of 2500W@4.5K refrigerator and 500W@2K refrigerator. For the small-scale cryogenic devices, most kinds of the regenerative refrigerators on miniaturization, low temperature, and large cooling capacity refrigeration in China have prototypes or products comparable to those of developed counties, but the G-M pulse tube refrigerators and dilution refrigerators in China still need to be further improved since most of which currently used are imported from abroad. In the field of the cryotherapy and cryopreservation, the application of cryotherapy still lags behind that of America, but we lead the way in research and equipment development. We are currently frontrunner in the development of biomimetic materials and the preservation of small-scale samples, but we still need to catch up with the world class level on the preservation of large-scale samples. And the construction of biobanks is rapid in China, but it is still necessary to learn from the advanced international institution.

5.4 Trends of Development

Overall, high-level original researches and products have emerged in this field, and the gaps with the international advanced level are becoming smaller and smaller. In the future, we will further enhance interdisciplinary integration, promote original innovation, develop key devices and basic technologies, break away from dependence on foreign countries, and better serve national strategies and economic construction.

As for the refrigeration technologies, breakthroughs in vapor compression refrigeration should be explored, including oil-free compressors, substitution of high-temperature heat pumps for boilers in industrial production, life cycle management of refrigerants, and deep integration of refrigeration and air conditioning technology with artificial intelligence technology. For the heat-driven refrigeration and heat pump, absorption heat pump technology should focus on developing new processes and systems for the flexible transformation of thermal energy. Sorption refrigeration and thermal pump technology should concentrate on the development of high-adaptability working pairs and multi-effect cycles and systems at variable heat sources and ambient temperatures. In addition, the thermoacoustic refrigeration system should enhance the level of sound field regulation, develop heat exchangers with alternating flow, and a new thermoacoustic refrigeration cycle that enhances mass transfer performance.

As for the solid–state refrigeration, thermoelectric cooling is expected to penetrate more into markets requiring thermal management of high heat flux, high temperature precision, and flexible devices. The three major caloric cooling technologies are expected to witness more breakthroughs in both materials and systems, while their application scenarios need further investigation.

In the application of refrigeration technologies, the residential building air conditioning is the most well–known. The future research focuses on reducing the carbon emissions of residential building air conditioning systems. As for high–temperature heat pumps, the future development mainly includes high–temperature heat and steam supply based on high–temperature heat pumps, and heat pump technology for on–site consumption and utilization of industrial waste heat. For heat pumps of vehicles, the future development should focus on green and efficiency, functional integration, structural modularization and intelligent control.

In the cryogenics technologies, for the air separation, to achieve the large–scale, energy–saving, and intelligent operation of cryogenic air separation systems, it is urgent to develop new processes, innovative subsystem designs, and novel operation control strategies. For the LNG technology, offshore natural gas liquefaction will be a focus of attention in the near future; research on carbon reduction in the LNG field is becoming a hot topic. The liquefaction of unconventional natural gas, especially natural gas containing hydrogen and ethane, may become a research hotspot; the new optimization algorithm for liquefaction process simulation is expected by the industry. Continuous efforts are still needed to research the localization of large gasifiers. In terms of the large–scale cryogenic devices, scientific research workers in our country continue to deepen on the existing basis. At the system level, the ten–thousand–watt liquid helium temperature zone refrigerator and the kilowatt superfluid helium temperature zone refrigerator are developing. As for the small–scale cryogenic devices, the developing directions of regenerative refrigerator are still on miniaturization, low temperature and large cooling capacity refrigeration, including the commercialization of G–M pulse tube refrigerator. The 4K J–T refrigerator needs to be further improved especially on the performance of its valved linear compressor. For the refrigerator below 1K, aiming to break international monopoly, great efforts on the research and productization of the domestic dilution refrigerator are still needed. As for the cryotherapy and cryopreservation, it is crucial to achieve precise regulation and control of energy and matter transfer in cryobiology and to optimize the construction level and thoroughly improve standardized laws and regulations of biobanks.

6 Artificial Intelligence

6.1 Introduction

Artificial intelligence is a science that uses computing technology to simulate intelligent behavior in the world. It utilizes knowledge computing, machine learning, deep learning and large models, natural language processing, computer vision, speech processing, information retrieval, multi-agent systems, embodied intelligence, adversarial technology, and other technologies to help or replace humans efficiently complete some intelligent tasks. To provide important technological and theoretical support for the development of human society in the era of intelligence.

6.2 The Latest Research Progress of Artificial Intelligence in Recent Years

Due to the continuous and rapid development of artificial intelligence, new research directions and topics are emerging one after another, and even the discipline itself is in the process of updating. Therefore, the following will showcase the latest developments in the entire field of artificial intelligence from some of the currently popular branches of disciplines.

6.2.1 Knowledge computing

Knowledge representation has been further extended from traditional symbolic knowledge models to implicit knowledge models. Recent research progress has mainly focused on knowledge representation of neural networks, large language models, extraction and expression of dark knowledge, and collection and construction of knowledge graphs.

6.2.2 Machine learning

The unsupervised training method represented by self-supervised learning has greatly improved the model's ability to utilize and process data, thereby guiding machine learning algorithms to continuously achieve breakthroughs in practical applications.

6.2.3 Deep learning and large models

Deep learning technology automatically learns features from a large amount of data and models complex tasks by simulating the neural network structure of the human brain. Large models have a large number of parameters and significant advantages in handling large-scale and complex tasks. The combination of the two has brought many innovations and breakthroughs to the field of artificial intelligence.

6.2.4 Natural language processing

Natural language processing has achieved significant results in pre-trained language models, sentiment analysis, machine translation, and text generation.

6.2.5 Computer vision

Computer vision is a technology that uses computers to simulate human visual functions, aiming to obtain information from images or videos and process and analyze it. Computer vision has made significant breakthroughs in object detection and recognition, image generation, enhancement, and super-resolution.

6.2.6 Speech processing

The introduction and application of deep learning algorithms have significantly improved the accuracy of speech recognition. In addition, significant progress has been made in speech synthesis technology, which can generate more natural and realistic speech through technologies such as generative adversarial networks, providing a more realistic and smooth experience for speech interaction.

6.2.7 Information retrieval

Information retrieval has made significant progress in areas such as deep learning based information retrieval technology, cross language information retrieval, and multimodal information retrieval.

6.2.8 Multi-agent system

Multi-agent systems have achieved results in collaborative control, such as convergence analysis of consensus algorithms and robust control strategies for delay and uncertainty. In terms of intelligent agent modeling, agents based on large-scale language models exhibit strong perception, decision-making, and action capabilities, providing a new direction for the implementation of artificial general intelligence.

6.2.9 Embodied intelligence

The perception and action fusion technology of intelligent agents has been able to achieve efficient environmental perception and precise control, and advances in cognitive modeling have enabled agents to simulate human adaptation and learning mechanisms to cope with complex and changing environments.

6.2.10 AI security

AI security technology has made significant research progress in adversarial attack defense, privacy protection, and model security evaluation.

6.3 Comparison of Research Progress in Artificial Intelligence at Home and Abroad

Artificial intelligence, as a strategic emerging technology leading a new round of technological revolution and industrial transformation, plays an important role in promoting scientific and technological progress, industrial upgrading, economic development, and is becoming a key area for countries to compete for development. There are significant differences in the application fields and development levels of artificial intelligence both domestically and internationally.

6.3.1 Knowledge computing

In terms of knowledge representation and modeling, China has gradually developed distinctive knowledge representation methods, continuously improving its ability to express and model knowledge. In terms of knowledge acquisition and inference technology, China has closely followed the pace of the big data era and made substantial progress, especially showing obvious advantages in data scale and processing speed.

6.3.2 Machine learning

Compared with foreign countries, China has made certain progress in algorithm optimization and traditional machine learning applications, especially in the fields of big data processing and pattern recognition, showing many advantages. However, there is still a gap in the innovation of machine learning theory and exploration of cutting–edge technologies.

6.3.3 Deep learning and large models

China has shown obvious advantages in algorithm optimization, model compression, and application implementation, but there are still shortcomings in the original research of large models, the scale and efficiency of model training, as well as the generalization ability and robustness of models.

6.3.4 Natural language processing

Compared with foreign countries, China still has gaps in theoretical depth, technological innovation, and cross language processing capabilities in natural language processing. In terms of application technology, China has successfully applied natural language processing technology to many fields.

6.3.5 Computer vision

China has conducted in–depth research on visual feature representation, sparse encoding, and visual attention mechanisms, and has achieved a series of important results. However, in the fields of basic theoretical research and some cutting–edge technologies, China still needs to further strengthen.

6.3.6 Speech processing

In terms of key technologies such as speech recognition, speech synthesis, and speaker recognition, China has reached an international leading level. However, there is still a certain gap in basic theoretical research on speech processing, innovative original algorithms, and adaptability across languages and multiple scenarios.

6.3.7 Information retrieval

Compared with foreign countries, China has obvious advantages in Chinese word segmentation, part of speech tagging, named entity recognition, and other technologies. However, there is still a certain gap in the basic theoretical research of information retrieval, innovative original algorithms, and adaptability across languages and fields.

6.3.8 Multi-agent system

China has made innovations in communication protocol design and information sharin-strategies, but there are still shortcomings in the basic theoretical research and experimental facility construction of multi-agent systems.

6.3.9 Embodied intelligence

China has shown a positive trend in the field of embodied intelligence, especially in the interactive applications of robot perception and control, virtual reality, and augmented reality, forming its own characteristics and advantages. However, there is still room for improvement in areas such as biomimetic materials and structural design, brain computer interface technology, and long-term autonomous learning and adaptive optimization of intelligent agents in complex environments.

6.3.10 AI security

China has achieved significant results in model security, multi-party secure computing, and differential privacy protection. However, there is an urgent need to further strengthen the construction of AI security basic theoretical system, cross domain security attack and defense research, and standardization formulation.

6.4 Development Trend and Prospect of the Discipline

Artificial intelligence has obvious characteristics of multidisciplinary integration and strong technical synthesis, and with the continuous progress of technology, artificial intelligence will play an important role in more fields.

First, the multi-modal model promotes the deep integration of image, text and video. From video generated videos to cultural and graphic generated videos, the development of multimodality emphasizes achieving richer AI generation results with less user input information.

Second, high-quality data has become a magic weapon for the value transition of large models. At present, algorithms with supervised learning have a much greater demand for training data than existing annotation efficiency and investment budget. Basic data services will continue to release their basic support value for algorithm models, while the value of high-quality data processing, data annotation services, and a comprehensive data collection and evaluation system will be further highlighted.

Third, artificial intelligence chips are developing in the direction of diversification and ecology.

Large models are not the ultimate form of AI, and the next wave of AI will be multimodal embodied intelligence in 10 years, which combines intelligent algorithms with robots to enable machines to perform tasks more intelligently.

Fourth, embodied intelligence has become a new form of AI development. At present, the interpretability of algorithms is still in the early stages of development, and how to make people understand and trust artificial intelligence systems has become one of the important directions for the future development of artificial intelligence.

7 Industrial Internet

7.1 Introduction

Industrial Internet (II) is a new infrastructure, application mode, and industrial ecology for the deep integration of new-generation information and communication technology and industrial economy. II builds a new manufacturing and service system covering the whole industrial chain and the whole value chain through the comprehensive connection of people, machines, things and systems. II provides a way to realize the development of industrial and even industrial digitalization, networking and intelligence, and serves as an important cornerstone of the Fourth Industrial Revolution.

The II discipline is a knowledge system centered around relevant theories, technologies, applications, and management methods. It encompasses various cutting-edge technologies, including the Internet of Things (IoT), cloud computing, big data, artificial intelligence (AI), 5G, and others. This interdisciplinary field combines knowledge and methods from multiple disciplines, such as manufacturing engineering, communication technology, computer science, and more. Its primary objective is to cultivate engineering, technical, and management professionals with expertise in Industrial Internet technology and applications. It aims to drive research and application of Industrial Internet-related technologies and promote the transformation and upgrading of the industrial manufacturing sector.

This report aims to comprehensively reflect the research status and development trends of the Industrial Internet discipline. It systematically examines and analyzes the development history, research status, application scenarios, and technological framework of the Industrial Internet discipline. The purpose is to provide strong academic and intellectual support for the development of the second phase of this project.

7.2 Recent Research Development in China

7.2.1 Industrial Internet perspective and architecture

General Electric (GE) has systematically introduced the concept of the Industrial Internet in its publication "Industrial Internet: Breaking the Boundaries Between Intelligence and Machines." In this publication, GE systematically presents the concept of the Industrial Internet, emphasizing that its purpose is to enhance industrial production efficiency and the competitiveness of products and services in the market. The German government views Industry 4.0 as the combination of production methods with advanced information and communication technologies to conduct production in a more flexible, customized, and sustainable manner, thus achieving intelligent production processes and new business models. The United States defines Advanced Manufacturing as improving innovation in the manufacturing process of existing products and utilizing advanced technologies to produce new products. Cisco, a globally renowned digital communication technology company, considers the Industrial Internet of Things (IIoT) as an ecosystem consisting of devices, sensors, applications, and related network devices that collectively collect, monitor, and analyze data from industrial operations. The Industrial Internet Innovation and Development Action Plan (2021–2030) points out that IIoT is a new industrial ecosystem, a key infrastructure, and a new application mode formed by the deep integration of next–generation information and communication technologies with the industrial economy.

Academician Hequan Wu of the Chinese Academy of Engineering believes that the Internet of Things is the foundation for the digitalization, networking, and intelligence development of industry. The second generation of information technology needs to meet the requirements of high security, ultra–reliability, low latency, large connectivity, personalization, IT and OT compatibility, etc., and requires the development of ICT technologies optimized for the second generation of information technology. Academician from the Chinese Academy of Engineering, Bohu Li, proposed that China's IIoT development has entered a new stage of "Intelligent Industrial Internet System–Industrial Internet 2.0", initiating intelligent connections for everything. Similarly, Chinese Academy of Engineering Academician Yunjie Liu predicts that by 2030, IIoT can upgrade and transform traditional industries in China, with an economic scale exceeding $5.6 trillion, driving the transition of the Internet from "consumer-oriented" to "production-oriented."

The architecture serves as the fundamental guideline and leading document for the development of the Industrial Internet, playing a vital role in its development. Under the guidance of the Ministry of Industry and Information Technology (MIIT), the China Industrial Internet Industry Alliance released the "Industrial Internet Architecture (Version 1.0)" and "Industrial Internet Architecture (Version 2.0)" in 2016 and 2020, respectively. In April 2015, Germany released the "Industry 4.0 Implementation Strategy," which includes the "Industry 4.0 Reference Architecture" jointly developed by the German Federal Ministry for Economic Affairs and Energy, the German Federal Ministry of Education and Research, and the Industry 4.0 Platform. This reference architecture aims to improve productivity, reduce costs, enhance flexibility, and improve quality. In the United States, where architectures are developed for commercial purposes, the Industrial Internet Consortium (IIC) released the "Industrial Internet Reference Architecture" in June 2015. Its value lies in promoting interoperability through broad industry applicability, positioning applicable technologies, and guiding the development of technologies and standards.

7.2.2 New achievements and new journals in the field of Industrial Internet

II patents play a vital role in the research and development, protection, innovation and cooperation of II technologies. Through retrieving with the keyword "Industrial Internet" contained in the patents'names in the State Intellectual Property Office while restricting the application date prior to December 31,2022, a total of 1,735 domestic II patents were found. The number of patent applications shows an overall rising trend, in which the China Academy of Information and Communication Research, Shenzhen Xuanyu Technology and Shandong Wave have applied for the largest number of Industrial Internet patents. By analyzing the categories of Industrial Internet-related patents, the report finds that the proportion of invention patents in Industrial Internet-related patents is the largest, up to 70%. Additionally, foreign patents were searched in the Derwent Patent Intelligence Database under the titles of "IIoT" and "Industry 4.0" with application date before December 31,2022, and a total of 67 foreign Industrial Internet patents were collected. There is a general upward trend in the number of foreign patents, in which engineering and computer science are popular topics.

In terms of Industrial Internet papers, with CNKI as the search platform, high-quality papers with the title "Industrial Internet" were searched, and the time range was limited to January 1,2018, to December 31,2022, resulting in the retrieval of a total of 258 Industrial Internet papers. The number of papers demonstrates a rising trend, and the research fields are mainly concentrated on two disciplines: information economy and industrial economy. Foreign papers were searched in the Web of Science core library, with the terms "IIoT", "Industria 4.0" and "Industrial 4.0" in the search title, limited to the time range between January 1,2018, and December 31,2022, and a total of 5,321 relevant papers were retrieved. The number of papers also shows an upward trend, and the foreign II research field is extensive, mainly focusing on the two major research directions of computer science and engineering.

In terms of the development of Industrial Internet journals, through the search of CNKI database, computer research and development and China engineering science are the journals with the highest impact factor in the column of Industrial Internet, respectively, and they have opened the column of "Topics of industrial Internet security technology", "Development of new–generation Industrial Internet security technology", and "Strategic research", respectively. The columns "Industrial Internet Security Technology Topics" and "Development of New Generation Industrial Internet Security Technology and Strategy Research" are respectively opened to carry out research on the security field of Industrial Internet. The column with the most papers published is "Industrial Internet System" under the Journal of Control Engineering, where 82 papers have been published so far. Their research areas encompass the architecture, technology and application of II; security, privacy and trust of II; data analysis, mining and intelligence of II; standards, specifications and testing of II, etc. Other journals with columns relating to II include Network Security and Data Governance, Information and Communication Technologies and Policies, Introduction to Software, Industrial Technology Innovation, and Standardization and Metrology of Instrumentation.

7.2.3 New methods and new technologies of Industrial Internet

In terms of 5G, China established the IMT–2020 (5G) Promotion Group in 2013, which started the research in 5G. Michae Gundall and other scholars analyzed the specific case of 5G in the industrial field, identified the main challenges and requirements of the communication network in Industry 4.0, and predicted that wireless communication technologies in future industrial systems will better meet the requirements of Industry 4.0. In the following year, Zhang Yunyong introduced the network characteristics of 5G for industrial development, pointing out that 5G will continue to consolidate the foundation of the Internet of Everything, fully empower II, and accelerate the integration of multi–industry Internetization. In 2022, the General Office of MIIT issued the guidelines for the construction of 5G fully–connected factories on August 25, guiding various regions and industries to actively carry out the construction of 5G fully connected factories. The guidelines also call for driving the development of the 5G technology industry, further accelerating the deep expansion of "5G+Industrial Internet" new technologies, new scenarios, and new modes into all fields and segments of industrial production, and promoting the quality, cost reduction, efficiency, green, and safe development of traditional industries.

Cloud computing and edge computing solve the difficulties of II in data processing. Mohammad Aazam and other scholars defined a middleware that is used in both cloud and edge middleware, fog, which can provide local processing support with acceptable latency for actuators and robots in manufacturing. The article also discusses how fog can provide local computing support in IIoT environments as well as the core elements and building blocks of IIoT. Luo Junzhou's team focuses on investigating and summarizing the current state of research and challenges faced by several important directions involved in the field of combining cloud and edge computing, and proposes a solution for a new type of cloud-converged II architecture and related key technologies. Li Hui's team discussed the II edge computing architecture and the core technology that promotes the development of Industrial Internet edge computing, and elaborated on the current situation and challenges of II edge computing.

In terms of II security, in 2016, Tao Yaodong and other scholars innovatively proposed overall recommendations and a PC4R adaptive protection framework to guide daily security operations based on existing II security challenges. In 2018, Du Lin and other scholars argued that the connection of the II opened up the industrial system and the Internet. Compared with the closed and trusted production environment of the traditional industry, in the era of the II, the network security and the industrial security risks are unprecedentedly complex. Thus, situational awareness will become an important technical means for safeguarding Industrial Internet security, and endogenous security defense will become a major trend in the future of Industrial Internet security protection. In the Industrial Internet Architecture (Version 2.0), II security is listed as an important part of the functional framework of the Industrial Internet, in which the key technologies include encryption algorithm technology, access control technology, privacy protection technology, intrusion detection technology, digital watermarking technology, digital signature technology, blockchain technology and so on. In the same year, foreign scholars such as Jayasree Sengupta mainly introduced the security problems and challenges in IoT and Industrial IoT. Firstly, various attacks were categorized and each attack was mapped to one or more layers in the IoT/IIoT architecture. Then relevant countermeasures and cases are discussed and the application of blockchain technology in solving the problems posed by centralized architectures is presented.

In 2017, Yang Zhen and other scholars believed that logo resolution technology is the key technology to solving the problem of "information silos", completing the convergence of industrial big data, and forming the information fusion and understanding based on this technology. In 2019, Ren Yuzheng and other scholars discussed the design principles and key supporting technologies of Industrial Internet marking resolution systems, including marking scheme, marking allocation mechanism, registration mechanism, resolution mechanism, data management mechanism and security protection scheme, etc. They also discussed the importance of identity resolution systems in the IIoT, provided a comprehensive survey of identity resolution systems that may be used in the IIoT, and compared them from the perspectives of IIoT requirements and technology selection.

Concerning AI, the industrial grand modeling represents a key recent application of artificial intelligence technology. It is an important and forward–looking technology for applications within the II domain, providing a comprehensive, real–time understanding of industrial systems by integrating large amounts of data, simulations, and analytics. Germany's research on the integration and development of II and AI is relatively thorough. The German Research Center for Artificial Intelligence (Deutsches Forschungszentrum für Künstliche Intelligenz, DFKI) has been working on the field of AI and II for a long time, and from 2006 to 2023, the DFKI has carried out a total of 145 completed or ongoing II–related projects. One of the newest projects for 2023 is titled "Capabilities and Skills for Reusable Robots Handling Parts in Manufacturing Environments", which aims to provide components for automated planning and control of grasp–release and contact–assembly tasks, and to propose innovative electro–adhesion–based gripping concepts to handle many different products. These components will make production lines more efficient, flexible, and reconfigurable. The main technological objectives are self–monitoring recognition and control of gripping strategies, learning and control of assembly techniques based on human demonstrations, development of AI–based multimodal perception for visual control and continuous monitoring during handling operations, and development of multifunctional and dexterous soft grippers with electrically active fingertips.

Furthermore, digital twins have also received plenty of attention. Digital twin technology allows organizations to create, simulate, and test virtual copies of their physical devices, systems, and processes in the digital world. Digital twin technology functions in the following ways: ① improve productivity: Digital twin technology allows manufacturers to simulate and optimize production processes, reducing waste and unnecessary costs in production; ② increase productivity by allowing problems to be detected in advance and improvements to be made through virtual testing; ③ reducing maintenance costs: Digital twin technology can create virtual copies of physical equipment, monitor its status in real–time, and predict potential failures. This helps to reduce equipment maintenance costs and minimize downtime due to equipment failure; ④ product design and innovation: Digital twin technology enables product design teams to test and optimize product prototypes in a virtual environment, accelerating the development cycle of new products and improving product quality; ⑤ real–time data analysis: Digital twin technology can integrate a variety of sensors and device data to provide real–time data analysis and prediction capabilities to help companies better understand their operations and make intelligent decisions.

7.3 Comparison of Chinese Domestic and International Research Developments

Induatrial Internet is the cornerstone of the Fourth Industrial Revolution and occupies the forefront of the II, helping our country to lead in the new global industrial transformation. Industrial Internet has been an internationally highly regarded field in recent years, with various countries listing it as a national strategy and increasing their investments. Notable examples include Germany's Industry 4.0 and the United States' "Advanced Manufacturing." Currently, China, the United States, and Germany are the three fastest-developing countries in the global II.

In terms of the number of II projects, the development trends of these three countries are similar, all having emerged in the last five years. This is partly due to strong policy support from their respective governments, such as China's "Internet Plus Advanced Manufacturing" initiative launched in 2017, Germany's well-known "Industry 4.0," and the United States' Advanced Manufacturing Partnership program in 2011. Furthermore, the rapid development of information technology and its timely integration with manufacturing have promoted the continuous renewal and iteration of the manufacturing industry.

Regarding the funding amounts for II projects, both China and the United States have made substantial investments in Industrial Internet development. However, on average, the funding scale and amount for II projects in the United States are higher than those in China. This is partly because the total budget of the U. S. National Science Foundation is larger than that of the Chinese National Natural Science Foundation. Additionally, the U. S. National Science Foundation not only supports scientific research but also funds educational and technology commercialization projects.

As for the keywords of II projects, different countries have commonalities and specific focuses. China, the United States, and Germany all emphasize the importance of data, considering it the lifeblood of the II. Each country also has its own emphasis, with China focusing more on networking, Germany on production and processes, and the United States on equipment and security.

As for the disciplinary background of II projects, China and the United States have similar disciplinary divisions, with a greater emphasis on computer and information science, supplemented by engineering and management science, among other fields. Germany, on the other hand, primarily focuses on mechanical and industrial engineering, with computer science as a secondary discipline, and involvement of other natural science departments in II applications. The main reason for the differences between China and Germany may be that China's II places more emphasis on the technological aspect, while Germany emphasizes practical implementation, aligning with the respective architectural designs of II in these countries. Germany has a wider distribution of disciplines compared to China, covering basic areas like physics, chemistry, and humanities, as well as application areas in earth sciences and agriculture.

In terms of project affiliations, the United States has a larger number of projects and affiliations, with George Mason University being one of the most prominent affiliations. In China, institutions like Beijing University of Posts and Telecommunications, Beijing University of Aeronautics and Astronautics, and Dalian University of Technology have received a significant number of projects.

7.4 Development Trends and Prospects

Firstly, the development of the II discipline is a crucial support for the II strategy. II has become a vital direction for industrial development in countries worldwide, and governments are continually advancing related policies and plans, aiming to achieve industrial upgrading and economic growth through II. International cooperation in the field of II is increasing, and competition is becoming more intense. The development of II discipline will continuously provide talent and technological support for II strategy.

Secondly, the development of the II discipline is an important foundation for driving the digital transformation of industries and enterprises. Digital transformation leads to the continuous emergence of new industries and the transformation and upgrading of existing ones. Through digital transformation, the efficiency and profitability of enterprises are continually improved, while also effectively enhancing their management efficiency and decision-making capabilities. The development of II discipline will provide multidisciplinary professionals for the digital transformation of industries and enterprises, serving as a crucial foundation for their digital transformation.

Thirdly, the development of the II discipline is a crucial area for nurturing talent in the era of new industrialization. With the rapid development and widespread application of II, talent for the new industrialization era needs to possess a wide range of capabilities. They must not only have a solid technical foundation but also exhibit innovation, interdisciplinary collaboration, and systems integration skills, among other comprehensive abilities. Traditional disciplinary education alone cannot meet the talent requirements of the new era. The development of the II discipline is an integral part of nurturing talent for the era of new industrialization, providing a platform for the next generation of students to become highly skilled individuals with II application capabilities and innovation spirit.

Looking at the development trends of II discipline in the next five years, several key characteristics become increasingly prominent: its foundational nature, comprehensiveness, practicality, and personalization. The cross-fusion of II discipline with other disciplines such as computer science, engineering, and various fields like energy, aviation, and chemical engineering will deepen and broaden. The establishment of diversified and international "digital craftsman" training centers will expand further. These developments will facilitate the synergistic development of the three sectors and the cultivation of talent with an innovative spirit, promoting the transformation of new industrialization.

8 Aeronautical Science and Technology

Commencing from 2016, under the steadfast leadership of the government and bolstered by sustained and stable funding, China has witnessed rapid advancements in aerospace science and technology, marked by several notable achievements. The realm of aeronautical science and technology spans a multitude of specialized disciplines. This report delineates the latest research progress made since 2016 in eight key areas: aerostat technology, advanced composite materials, aerospace engine development, maintenance engineering, the field of reliability engineering, innovations in rotorcraft and their systems, unmanned aerial vehicles (UAVs) and their systems, along with advancements in management science.

8.1 Aerostats

The "Dream" airship developed by Beihang University and the stratospheric airship technology research under the Honghu Pilot Project "Near-Space Scientific Experiment System" led by the Chinese Academy of Sciences Aerostat Center have both continuously carried out related flight validations, achieving long-duration controlled flights across day and night. Several key technologies have made breakthroughs, marking a high level of maturity in critical technologies and laying a solid technical foundation for the early application of near-space platforms in engineering. This holds groundbreaking significance. The "Ji Mu Yi Hao" tethered balloon, developed by the Chinese Academy of Sciences Aerostat Center, ascended to 9,032 meters in the Qinghai-Tibet Plateau scientific expedition, setting a new record for scientific observation.

8.2 Composite Materials

Composite materials account for 12% of the domestic large passenger aircraft C919. The C919's composite material wing has successfully completed typical box section static and damage tolerance tests. The long-range wide-body passenger aircraft composite material fuselage curved stiffened wall panel process verification piece has been developed, and the wide-body aircraft composite material forward fuselage full-size barrel section has smoothly come off the assembly line. Additionally, the development of several models of all-composite general aviation aircraft, such as the Shanhe SA160L and ZA800, has been successful.

In terms of functional aerospace composites, structural absorbing composite materials were first applied in stealth integration assessments of third-generation aircraft and have now been extensively used in fourth-generation aircraft. Structural transparent composite materials have evolved to the current low moisture absorption, low dielectric, low density, and high integrated mechanical performance, expanding their application and supporting the enhancement of the detection capabilities of aerospace equipment.

8.3 Aerospace Engines

China has entered a favorable development path primarily based on indigenous R&D in aerospace engines, achieving comprehensive domestic coverage across all types of aviation engines. In 2015, substantial progress was made in the field of aviation engines and gas turbines. In May 2016, the establishment of Aero Engine Corporation of China (AECC) marked a significant shift in the nation's approach, elevating the status of aviation engines to be on par with aircraft and forming the "separation of flight and engine" system. This move also facilitated the implementation of the "Two Engines Project".

Significant technological and product breakthroughs have been achieved in military aviation engines. The successful development of the "Kunlun" engine ended the long-standing dependence on foreign technologies for aviation engine manufacturing in China. Technologies like the WS-10 "Taihang" series, representing third-generation military turbofan engines, have matured and are extensively deployed, marking a historic leap for China in aviation engines from turbojets to turbofans, from medium to high thrust, and from third to fourth generation. The development of fourth–generation military turbofan engines, such as the WS–15 "Emei" and WS–20, is in the final stages. Post the "Two Engines Project", indigenous civil aviation engines have become a new focus in China. The development of large aircraft engines like the CJ1000 and CJ2000 is progressing well, achieving significant milestones.

In the realm of ultra–high–speed propulsion technology, multiple hypersonic wind tunnels have been constructed and put into use, laying a solid foundation for scramjet engine research and testing. Northwestern Polytechnical University's "Feitian Yi Hao" kerosene–fueled rocket–scramjet combined cycle engine has completed flight tests, verifying its multi–mode stable transition and broad–spectrum integrated capabilities. Tsinghua University's "Qinghang Yi Hao" novel rotary detonation scramjet engine successfully completed its flight demonstration test, positioning China at the forefront of new aerospace propulsion technology. The Aerospace Science and Industry Corporation's "Tengyun Project" has completed China's first flight validation of mode transition in a liquid rocket–scramjet combined cycle engine, achieving a major breakthrough in combined propulsion technology for space flight.

8.4 Maintenance Engineering

Domestic civil aviation maintenance in China adheres to global advanced models in maintenance management and production, overall ensuring safe and economically effective operation of civil aviation flights. The main progress in the military aviation maintenance sector includes extensive exploration and application in Condition–Based Maintenance (CBM) and on–condition maintenance, new advancements in maintenance technology, and fresh explorations in digital maintenance.

8.5 Reliability Engineering

Applying systems engineering theory and methods, reliability engineering has evolved into reliability systems engineering. A series of standards and technical specifications for aerospace equipment products have been formulated, establishing a reliability technology pathway that spans the entire lifecycle of aerospace equipment products, from fault prevention, prediction, diagnosis, to treatment. A foundational theory of "Integrated Reliability" has been developed, with application technologies such as "Model–Based Reliability Systems Engineering" and "Fault Diagnosis and Prediction" continuing to be practically applied in representative aerospace equipment products like the Y–20, J–20, Z–20, C919 large aircraft, and new types of military/civil UAVs.

8.6 Rotorcraft

Significant advancements have been made in the development of the civilian helicopter AC series, with models such as the AC312E, AC332, and AC313A successfully completing their maiden flights. The AC352 has been awarded a type certificate by the Civil Aviation Administration of China. The new military general–purpose transport helicopter Z–20 has completed its development and is now being mass–produced for the armed forces, marking a monumental leap in China's helicopter technology from the third to the fourth generation. The eVTOL (electric Vertical Take–Off and Landing) aircraft has emerged as a new popular direction in rotorcraft. In 2023, domestic companies such as Fengfei, Wofei Changkong, Wolante, Shide, Yufeng Future, and Zero Gravity Aviation successively achieved maiden flights of their full–scale prototypes. In October 2023, the Civil Aviation Administration of China issued a type certificate for the EH216–S unmanned aerial vehicle system to EHang Intelligent. This demonstrates that China's eVTOL technology and industry development are now in sync with leading countries worldwide.

8.7 Unmanned Aerial Vehicles (UAVs)

There has been a surge in enthusiasm for UAV development in China. Military groups, state–owned enterprises, and private capital have established specialized UAV companies, developing various types of drones. Currently, there are thousands of UAV enterprises in China. Through intense competition, UAV technology has made significant progress, continuously improving in performance and leading to a rich and diverse UAV system. Various types and functions of UAVs have been put into service.

8.8 Management Science

Key enterprises are vigorously advancing the construction of operational management systems. Aviation Industry Corporation of China has established the AOS system, Aero Engine Corporation of China has built the AEOS system, and Commercial Aircraft Corporation of China has developed the COMAC management system. These advancements have enhanced management capabilities, ensured strategic implementation, and improved management efficiency. Research progress is mainly reflected in strategic planning management, end–to–end process management, innovative R&D management, evolution of enterprise management information architecture, quality management, and comprehensive risk management in aviation enterprises.

Building upon the foundation of the latest developmental research in various domains, we have undertaken comparative studies against the backdrop of internationally recognized advanced standards. Additionally, we have engaged in forward–looking assessments to chart the future trajectory of these fields.

9 Aerospace Science and Technology

9.1 Introduction

Based on the development of aerospace science and technology disciplines in the past five years, and in accordance with the principles of major landmarks, key urgency, and technical feasibility, the research team selected 15 major disciplines to conduct comprehensive research. This report summarizes the latest research progress of 15 aerospace science and technology disciplines in China, as well as evaluates gaps between domestic and international development status, and forecasts future development trends and prospects.

9.2 Latest Research Progress

In the past five years, China's aerospace industry has made a series of brilliant achievements, and the aerospace disciplines have made rapid and innovative progress:

(1) Aerospace transportation system has accelerated the updating and upgrading, driving the development of a series of key technologies.

(2) Communication satellites, navigation satellites, remote sensing satellites, and space science and technology test satellites have been improved steadily. For example, China's launched first high-throughput communication satellite, ChinaSat-26, with a transfer capacity of 100 Gbps. The completion and operation of the 30-satellite BeiDou Navigation Satellite System represents the successful conclusion of its capacity to serve the world.

(3) Due to the great progress of small satellite technology, the low orbit large-scale constellation has shown great value in a series of application fields such as space-based global communication and remote sensing, and has set off a boom in the development of low orbit giant constellation. The small satellite industry has changed from a "trend" of rapid development to a "normal" development.

(4) The three-step process of the Chinese manned spacecraft project has been successfully realized, and major progress has been made represented by the completion of the assembly and construction of China Space Station. It made comprehensive breakthroughs in a series of key professional technologies in the design, construction and operation of space station. China manned spaceflight has entered the space station era.

(5) China has successfully completed the last step of the three-step lunar exploration program ("orbit, land, and return"), and carried out a smooth interplanetary voyage and landed beyond the earth-moon system by Tianwen-1, followed by the exploration of Mars. It has mastered a series of deep space exploration key technologies, and established space infrastructure and capabilities. A large number of scientific achievements have been obtained.

(6) Outstanding progress has been made in fundamental technology fields, such as aerospace propulsion, space energy, aerospace guidance, navigation and control (GNC), aerospace detection and guidance, aerospace intelligent detection and recognition, aerospace telemetry, tracking and command, space remote sensing, advanced aerospace materials, space biology and medical payload technology, spacecraft recovery and landing technology, etc. It has enriched the content of aerospace disciplines and set up a solid technical foundation for accelerating aerospace development.

9.3 Gaps between Domestic and International Development

In the past five years, China's aerospace industry has achieved a series of brilliant achievements, accelerating China's step towards a strong space presence. Space vehicles have broken through the bottleneck of high-frequency and normalized launch, and speeding up the development towards non-toxic, pollution-free, modular and intelligent direction. The technological level of satellite systems has steadily improved, supporting and constructing national space infrastructure with multiple functions and orbits; the space station system has broken through a series of key technologies in space station design, construction, and operation; the field of deep space exploration has broken through a series of key technologies; significant progress has been made in numerous fundamental technical fields. However, compared with the world's aerospace powers, China still has many gaps in the fields of key and fundamental aerospace disciplines.

9.3.1 Main gaps in key aerospace disciplines

In the field of space vehicle technology, compared with the international mainstream level, there is still a gap in the core performance indicators such as the carrying capacity index; the carrying capacity of China's new generation of rockets has reached the level of large launch vehicles, which is equivalent to the United States' Delta IV. However, it still cannot support the demand for manned moon landing and large-scale space infrastructure construction; China's Long March series of launch vehicles do not yet have the ability to be reused.

In the field of satellites and applications, compared with international in-orbit and under-development satellites, there is a gap in the system capacity of China's communication satellites in terms of load weight. Taking the ChinaSat-26 as an example, its communication capacity is 100 Gbps, and the system capacity per unit weight is 12 Gbps/100 kg, while the system capacity of ViaSat-3 has exceeded 1Tbps, and its system capacity per unit weight is 50 Gbps/100 kg. There is a significant gap in the quantitative detection capabilities of remote sensing satellites, and there are still certain deficiencies in detection elements. For example, the detection methods for gaseous substances such as PM 2.5 and ozone are still blank.

Compared with SpaceX, there are many gaps in small satellite fields, such as the development of small satellite technology iterations and breakthroughs, system cost reduction and efficiency improvement, batch launch deployment, etc.

9.3.2 Main gaps in fundamental aerospace disciplines

In the field of aerospace propulsion, the development of solid propulsion technology is slow, and there is gap of generations in terms of performance and scale. As for cutting-edge propulsion technologies such as nuclear propulsion and tethered propulsion, China is still in elementary stage.

In the field of space energy, in terms of high-specific energy long-life batteries, the specific energy of domestic batteries has exceeded the international technical level of similar products; there is a significant gap in terms of integrated development and batch preparation of power generation, energy storage, and energy control products.

In the field of aerospace intelligent detection and recognition, China lags behind international research on brain-inspired theories such as neuron models and pulse spatiotemporal dynamics, and development of hardware such as brain-inspired chip and neuromorphic sensor.

In the field of aerospace telemetry, tracking and command, China has shortcomings in theory and standards; full–area coverage and ubiquitous interconnection capabilities are insufficient.

In the field of advanced space materials, the system of metallic materials and non–metallic materials is not yet complete; basic research in the field of advanced functional composite materials is relatively weak.

9.4 Development Trends and Prospects

Driven by national strategic needs and the market, the capacity and performance of space transport system will be continuously improved, towards the directions of reusability, new model development, heavy–duty, and diversified applications. It will make space entry and exit more efficient. China will continue to improve its space infrastructure, and integrate remote–sensing, communications, navigation, and positioning satellite technologies. Besides improving the performance of satellites, satellite system technology will be further developed towards systematization and integration. While large–scale low–orbit constellation showing an explosive development trend, the main direction of small satellite technology and applications development will be satellite–ground integration, and communication–navigation–remote sensing integration. Towards reusability, low cost, long–term on–orbit operation, and manned deep space exploration, manned spacecraft technology will focus on the plan for a human lunar landing, develops new–generation manned spacecraft, and research key technologies to for exploring and developing cislunar space. Deep space exploration will promote a series of breakthroughs in core key technologies through national aerospace major missions and produce fruitful space science achievements. Related fundamental technologies such as advanced space materials, aerospace GNC, space energy, and space biology will achieve further breakthroughs. As cutting–edge technologies such as intelligence, multi–domain integration and collaboration of air–space–ground are increasingly used in aerospace engineering, the multidisciplinary fusion, boundary crossing and knowledge integration of cutting–edge academic achievements in both non–aerospace disciplines and aerospace disciplines will become an important direction.

10 Modern Advanced Damage Technology and Effects Evaluation

10.1 Introduction

Damage is the final link in the chain of military strikes, which affects the course of the battle and determines the end of the war. The scientific essence of damage is "the process of releasing and transferring energy to the target", and when the energy density used for the target exceeds a certain threshold, it will lead to the destruction of target materials and structure or loss of function, so as to achieve the combat purpose of target damage. The energy used by conventional weapons to destroy a target is derived from the chemical energy released by the explosion of propellants and explosives, which is very limited. It is difficult to achieve "one-strike destruction", with a greater degree of randomness and significant probability, so that is referred to as the "probabilistic nature of conventional destruction". The basic scientific problem of this discipline is "high-density energy storage, harnessing, efficient utilization and evaluation", which is a cutting-edge cross-cutting field integrating many disciplines, such as mechanics, engineering, material science, chemistry, explosion science, condensed matter physics, computational science and so on, and it has typical cross-scale characteristics. In this report, the research progress and future development trends of modern advanced damage technology and effects evaluation are discussed regarding the energy enrichment and creative technology, energy release and control technology, energy efficient utilization technology, and energy utilization effect evaluation technology.

10.2 Latest Research Progress

In terms of energy enrichment and creative technology, this discipline need to mainly focus on solving the problem of damage energy sources by using the energy-containing materials as carriers to realize energy enrichment, research on energy storage and stabilization pathways, master the mechanism of energy excitation and transformation and then expand the energy spectrum of energy-containing materials. At present, the researches on the second, third and new generation of energy-containing materials are very active, showing three major development trends: Firstly, the second and third generation of energy-containing materials are developed in the form of systematization, and the related technological research is also being deepened. Secondly, the energy-containing materials are developed from the traditional single nitro energy storage unit to the direction of combining multiple energy storage units. Thirdly, the energy-containing materials are mainly developed from chemical bond energy storage to the direction of combining high tension bond and chemical bond energy storage.

In terms of energy release and control technology, focusing on solving the problem of high-density energy utilization, the traditional carbon-hydrogen-oxygen-nitrogen (CHON) type of energy-containing materials are close to the energy limit of nitro. So it is time to pay attention to the reaction zone of the propellants and explosives detonation, and enhance the energy level through dynamics regulation.

In terms of energy efficient utilization technology, precise controlling of the power field, such as the strengthening of the thermal effect of the charge blast, the superposition of multi-domain energy coupling, and the modulation of the destructive effect, which can achieve the efficient use of high explosives.

In terms of energy utilization effect evaluation technology, focusing on solving the problem of evaluating the efficiency of high-density energy utilization. It has constructed a target system based on damage, an analysis of the damage characteristics of the target and the damage standard, the damage criterion of a typical target and an equivalent target design method. Then the valuation of energy utilization efficiency is carried out by combining target vulnerability characteristics, power field distribution and environmental factors.

10.3 Comparative Analysis of Research Progress at Home and Abroad

10.3.1 Energy enrichment and creative technology

The United States is attempting to realize new technologies with more than 10 times TNT destructive effectiveness by 2040, focusing on the development of metastable nanomaterials, high-energy hydrogen storage materials, high-tensile bond energy-releasing materials and other subversive energy-containing materials, with the aim of raising the destructive capability of conventional weapons to 6 times, 15 times and 100 times TNT equivalent, thus completely changing the performance of weapons and equipment and subverting the shape of war. Comparatively, China has mastered high-quality, low-cost preparation technologies for second-and third-generation energy-containing materials. Its application level continues to rise, and the research progress of the new generation of energy-containing materials has accelerated significantly, catching up with international development, with some of the research results ranking among the world's top.

10.3.2 Energy release and control technology

The United States has deepened the explosive reaction dynamics research, mastered the shear ignition, hot spot growth, ignition suppression and other mechanisms, realizing the reaction dynamics of aluminum powder, detonation reaction dynamics and other technologies. Compared to them, China's propellants and explosives technology has broken the technical bottleneck of single formula design method, energy output structure and other technical bottlenecks, so that the energy density has been greatly improved, and the technical indicators have reached the advanced level of similar products in the international arena.

10.3.3 Energy efficient utilization technology

In the area of anti-armor, France has adopted the multi-purpose tandem armor-breaking technology, and the penetration depth of rolled homogeneous armor is 30% higher than that of existing anti-tank missiles of the same class. In the area of countering solid targets, the United States has developed a 13.6-ton giant drill bomb, with a penetration capacity of more than 8 meters on super-strength reinforced concrete. Correspondingly, China has established new principles and methods of energy utilization, such as adjustable power/effect selectivity, Low-Collateral, high light, electromagnetic pulse, etc., and the destructive performance has reached the technical level of similar equipment in Europe and the United States.

10.3.4 Energy safety applied technology

The proportion of safety ammunitions in service in the United States has exceeded 50%, and the proportion of safe munitions in its Navy and Air Force has exceeded 70%, which indicates that the level of energy release control of foreign propellants and explosives is constantly improving, and the risk of large-scale chain explosions triggered by munitions safety problems has been basically eliminated. China has mastered the preparation technology of many kinds of insensitive energy-containing materials, synthesized a variety of heat-resistant compounds with decomposition temperatures exceeding TATB, and put them into engineering applications; mastered the technology of quantitative testing and characterization of reaction intensity, as well as has formed model, methods and specifications for testing and evaluating the ammunition safety.

10.3.5 Energy utilization effect evaluation technology

The United States has accumulated a large amount of target vulnerability test data, formed generalized assessment methods and software, and established perfect target vulnerability assessment methods and databases, such as the Modular Unix-based Vulnerability Estimator Suite (MUVES). Correspondingly, China has established methods for evaluating the vulnerability characteristics of common military targets and for designing damage-equivalent targets, and developed a variety of damage evaluation software systems and databases. The research on energy utilization effectiveness evaluation technology has shifted to the study of dynamic power distribution law and the evaluation of multiple damage effects in the complex conditions.

10.3.6 Energy enrichment and utilization of the whole process of simulation capability molding

The United States and European countries have developed a variety of specialized series of simulation software, such as LS-DYNA, AUTODYN, Material Studio, Cheetah (USA), which can achieve rapid design and performance prediction, saving the research cycle and funding, and improve the efficiency of research and development and evaluation. Compared with the United States and European countries, China has established a database on the properties of propellants and explosive materials and testing, formed a physical model of high-density energy release and an engineering model of the damage effects of energy utilization; mastered the solution algorithm and a method of accurate analysis and simulation for the whole process, including high-density energy enrichment, output, transformation and utilization.

10.4 Development Trends and Prospects

Due to the application of new technologies, in the future, the warfare will gradually become intelligent, all-domain operation, precise, diversified, and autonomous. Accordingly, the weapons systems must have high technical performance, such as advanced destruction, long-range suppression, precision strikes, high-speed mobility, rapid response, battlefield survival, and round-the-clock operation. Modern advanced damage technology and effect evaluation need to focus on energy merging and utilization, damage precise control and characterization, damage effectiveness prediction and verification, target damage characteristics, and so on. Meanwhile, the following methods including energy enhancement and regulation of explosives, energy excitation and transformation mechanisms, regularity of coupling effect between energy and target, all-domain precise control of explosion and combustion effects, and damage energy merging should be possessed. The key critical technologies, such as, multi-domain energy coupling and new destruction mechanism, need to be developed desperately. On the basis of the above-mentioned, we will gradually build an efficient damage technical system with complete spectrum, comprehensive technology, and advanced level, realizing adjustable weapon force and power, improving the application level and forging efficiency of advanced weapons and equipment, and then driving the development of modern advanced damage technology and effects evaluation.

In conclusion, the independent innovation and development of advanced damage technology and effect evaluation will promote the change of combat mode, give birth to the new quality combat force, and enable our country to grasp the initiative of cutting-edge scientific and technological competition; at the same time, it will also drive the rapid development of basic disciplines such as materials science, high-pressure physics, computational science, etc., and the results obtained will be widely applied to new energy, new materials, high-end manufacturing and other related fields, so as to enhance the scientific and technological innovation capability and industrial manufacturing level, and promote the overall improvement of national scientific and technological strength.

11 Mining Engineering

11.1 Introduction

Mining Engineering is a key discipline supporting the sustainable development of China's national economy. It is a practical fundamental discipline aimed at the safe, efficient, and environmentally friendly extraction of mineral resources, and it is a second-level discipline categorized under the first-level discipline of Mining Industrial Engineering. The Mining Engineering discipline has a long history. After the reform and opening-up, China's mining industry has gained remarkable achievements, and the Mining Engineering discipline has entered a new era of robust development, ensuring the increasing demand for mineral resources in support of China's national economic growth.

In recent years, China's mining engineering has made significant progress in green mining, deep mining, and intelligent mining. China's production of ten non–ferrous metals has ranked first in the world for consecutive years. In 2022, the national coal output reached 4.56 billion tons, exceeding 54.7% of the global coal output. However, compared with international status, there are still significant differences in terms of high–quality development, mining depth, and mining intelligence. In the future, carbon–negative and efficient backfill mining, co–exploitation of mineral and geothermal resources, in–situ fluidized mining of deep solid resources, exploration and exploitation of deep space resources, and mining of deep–sea resources are important trends in the development of mining engineering in China.

11.2 Latest Research Progress in Mining Engineering Discipline in Recent Years

11.2.1 Latest advancements in research on mining science and technology in metal mines

11.2.1.1 Theory and key technical equipment for paste backfill in metal mines

Paste backfill technology is the key technology to realize green mining. The rheological behavior of full tailings in deep thickening, the rheological behavior of paste in mixing, the rheological behavior of paste in transportation, and the rheological behavior of the filling body have been thoroughly studied, and a rheological framework system for paste in metal mines has been constructed. A complete set of paste backfill technology and equipment has been developed: ①technology and equipment for full tailings paste deep thickening; ②technology and equipment for paste homogenization mixing; ③technology and equipment for long–distance and steady–state transportation of paste; ④Intelligent control technology and equipment for the entire process of paste backfill. The National Standards, such as Technical Specification for The Total Tailings Paste Backfill and the Technical Specification for Total Tailings Paste Production and Disposal, were formulated.

11.2.1.2 Theory and technology of mine slope disaster prevention and control in complex environments

The induced mechanism of rock fracture was elucidated, and the rock strength criteria were constructed. An innovative deterministic solution was presented to address the uncertainty of safety factors in slope design, and the Specification for Rock Slope Engineering Design of Open–pit Mine was formulated. A slope deformation monitoring equipment with completely independent intellectual property rights, S-SAR synthetic aperture radar, was developed; the "double-body catastrophe mechanics theory" for landslides, based on Newton force change measurement, was put forward for the first time; the "reinforcement-monitoring-early warning" integrated prevention and control technology for landslides was formed; and two diaphragm wall structures for water containment and slope stabilization were put forward, namely single diaphragm wall structure and anchored diaphragm wall structure.

11.2.1.3 Cloud platform for monitoring and early warning of mining disasters in metal mines

An intelligent mesh division algorithm for structured hexahedron with hundreds of millions of degrees of freedom, elastic-brittle damage constitutive model, and a dynamic correction algorithm for temporal and spatial variability of rock mechanics parameters driven by microseismic data were proposed, forming a dynamic simulation cloud service propelled by monitoring data and a disaster prediction and early warning method combining monitoring and simulation. An algorithm for constructing an early warning index system based on case mining was proposed; finally, a "four-in-one" dynamic evaluation method of mine disaster risk was formed, including geological disaster case base, field monitoring big data mining, simulated cloud computing analysis, and expert system evaluation; and a cloud platform for monitoring and early warning of geological disasters in metal mines was built.

11.2.1.4 Efficient and intelligent collaborative technology for the entire activity chain in underground metal mine production

The core challenges, such as "continuous, accurate, and rapid acquisition of mine production operation data and integrated management of multi-source and heterogeneous data", "automatic driving of underground trackless equipment", "visual integrated management and control of the underground environment and working conditions", and "real-time scheduling of production process throughout the entire activity chain", have been studied and overcome, realizing efficient and intelligent cooperation throughout the entire activity chain in underground metal mine production, and achieving intelligent mining of mines. The operational mode in underground metal mines has been innovated. The efficient and intelligent collaboration throughout the entire process of the activity chain in underground metal mine production has been realized. A typical metal intelligent mine demonstration has been established.

11.2.1.5 Key technologies for intelligent management and control of open-pit metal mines

A new comprehensive intelligent production management and control system for unmanned mining in open pit metal mines has been established, and the key technologies for intelligent management and control have been overcome: ①open architecture design and integrated control technology for unmanned driving vehicles; ②unmanned autonomous operation and obstacle avoidance technology in the complex environment of open-pit mining area; ③all-factor intelligent fine ore blending technology for multi–metal and multi–target open–pit mines; ④data–driven unmanned multi–vehicle collaborative intelligent scheduling technology for open–pit mines; ⑤integrated management and control platform for unmanned mining in open–pit metal mines under cloud services.

11.2.2 Latest advancements in research on mining science and technology in coal mines

11.2.2.1 Mining stress and rock mechanics theory

The distribution and rotation regulation of mining stress in the deep long–wall working face with ultra–large length are studied, a fractional order model is proposed to describe the mining stress attenuation mode, and the mechanism of mining–induced fracture propagation driven by rotational mining stress is revealed. The optimization principle of working face advancing direction is proposed based on the rotation trajectory of ground stress and mining stress. The relationship between the principal stress rotation around the roadway and the asymmetric failure of the roadway is expounded, and the butterfly failure theory is developed. The hydraulic fracturing technology in the overlying thick and hard roof area of the coal seam is developed, and the destressing method of rock burst prevention named "artificial liberation layer" is put forward.

11.2.2.2 The basic theory and key technical equipment of thick coal mining

Based on the combined effects of top coal fracture distribution, stress loading and unloading, and stress rotation, a prediction model of top coal broken degree is constructed, and a "cantilever beam–articulated rock block" structure model of overlying rock in fully mechanized mining caving face with large mining height is established. The relationship between coal wall stability and roof pressure, coal wall strength, coal wall height, support resistance, and other parameters are obtained, and the timing rule of coal–gangue–rock caving flow is revealed. The technology of fully mechanized caving with large mining height of extra–thick coal seams and top coal caving of steep thick coal seams have been developed, and breakthroughs have been made in intelligent fully mechanized caving and intelligent top coal caving technology and equipment.

11.2.2.3 The Key theory and technology of strata movement and surrounding rock control

A thorough rock strata control theory and technology system has been formed. The intelligent control technology framework of "multi–parameter sensing, analysis pattern discrimination, equipment independent decision–making, quick execution of processes, dynamic evaluation of effects" is established in mining face, and intelligent control technologies such as intelligent mining system, adaptive cruise of equipment and in–situ sensing of rock formation movement are innovated. Intelligent underground sorting and in–situ backfilling technology has been developed. The integrated safe, green, and efficient mining of "mining, selecting, backfilling + X" are realized. A method of far and near field cooperative pre–control of hard roof combines surface drilling fracturing with underground roof pre–cracking is proposed.

11.2.2.4 Coal mine intelligent mining equipment and technology

The three major equipment of "mining, supporting and transporting" in the intelligent working face has made considerable development and progress. With the help of CAN bus and industrial Ethernet technology, the shearer realizes the detection, control, and protection of oil level, frequency converter, motor, sensor, and other parameters. The maximum support height of the full–height hydraulic support has reached 10.0m, the maximum support height of the fully mechanized top coal caving hydraulic support has reached 7.0m, and the weight of a single hydraulic support has reached 100t. A new idea of intelligent mine construction based on "digital twin + 5G" is proposed, and a smart mine platform with global perception, edge computing, data–driven, and auxiliary decision–making is constructed.

11.2.2.5 Green mining theory and technology

The pressure behavior law, key strata movement, and surface deformation characteristics of backfilling mining are studied, and the key strata filling control theory aimed at controlling the bending deformation of key strata is constructed. Under the engineering requirements of groundwater protection, low surface damage, and near–zero discharge of gangue and gas, a green mining mode of "mining separating and filling + X" was formed, and key technologies such as "mining separating and filling + control", "mining separating and filling + retention", "mining separating and filling + pumping", "mining separating and filling + prevention", "mining separating and filling + protection" were developed, which promoted the application of underground separation and in–situ filling in coal mines.

11.3 Comparative Analysis of Domestic and International Research Advancements in Mining Engineering Discipline

By promoting carbon peak and neutrality strategies, countries globally are expediting the shift toward intelligent production, information–driven management, and cleaner coal utilization while ensuring a secure and stable energy supply foundation. The forefront of international research includes automation and intelligent technology, cost–effective mining techniques, green mining, machine learning and data analysis, and deep–sea mining. Combining the latest research advancements in this discipline in recent years in China, a comparison is made between the domestic and international research progress in this field as follows.

The high–quality development of the coal industry, both domestically and internationally, reveals significant disparities. After over 40 years of development, China's sub–level caving mining technique has achieved a globally leading position. However, imbalances persist in the efficiency and technological levels of coal mine exploitation, market structures, and the cleanliness and quality of coal utilization in China. Simultaneously, the coal industry in China is not fully developed in terms of safety, green, low-carbon, human resources, international cooperation, and enterprise transformation and upgrading.

Significant differences exist in the mining depth of metal mines between China and foreign countries. Currently, the maximum mining depth in metal mines in China is 1990m, which has not yet reached 2000m. In contrast, several countries, including South Africa, India, Canada, the United States, and Chile, have already achieved or exceeded depths of 3000m in metal mine exploitation. South Africa, in particular, has more than ten metal mines with depths reaching or exceeding 3000m, with the deepest reaching 4800m. It is evident that there is a substantial gap between the depth of metal mine exploitation in China and that in developed mining countries abroad.

A considerable gap exists in the deep intelligent mining of metal mines in China. Currently, most mines in China have established underground optical fiber backbone communication networks, and most are equipped with environmental sensing devices, automation control systems, disaster monitoring systems, mine management software, etc. However, compared to foreign mines, the development of intelligence is relatively lagging: weak intelligent sensing technology for mining environments; lack of efficient mining technology under deep high-stress and high-temperature conditions, high mining costs, and slow progress in shaft and roadway engineering; poor matching of mechanized equipment in mines and low manufacturing levels of large underground mining equipment; and a relatively backward development of integrated information platforms for mining production control.

11.4 Development Trends and Prospects of Mining Engineering Discipline

11.4.1 Theory and technology for carbon-negative and efficient backfill mining in coal mines

Innovative theories have been developed for the structural topology and strength of carbon-negative high-porosity backfill materials as well as theories on carbon sequestration in backfill bodies, rapid reaction kinetics for quick-setting cemented materials, and conceptions of carbon-negative efficient backfill theories, including backfill mining for preventing rockbursts in mining areas. The key technology system has been developed, including the preparation of rapid and efficient cemented high-porosity backfill materials from gangue, green and efficient preparation technology for quick-setting cemented materials, carbon–negative and efficient backfill mining technology, multi–side simultaneous backfill mining technology and process, and full–cycle three–dimensional efficient backfill mining and impact prevention technology.

11.4.2 Theory and technology for co-exploitation of mineral and geothermal resources

To strengthen exploration of mineral–thermal coexistence zones and promptly initiate mining–thermal resource co–mining pilot projects. To develop rock breaking and tunneling technology in deep, high–temperature, and hard rock stratum. To strengthen the theoretical and experimental study of rock mechanics in deep multi–field coupling environment. To establish a classified utilization system of thermal energy for co–exploitation of mineral–thermal resources.

11.4.3 Theory and technology for in-situ fluidized mining of deep solid resources

To overcome the limit of critical depths in solid mineral resource extraction and to construct a fluidized mining technology system that realizes unmanned and intelligent mining–processing–backfill and thermal–electricity–air transformation in underground coal mines in situ. To develop theories and technologies for in–situ leaching and integrated mining, processing, and backfill in deep metal mines.

11.4.4 Exploratory research on exploration and exploitation of deep space resources such as the Moon and Mars

To study the remote sensing detection and acquisition technology for lunar and Martian resources, to innovate the in–situ true–core sampling technology for large depths on the Moon and Mars, to study the concept and technical implementation scheme for the utilization of underground steady–temperature layers on the Moon and Mars, and to develop supporting technologies such as underground activity space, underground heat storage, thermoelectric power generation, and mineral mining structures on the moon and Mars, thus forming new theories, technologies, and schemes for the development and utilization of the underground space of the moon and Mars.

11.4.5 Theory and methods for the mining of deep-sea resources

To study the hydrodynamics of deep–sea mining systems, to explore the coupled response characteristics of deep–sea mineral rocks and mining stripping equipment, to establish a joint system of lithology detection reporting and online numerical analysis, to develop intelligent decision–making and intelligent mining stripping methods of mining equipment, and to establish response mechanisms and models of the deep–sea mining system to changing ocean environments.

12　Seed Science

12.1　Introduction

The country is based on agriculture and prioritizes seed production. Seeds are the "chips" of agriculture, which are irreplaceable, basic, and important means of production in agricultural production. They are also the foundation of human survival and development, and play an irreplaceable role in the entire agriculture and national economy. Seed science is the study of the characteristics, physiological functions, and life activities of crop seeds, serving agricultural production and solving various scientific and technological problems in seed production. Seed science is an ancient and young discipline that has been systematically studied for over a hundred years. Early seed studies mainly included seed biology and seed physiology. In the 20th century, with people's recognition of the important position of seeds in agricultural production, the development of seed science research was promoted. In recent years, with the rapid development of the seed industry and the process of seed industrialization worldwide, seed science has expanded into seed science and technology on the basis of traditional and emerging disciplines such as molecular genetics, molecular biology, and genetic engineering. Research has expanded from population to individual, and from cellular to molecular levels. Seed science is a discipline that closely combines basic research and application. Seed research and production need to be closely integrated to serve agricultural production on a broader scale.

This section reviews and summarizes the development of seed science in China in the past five years. It reviews and summarizes the current development status of new theories, technologies, methods, and achievements in seed science in China in recent years. Combined with major agricultural projects since 2022, it summarizes key technological advances related to crop breeding and briefly introduces the progress made in the field of seed science research, and analyzes and compares the latest research hotspots, cutting–edge trends, and development trends of this discipline internationally. Based on the current development status of seed science in the past five years, compare the gap between domestic and international agricultural technology development, analyze the development strategy and key development directions of seed science in China in the next five years, and propose relevant development trends and strategies.

12.2 The Latest Research Progress in this Discipline in Recent Years

12.2.1 The latest theoretical and technological research progress in seed science

12.2.1.1 Seed and crop yield and quality

As the "chip" of agriculture, seeds are an important guarantee for increasing agricultural production and food security. Good varieties play a crucial role in increasing and stabilizing agricultural production. It is estimated that the breeding and promotion of good varieties contribute more than 50% to the increase in yield per unit area. For example, due to factors such as breeding and cultivation, the yield level of soybeans and corn in China is only about 60% of the world's advanced level. Behind the yield gap is the difference in density and stress resistance of varieties. In the past decade, the country has continuously introduced a series of policies to promote grain production, and China's seed industry has steadily developed and continued to strengthen, with significant results in high-quality agricultural development. The compound growth rate of the number of seed related patents in China has reached 28.8%, and the contribution rate of high-quality seeds to grain production has exceeded 45%. With the improvement in living standards, people's demand for high-quality grains has increased, and the direction of scientific research and breeding has gradually shifted from focusing on yield to emphasizing both yield and quality. The transition from eating full to eating well, and then to eating nutritious and healthy, has put forward more demands for the breeding of good varieties.

12.2.1.2 Seed production

Seed production is the first step in agricultural production and has made significant contributions to ensuring food security and sustainable agricultural development. The self-sufficiency rate of seeds in China's main agricultural products is at the forefront of the world. Currently, China has established an efficient breeding technology system for super rice, dwarf wheat, hybrid corn, and other crops. The area of independently selected crop varieties accounts for over 95%. In recent years, with the continuous progress of technology and the continuous development of agricultural production, new progress has also been made in seed production, including gene editing technology, haploid breeding, molecular design breeding, intelligent agriculture, and plant factories. The application of these new technologies and models brings broader development prospects and more opportunities for seed production. The progress of technology has empowered innovation in the seed industry, achieved changes in breeding models and processes, and played an important supporting role in ensuring food security in China.

12.2.1.3 Seed storage and processing

Seed storage technology is the process of preserving seeds under suitable environmental conditions to extend their shelf life and ensure their quality and availability. The key to seed storage technology is to control temperature, humidity, and oxygen concentration. The newly built National Crop Germplasm Resources Database in September 2021 has been officially put into operation, basically achieving ultra-low temperature preservation of seeds, as well as the preservation of test tube seedlings and DNA. The entire preservation process has achieved intelligence and informatization, and the seed storage life can reach 50 years. The application of seed storage technology can not only improve the efficiency of agricultural production, but also reduce resource waste and environmental pollution.

Seed processing technology refers to the treatment of seeds to improve their quality and adaptability. In recent years, China's attention to research on seed processing equipment has significantly increased. China's agriculture has basically achieved full mechanization of operations, and equipment such as dryers, cleaning machines, selection machines, and coating machines have basically achieved digital control. Under intelligent control, sensor measurement is achieved, improving the accuracy of seed processing. At the same time, new technologies such as wind balancing cleaning, self balancing gravity selection, optical color selection, and precise granulation treatment have gradually been widely applied, and the quality of processed seeds has been further improved.

12.2.1.4 Collection and protection of germplasm resources

Crop germplasm resources are the source of innovation in crop seed industry. Germplasm resources are carriers of genetic information carried by organisms, with actual or potential utilization value. Their forms include seeds, plants, stem tips, dormant buds, pollen, and even DNA. During the 13th Five Year Plan period, China accelerated the construction of a germplasm resource protection and utilization system, and established and improved a national crop germplasm resource protection system consisting of one long-term database, one duplicate database, 10 mid-term databases, 43 germplasm nurseries,205 original habitat protection points, and a germplasm resource information center. As of the end of 2022, the total amount of collected and preserved resources in China has exceeded 540000, protecting a large number of rare and endangered resources; every year, more than 100000 copies are effectively utilized for scientific research, breeding, and production, providing strong support for crop breeding and agricultural technology innovation in China.

12.2.2 The application of seed science in the seed industry

The seed industry is a systematic engineering that mainly includes five systems: breeding of excellent varieties, production and reproduction of seeds for field production, processing and coating of high-quality seeds, seed quality inspection, packaging, storage, circulation and sales, and seed industry management. After years of development, China has made significant breakthroughs in the research of major crop seeds, greatly promoting the development of the seed industry. In the identification, evaluation, and innovation of germplasm resources, a series of achievements have been made in seed development, seed dormancy and germination, seed deterioration and storage tolerance theory, new variety breeding theory and technology, seed breeding technology, key technologies for high vitality seed production, hybrid seed production technology and key equipment, and seed processing. During the 13th Five Year Plan period, significant achievements have been made in scientific and technological innovation in China's seed industry. Firstly, large-scale precise phenotypic identification and whole genome genotype identification of crop germplasm resources have been carried out, providing an important material basis for new variety breeding and basic research; secondly, a number of favorable genes for important traits have been precisely mapped and cloned, providing important genetic resources and pathways for crop molecular directed design and breeding; thirdly, significant breakthroughs have been made in core key breeding technologies such as crop gene editing and haploid breeding, accelerating the rapid application of efficient crop breeding systems; the fourth is to comprehensively utilize plant molecular design, chromosome cell engineering, mutagenic biotechnology, and hybrid advantage utilization technologies to cultivate a batch of high-quality rice, drought resistant and water-saving wheat, high-yield machine harvested corn, early maturing and high-quality vegetables and other new crop varieties. China's seed industry enterprises have become one of the top ten in the global seed industry through technological and product innovation, providing strong technological support for ensuring national food security and promoting green agricultural development.

12.3 Comparison of Research Progress in this Discipline both Domestically and Internationally

In recent years, seed science has achieved significant success in both basic and applied research that promotes the development of the discipline itself, as well as in the cross development with related disciplines and the expansion of new application fields. This has had an important impact on promoting the independence, self-reliance, and controllability of seed sources in China's seed industry technology. The level of crop breeding, excellent varieties, and supply capacity in China have significantly improved, with over 95% of the planting area of independently selected varieties, achieving the goal of "China's grain mainly uses Chinese varieties". The self-sufficiency rates of pigs, cows, sheep, and some characteristic aquatic core species have reached 75% and 85%, respectively, providing key guarantees and support for the stable production and supply of food and important agricultural and sideline products. At present, China's seed industry market size has ranked second in the world, second only to the United States. However, from the perspective of the seed industry itself, the protection and utilization of germplasm resources are still far from enough, and the independent innovation ability is not strong, especially in the theory and key core technology of breeding, which is still far from cutting-edge countries.

In terms of excavation and identification of germplasm resources, although the total amount of preserved germplasm resources in China exceeds 520000, ranking second in the world, there are still many shortcomings in China's existing germplasm resources. The structure of existing germplasm resources is relatively single, mainly domestic resources, with a small proportion of foreign resources, poor resource diversity, and many disease resistant, stress resistant, and high-quality breeding materials have not yet been collected. In terms of basic theoretical research in seed science, China has made significant progress in biological breeding technology innovation and patent inventions. In some sub fields, the gap with developed countries represented by the United States has significantly narrowed, and in some aspects, it has even surpassed developed countries. However, there are few discoveries of important functional genes, and the biological basis and regulatory mechanism for the formation of important economic traits, quality, and Research on the coordinated improvement of yield and resistance, as well as the interaction mechanism between abiotic stress and crop development, is weak. In terms of key breeding technologies, there is a significant intergenerational gap between China's breeding technology and the international advanced level. International breeding technology has basically entered the "4.0 era" of intelligent breeding, which deeply integrates life science, information science, and breeding science. A smart breeding research and development system based on genotype phenotype environment data collection and simulation analysis has been established, and it has expanded from greenhouse phenotype technology to field phenotype technology. The data processing capacity of research and development is increasing exponentially every year. However, most breeding in China is between the 2.0 and 3.0 eras, where hybrid breeding and molecular assisted breeding are the main methods. There is still a significant gap in breeding quality, efficiency, and regional adaptability compared to the international advanced level. In terms of research tools and equipment used in the field of seed science, China is currently highly dependent on imports, and some technical equipment is still unable to be obtained from abroad. The lack of advanced scientific research tools and equipment seriously affects the development of seed science research in China.

12.4 Development trends and prospects of this discipline

The world today is experiencing unprecedented changes, with a new round of technological revolution and industrial transformation advancing rapidly. The paradigm of scientific research is undergoing profound changes, and interdisciplinary integration is constantly developing. The new round of technological revolution and industrial transformation is reshaping the global economic landscape, and international forces are deeply adjusting. The international environment is becoming increasingly complex, and ensuring food security has become a common challenge faced by the world. To carry out the research and development of seed source bottleneck technology and aspire to fight a revolutionary battle in the seed industry, China's agricultural biological breeding technology research and industrialization development has entered a new stage of self–reliance and leapfrog development. However, there are still a series of constraints in the research and development of biological breeding, including weak original innovation, lack of key technologies, and disconnection in the innovation chain. Therefore, the development of seed science is particularly important. Only by vigorously developing technological innovation in the seed industry can we accelerate the research and industrialization process of biological breeding technology in China, enhance the core competitiveness of modern agriculture, achieve technological self–reliance and self–improvement, and ensure national food security, ecological security, and national nutrition and health.

With the development of technologies such as artificial intelligence and big data, the field of seed science research will gradually achieve intelligence and precision, with the main research goal of improving agricultural production efficiency and reducing resource waste. Through advanced bioinformatics data analysis, molecular biology technology, and artificial intelligence, the performance of seeds will be rapidly improved to improve their response to the challenges brought about by climate change and environmental degradation. In terms of seed science talent cultivation, the advent of modern seed industry and digital information era has led to an increasing trend towards applied and interdisciplinary talents in seed industry disciplines. In recent years, various universities have established new technology interdisciplinary majors to reshape the knowledge cultivation system of seed industry talents, providing a good fertile ground for modern seed industry talent cultivation. In terms of the construction of basic conditions for seed science research, we will build an internationally first–class seed industry technology innovation platform, and create a fertile ground for seed industry technology self–reliance and self–improvement through open sharing support and guidance of basic research in the field of seed science.

National policies play a guiding role in the development of the seed industry. Therefore, there is a need for more support and tilt at the national policy level. By continuously strengthening the strategic significance of the development of biological breeding, improving the national innovation and development system of biological breeding, and promoting scientific and technological innovation in the field of seed science research, we can promote China's rapid transformation from a major country in the biological breeding industry to a strong country, ensuring the independent and controllable control of key common technologies, and "China's food in a bowl". In addition, seed research should only focus on the production problems that urgently need to be solved in the development and innovation of China's seed industry, with more original innovations at the technical level, and further comprehensive development and promotion. In response to the significant needs of China's agricultural and rural modernization for food security, green development, healthy living, extreme climate response, and the development of strategic emerging industries, we will precisely cultivate and create new agricultural resources that increase production and quality, reduce investment and efficiency, and reduce losses to promote stability, achieve leapfrog upgrading of existing varieties, and lead the development of precision agriculture.

13 Materials Science

13.1 Introduction

All materials were strategic in space and military technologies, when the "materials" first appeared in the language of science policy makers. The discipline of Materials Science and Engineering (MSE) emerged from metallurgy and solid-state physics, and then combined math, physics, mechanics, chemistry and informatics principles. The intermingling of science and industry has launched to solve real-world problems associated with nanotechnology, energy, mechanical manufacturing, biotechnology, information technology and engineering disciplines. A large amount of research on electronic and magnetic materials, and advanced materials for batteries has been conducted under the umbrella of physics or within solid-state-chemistry departments.

Materials Science and Engineering aims to clarify the interaction of structure, properties, performance and process. A tetrahedron can be visualized them at the four vertices, with emphasis on their mutuality. A wide range of research fields can include metals, inorganic non–metallics, organic polymer, composites, biomedical materials, energy materials, environmental materials, electronic information materials, nano–materials, material genetic engineering, material surface and interface, material failure and protection, material testing and analysis technology, material synthesis and processing.

Here we review the eight frontiers of Materials Science and Engineering over the last years, including quantum materials, two–dimensional semiconductor, energy materials, metamaterials, biomaterials and nanomedicine in space environments, materials for extreme environments, materials genome engineering, as well as materials fabrication and characterization.

13.2 Frontiers of Materials Science and Engineering

13.2.1 Quantum materials

The new generation of spin quantum materials is related to ultra–high density, large capacity, nonvolatile magnetic storage and logic storage integrated devices, which is expected to become one of the main directions of the development of information industry in the "post–Moore era". Magneton spintronics can be used to, detect and control magnetons, as well as the mutual transformation between magnetons and various quasi–particles such as conduction electron spins. The excitations of spin lattice can form spin waves and their quantized elementary elements–magnetons in magnetic materials or insulators. This sort of magneton (spin wave) has wave–particle duality like photon (light wave) and phonon (sound wave), which can be used for directional and long–distance transmission of spin information, due to its electrical neutrality without Joule heat, less energy consumption, in order to realize the functions of data storage and non–Boolean logic operation. As an information carrier, magnetons have unique application prospects in new logic computing, neural brain–like computing, and even quantum computing. Therefore, the magnetic metals, ferromagnetic semiconductors, ferromagnetic semimetals, magnetic topological materials, spin–wave materials and antiferromagnetic semiconductors can contribute to the spin devices.

13.2.2 Two-dimensional semiconductor

Two–dimensional semiconductor has dominated the advanced materials and electronics in recent years, especially transition metal dichalcogenides. The better properties have been achieved in the atomic–scale ultra–thin layer, high carrier mobility, layer–dependent tunable band gap, spin–valley electronics, ultra–fast response speed and easy back–end heterogeneous integration.

The main challenge can break through the physical limitations of the mainstream silicon-based complementary metal oxide semiconductor (CMOS) chip technology in further miniaturization, such as short channel effect, which is one of the candidate chip materials to replace silicon in the post-Moore era. The large-scale wafer-level materials and multi-layer structure heterojunction can be fabricated for the high-density low-power storage, high-efficiency photovoltaic, high-sensitivity photodetection, ultra-short channel, and ultra-fast computing. Two-dimensional semiconductor can be expected to serve in the wearable electronics, sensors, and biomedicals.

13.2.3 Energy materials

Energy materials are ever increasing in the global demand for sustainable energy. The better performance, sustainability and environmental friendliness can be optimized the synthesis and preparation of electrode materials using various advanced microscopies and spectroscopies.

13.2.4 Metamaterials

Metamaterials are macroscopic dielectrics that composed of artificial microstructure units. These artificial metaatoms can be tuned by the size, structure and spatial periodic or aperiodic arrangement. The metamaterials can be obtained to alter the macroscopic electromagnetic parameters (including dielectric coefficient, permeability coefficient, refractive index coefficient, and absorption coefficient). Different propagation properties of electromagnetic waves can also be achieved, including refraction, reflection, transmission, absorption, and wave vector dispersion. A variety of novel and exotic electromagnetic applications have been realized, including negative refraction, perfect lens imaging, perfect absorption of invisibility cloak, and radiation cooling. Metamaterials are originated from microwave and visible light wavebands. The current research focuses on the electromagnetic wave frequency ranges such as terahertz, infrared and extreme ultraviolet. Metamaterials are beginning to be applied in the fields of national defense and economic construction, such as deep space, deep sea and deep earth exploration, high directional electromagnetic countermeasures, 5G/6G wireless communication, and green energy.

13.2.5 Biomaterials and nanomedicine in space environments

Medical micro-nano material is one of the important strategic frontiers of nanoscience and nanotechnology. Physical, chemical and biological properties can contribute to the disease prevention and diagnosis, treatment and prognosis monitoring in life science. Medical micro-nano materials have been widely used in disease diagnosis, targeted drug delivery and controlled release, tissue repair and regeneration, intelligent biological devices in medicine and health, and medical devices for life support of future space survival.

13.2.6 Materials for extreme environments

Extreme environment service materials are mainly used in many strategic national defense fields, such as extreme high pressure, extreme temperature, extreme radiation, extreme impact and strain rate, extreme energy and combustion reaction, extreme corrosion and hydrogen environment in deep space, deep sea and deep earth, where metal alloys in the conditions of extreme impact and high strain rate, while advanced ceramics at high strain rates.

13.2.7 Materials genome engineering

The generative AI ChatGPT can be extended to the high–throughput dynamics with data–driven technologies (such as machine learning and artificial intelligence), in order to accurately and rapidly screen thousands of microstructures and/or chemicals. Data and theory can be coupled to drive science. Current research focuses on machine learning that are trained on synthetic datasets to speed up computer simulations. Emerging data–driven methods are used for material synthesis, spectral interpretation and experiment optimization.while positive data–driven methods can promote the experimental chemistry and materials science.

13.2.8 Materials fabrication and characterization

The controllable preparation of materials is one of the cutting edges in new materials. The nucleation and growth of crystals can determine the preparation of solid materials. In–situ transmission electron microscopy has promoted the spatiotemporal resolution and high–throughput data analysis. Spherical aberration correctors with in–situ tensile function have equipped for the direct observation of the microstructural defects under strain at the atomic scale. Since the properties of metal materials are controlled by lattice defects, an automatic atomic column tracking can be used to characterize the grain boundaries sliding of bicrystals in real time during strain using in–situ electron microscopy.

13.3 Academic Discourse of Chines Materials Science and Engineering

The research and development (R&D) field of Materials Science and Engineering has achieved in the great leap in China over the past decades. New materials science, technology and industry have shifted into the high–quality development to fully meet the national strategic needs in the international competitiveness. The rapid and sustainable development of New materials science and technology need to improve the knowledge innovation system of materials, and then to promote the mechanism of discipline system and R&D innovation, finally to strengthen personnel training and international exchanges and cooperation.

13.4 Concluding Remarks

Artificial intelligence (Generative large language models) can greatly contribute to the reverse design of new materials, even new paradigm of material design. Green recycling and sustainable regeneration of new energy sources have also delineated the inter–discipline boundaries for materials science. The development trend of materials science can be promising potential to the integrated design of material structure, function and performance, the integrated manufacturing of material synthesis, components and devices, the normalized prediction of material service performance and damage failure, and the integrated research of green manufacturing and life cycle of materials.

The R&D strategy can emphasize the central engineering disciplines of materials science and engineering. The frontier areas can be focused on, such as the inter–disciplines of physics, chemistry, mechanics and life sciences. The decision making of materials industry can be expected to implemented the scientific and technological innovation, in order to coordinate the respective advantages of industry, university and research funds for the new material application scenarios. A new AI design paradigm is urgent for the computational materials and material characterization.

Disciplines of materials science and engineering can be extended to practical engineering training in the primary and secondary education system, such as the multi–level talent training. The academic surrounding can be given to foster the knowledge system of materials science and engineering, covering the scientific spirit, scientific ideas for the scientific quality–oriented education in the academic discourse, national science and technology.

14 Grain and Oil Science and Technology

14.1 Introduction

In the past five years (2019—2023), grain and oil science and technology discipline of China overcomes multiple scientific and technological problems. Fruitful achievements in scientific and technological innovation of subfields empowered the development of the industry and improved the overall level of discipline construction, which has made an important contribution to ensuring national food security.

In the next five years, the discipline will actively adapt to the new round of scientific and technological revolution and industrial transformation, and the strategic deployment of building an agricultural power. It will focus on the main problems and needs of scientific and technological innovation and industrial development, strengthen the basic research and key technical research in key areas, and constantly improve the capability of independent innovation, and support and lead the high-quality development of industries with the support of disciplinary construction to ensure national food security.

14.2 The Latest Research Progress in the Last Five Years

14.2.1 Steady increase in the level of disciplinary research

14.2.1.1 Significant improvement in scientific and technological innovation capacity for grain and oil storage

A number of storage grain application systems have been established and applied. The molecular regulation mechanism of reproductive and hypoxia adaptability of stored grain pests, pheromone biosynthesis of the main storage pest, the phosphine resistance of stored grain pests and controlling postpartum loss of rice by infrared and microwave drying were studied, and the intelligent monitoring system of cooling and ventilation was developed.

14.2.1.2 Significant improvement in grain processing technology and equipment level

The moderate processing technology system of rice and wheat was established, which realized the flexible rice milling and rice brushing/polishing and reduced the crushing and electricity consumption in the rice milling process. The smart wheat milling system can significantly improve the yield of wheat flour products and the retention rate of nutrients. The improvement of technology has effectively solved the outstanding problems in corn processing progress. New breakthroughs have been made in flour products, high activity yeast and coarse grain finishing equipment, etc.

14.2.1.3 Comprehensive advancement of oil and grease processing technology and equipment

Precision and moderate processing leads scientific and technological innovation. The processing technology of food specific oil and fats has realized full substitution of partial hydrogenated oils. Functional oils and fats products has been marketed, and protein products of vegetable oil tend to be serialized. New Breakthroughs have been made in the research on comprehensive utilization of oil and fats. Intelligent and digital technology has been applied in oil processing and oil equipment manufacturing enterprises.

14.2.1.4 Overall improvement of grain and oil quality and safety standards and evaluation techniques

Significant progress has been made in the standardization system, the institutional mechanism, the orientation of standards and international standardization work. The technology for evaluating the physical and chemical characteristics of grain and oil, for evaluating the quality characteristics of grain and oil, and for evaluating the safety of grain, oil and food have developed rapidly.

14.2.1.5 Digitalization and intelligence shaping the new industry of grain and oil logistics

Research has been carried out on digital empowerment of food supply chain innovation, new business forms and models of food logistics, optimization of inter–provincial distribution network layout, emergency logistics in the context of COVID–19etc. The new technologies have been applied in intelligent packaging, unmanned forklifts and robotic arms, etc.

14.2.1.6 Feed processing technology and equipment continue to innovate and promote high-level development of the industry

The basic research on feed processing has been constantly deepened. The manufacturing quality of feed processing equipment and facilities has been significantly improved. Various feed processing technologies focus on the entire industrial chain and sustainable development.

14.2.1.7 New achievements in basic research on grain and oil nutrition and food development

The database on nutritional and functional components of grain and oil has been improved. A new model of moderate processing of healthy grains and oils has been established. Whole grains and staple foods with a high content of grains and nutritious convenience foods have been industrialized, and the development of personalized food and oil products has tended to diversify.

14.2.1.8 Achieve a high level of basic and applied research in the discipline of grain and oil information and automation

The information and automation of grain and oil storage and processing have been significantly upgraded. The logistics information and automation technology of grain and oil gradually mature providing new models and paths for grain circulation, and a series of information platform have been established.

14.2.2 Fruitful development of disciplines

14.2.2.1 Excellent scientific research results

(1) Science and technology innovation gives new momentum to industrial development. ① 4 second prizes for the State Science and Technology Progress Award; several provincial awards; 2-second prizes for the State Technology Invention Award; 4 special prizes and 25 first prizes for the Science and Technology Award of China Cereals and Oils Association; ② applied 17159 patents and authorized 1946 patents; ③ about 25910 published papers; and multiple monographs; ④ initiate or issue over 190 national standards and 164 group standards; ⑤ a series of new grain and oil products and equipment, and grain logistics equipment.

(2) Undertake national science and technology projects to enhance innovation capabilities. During the 13th Five Year Plan period, multiple national key research and development program projects passed performance evaluations; during the 14th Five Year Plan period, multiple national key research and development projects have been successfully undertaken.

(3) The construction of scientific research bases and platforms continues to deepen.20 national and ministerial scientific research bases and platforms, and four original national engineering laboratories have been successfully integrated into the new sequence management of the National Engineering Research Center. The gap in research and development capabilities has significantly narrowed with the world's level.

14.2.2.2 Discipline construction strengthens the foundation and lead stable and far-reaching future

In the past five years, the grain and oil discipline has formed a relatively perfect discipline development system with coordinated development of basic science, applied science and engineering technology application. Formed a complete vocational education, bachelor, master and doctoral personnel training system; increasingly various active academic exchanges; group standard construction leads industry development, and science and technology awards promote achievement transformation; continuous strengthening of team building; a series of professional textbooks, monographs and academic journals have been formed; popular science education and publicity were carried in various forms.

14.2.3 Major achievements and applications of the discipline in industrial development

Several major achievements in scientific and technological innovation of grain and oil have been promoted and applied, generating significant economic and social benefits. The representative projects include: The successful pilot construction of grain gas film reinforced concrete dome warehouse; the key technology and system of real-time monitoring of grain inventory quantity network; the integration and demonstration of paddy low-temperature storage and rice moderate processing industrialization technology and equipment; construction of oil processing integration, data management system; precision and moderate processing technology of grain and edible oil has been widely promoted; construction of the world's largest milk fat database; and so on.

14.3 Comparison of Domestic and International Research Progress

14.3.1 Current status of foreign research

(1) Both basic and technological research are emphasized in grain storage, forming a relatively complete research system for grain harvesting, storage, and transportation technology.

(2) Equipment manufacturing is more refined in grain and oil processing, new products and leading technologies are constantly emerging, quality testing equipment and testing technology are leading, and resource development and high-value utilization of processing by-products are fully utilized.

(3) The quality and safety standards and evaluation systems for grain and oil are relatively complete, and advanced, efficient and non-destructive testing equipment ensure that testing capabilities continue to improve..

(4) Efficient and intelligent innovative equipment for grain logistics is constantly emerging, and the application of digital technologies for logistics systems effectively improve the efficiency of grain logistics system.

(5) The health effects, dietary patterns and functional factors of grain and oil foods have been intensively studied. Functional factors have been studied more deeply, and their diversified influences have been mined.

(6) The grain and oil collection and storage information and automation focused on the combination

of block chain, big data and cloud computing, and deepen with the supervision information in the grain field.

14.3.2 Gaps and reasons for domestic research

The existing gaps are mainly: Compared with developed countries, the combination of pre–and post–production of grain and oil in China is not close, and the weak links in the whole industrial chain are obvious; the difference between Chinese and Western food culture and structure was ignored when tracking the international frontier, just imitates or copies, and does not focus on the study of traditional Chinese food; the basic theoretical research is not deep enough, the high–value utilization of processing by–products and the kinetic energy of technological innovation are insufficient, and the types of high value–added products are few; the integration depth of fork disciplines is not enough; product homogeneity is serious; the standard system of grain and oil is not perfect enough; and so on.

Reasons for gaps:

The mechanism for cultivating high–level scientific and technological innovation talents is imperfect and team building is relatively lagging. The foundation and the integration between theoretical research and practical application is weak. The investment in scientific research funding is limited and scientific research is disconnected from the transformation of actual achievements. The level of standardization of data sharing and exchange in the industrial chain is low, and the degree of aggregation of grain information is not high.

14.4 Development Trends and Prospects

14.4.1 Strategic needs

The construction and development of the discipline will focus on the national strategies needs, take national food security as the purpose, accelerate the transformation of the economic development mode, support the major national strategic goals, and actively promote the high–quality development of the grain and oil industry.

14.4.2 Research directions and R&D priorities

(1) The discipline of grain storage. Strengthening basic and applied basic research. Focusing on the research and development of new green grain storage technologies and equipment and integrate grain storage technologies. Carrying out typical regional application demonstrations.

(2) Grain processing disciplines. Strengthening the intelligent development of equipment and

expanding the functionality of by–products. Promoting graded and precise processing of corn. Carrying out research on processing and preservation technology for products. Construction of yeast strain resource libraries. Research on key technologies for basic and high–value processing of miscellaneous grains.

(3) Fats and oils processing disciplines. Focusing on the traditional scientific research contents, combined with new disciplines to promote the interdisciplinary and integrated development.

(4) Grain and oil quality and safety disciplines. Improving the standard system of grain and oil; developing and improving rapid, intelligent testing technology; improving the grain and oil quality and safety database and early warning model.

(5) Grain logistics discipline. Carrying out research on the layout of grain logistics, the development strategy in the context of the "Belt and Road" Initiative and the new technology and equipment in grain logistics, and so on.

(6) Feed processing disciplines. Strengthening the applied basic research on feed and tackling the bottlenecks, the development and industrial application of feed resources and new feed additives Enhancing the intelligent level of feed processing equipment and technology.

(7) Grain and oil nutrition discipline. Focusing on the applied basic research and established a platform for building and sharing big data on grain and oil nutrition and health. Tackling key technologies and striving to enhance the development capacity of healthy food. Improving the standard system of nutritional and healthy grain and oil of China, and leading the healthy development of the industry.

(8) Grain Information and automation discipline. Establishment of an automated system for the entire process of grain and oil collection and storage. Construction of grain and oil processing intelligent factory and intelligent warehouse control system. Building a data center for the entire grain process to promote the full chain traceability of grain quality and safety.

14.4.3 Development strategy

First, continuously deepen the reform of the grain system, optimize the collection and storage system and strengthen pricing capacity. Second, actively participate in global grain governance. Third, build a major platform for grain science and technology innovation and fully leverage its leading and supporting role. Fourth, continuously carry out the "Five Excellence Linkages" and "Three Chain Synergy" in the grain industry, and deeply promote the "Six Improvement Actions" and the "Quality Food Project".

15 Basic Agronomy

15.1 Advances in Crop Cultivation and Farming System

Crop cultivation and farming system, a fundamental discipline in agricultural science, is pivotal in enhancing agricultural efficiency, productivity, and sustainability. This report highlights recent developments in theoretical and technological innovations within this field in China, emphasizing its role in boosting farmer incomes and fostering sustainable practices. Additionally, it reviews the historical progress of crop cultivation and farming to advance the discipline further.

During recent years, the crop cultivation and farming system discipline mainly focused on the following six aspects and has achieved great progress. ① Green–high yield and high–quality and high–efficient cultivation on the three major cereal crops, rice, wheat and maize; ② transformation of crop production driven by full mechanization, especially on the maize mechanical kernel harvesting technology; ③ smart crop production based on the modern information technology; ④ innovative multiple–cropping systems construction, including cereal–legume rotations and the modern intercropping systems; ⑤ development of the theories and technologies of conservation tillage systems; ⑥ climate resilience and low–carbon green farming constructions, especially the development of climate smart agriculture.

In general, seven major progress/landmark achievements were achieved in the crop cultivation and farming system discipline during the recent years. ① Chinese Academy of Agricultural Science acquired landmark achievement on the physiology and molecular mechanism of crop high yield with the results published on Science in 2022; ② water saving, fertilizer saving, high yield and simplified cultivation technology for the wheat–maize double cropping in the North China Plain; ③ integration and application of key technologies of multi–cropping system in China; ④ optimization of crop cultivation techniques adapting to climate change; ⑤ new zoning of farming system based on the big data platform; ⑥ high yield and high quality cultivation techniques of regenerative rice;

⑦ theory and technology of corn–soybean strip compound planting. These achievements contributed greatly to ensure China's food security and agricultural green development with higher environmental and climatic pressures.

Compared to leading crop–producing countries, China's crop cultivation discipline lags in large–scale efficient production, specialized quality, integrated management, and diverse market supply based on high, stable yields. The strategic focus for the coming years should include mechanization and precision in all crop cultivation processes, new farming systems for resource–environmental protection and green development, and low–carbon green crop production management models. Key areas and priorities are precise and intelligent crop production technology, composite planting modes and technologies for grain and legume productivity improvement, mechanisms and technologies for enhancing crop yield and quality, precise zoning for farming systems, climate–smart cultivation technologies, and sustainable techniques to boost farmland ecological function. Strategic measures to advance this discipline include adapting to new production demands with modern technology, strengthening interdisciplinary integration, expanding research areas, and enhancing training to align with international standards.

15.2 Advances in Plant Protection

Crop productions are often threatened by crop insect pests and diseases worsening the insecurity of food. Globalization has rapidly increased the introduction and threats of invasive pests. Climate change results in a changed suitability of landscapes to pests, further increasing the threat and uncertainty of their impact.

China faces numerous complex crop pests, posing significant threats to food security and safety. Over the years, plant protection research has evolved from basic methods to integrated, eco–friendly strategies for effective pest control, leveraging improved understanding of pest behaviors, crop damage mechanisms, and advanced disease control technologies. As agriculture evolves during the 14th Five–Year Plan period, plant protection is adapting to new technologies and changing scenarios. Future research will focus on: ① study on the new regularity and countermeasures of crop insect pests and diseases adapting to the new situation of agricultural production; ② study on new theory and method of plant protection adapting to new development of modern science and technology; ③ research on detection, monitoring and early warning technology of insect pests and diseases to meet the requirements of large–scale and long–term effect; ④ research and development of new technologies and new products for prevention and control of insect pests and diseases to meet the safety requirements of agricultural products; ⑤ research and development of intelligent new equipment and application technology for plant protection to meet the requirements of automation and intelligentization.

Future plant protection aims to ensure national food security, revitalize rural areas, and modernize agriculture. Advancements in science and technology will help prevent new pest outbreaks, control agricultural losses from major pests and diseases, and implement eco-friendly pest management strategies. Ultimately, developing a comprehensive national plant protection system will support food security, environmental safety, and contribute to global sustainable development goals.

15.3 Advances in Agricultural Information Science

Agricultural Information Science is an emerging discipline formed by the intersection and integration of agricultural science and information science. It is based on agricultural science theory, uses information technology as a means, and takes agricultural related activity information as the object to study the acquisition, processing, analysis, and application of agricultural information. Agricultural information technology has provided a strong driving force for the construction of modern agriculture and new impetus for the progress of agricultural technology. With the continuous progress of agricultural information technology, information, like human capital, land resources, agricultural inputs, etc., is playing a decisive role in agricultural efficiency and industrial competitiveness. According to different dimensions, agricultural information technology research has different classification methods. According to the agricultural information workflow, it can include aspects such as agricultural information acquisition, agricultural information analysis, agricultural information management, and agricultural information services.

The agricultural field is increasingly adopting technologies like the Internet of Things, big data, intelligent equipment, and advanced artificial intelligence, marking a widespread application phase in agricultural information technology. In information acquisition, intelligent search engines and integrated remote sensing technologies have gathered extensive agricultural data, alongside breakthroughs in specialized sensors and machine vision, enhancing digital transformation. Progress in information analysis includes advancements in agricultural production models for understanding growth mechanisms, individual identification, and decision-making, as well as improvements in analyzing and predicting product demand, consumption patterns, and behaviors. Agricultural monitoring and warning systems have also advanced in theory, technology, and application. AI developments in machine learning, intelligent systems, and agricultural robots are notable. In information application, key technologies in agricultural robots have led to unmanned farms and widespread facility factory production, like plant factories. Significant strides have been made in the intelligent transformation of animal husbandry and in enhancing breeding efficiency through digital crop breeding technologies.

The development of agricultural informatization based on the new generation of information technology has become a new driving force for promoting high-quality agricultural development. In the coming years, the key areas and priority development directions of agricultural information science include new high-performance agricultural sensors, basic research and core technologies of agricultural big data, animal and plant growth models and algorithms, and agricultural information services based on artificial intelligence. We need to focus on the major needs of improving agricultural quality, efficiency, and competitiveness, concentrate our efforts on tackling major scientific issues and key technical challenges such as agricultural information acquisition, processing, analysis, and application services, and comprehensively promote the "replacement of human resources by machines", "replacement of human brains by computers", and "replacement of imported technologies by independent technologies" in agriculture.

15.4 Advances in Agricultural Resources and Environment

Agricultural resource and environment studies focus on the interaction between agricultural production and factors such as soil, water, climate, and organism. It aims to maximize the efficiency of these resources and minimize their environmental impact. Key areas include arable land, water, climate, biological resources, agricultural waste management, and non-point source pollution control. The integration of informatics and biotechnology with this field is set to enhance the transformation towards green, low-carbon, and high-quality agricultural development.

This report summarized the development history, development status, significant achievements, and dynamic progress of agricultural resource and environment of China during the period of 2022-2023. This report also concentrated on the representative achievements, such as "soil fertilization and improvement of cultivated land, and rehabilitation of degraded farmland", "water-saving and quality regulating irrigation of crops and drought resilient and rain-adaptive dryland agriculture", "response mechanisms and adaptation strategies of agriculture to climate change", "comprehensive prevention and control of agricultural non-point source pollution throughout the entire process", "resource utilization and high-value utilization of agricultural waste", and "collaborative regulation of biodiversity and multi trophic level biological interactions by nutrient resources". This report also compared with similar foreign disciplines from six aspects (including quality improvement of arable land, efficient water-saving and circular utilization in agriculture, adaptation and resilience of agricultural environment to climate change, prevention and control of agricultural non-point source pollution, and high-value utilization of agricultural waste and protection of biodiversity), and clarified the overall research level, technical advantages and gaps of the disciplines in the world. Aiming at the development requirements of the agricultural resource and environment discipline in the future, this report confirmed the key areas and priority development directions for six aspects of this discipline. This report also proposed the strategic thinking and countermeasures for discipline development in the next few years: to strengthen the top-level design and systematic strategic layout of the cross integration of agricultural resources and environment; to integrate the advantages of agricultural resources and environment, and build a new technological innovation system that integrates resources and environment; and to strengthen the capacity building of agricultural resources and environment discipline and build national strategic science and technology forces.

15.5 Advances in Agricultural Biotechnology

Agricultural biotechnology, encompassing genetic, fermentation, cell, and enzyme engineering, aims to enhance agricultural organism production, breed new varieties, and produce biopesticides, fertilizers, and vaccines. With advancements in genomics, systems biology, synthetic biology, and computational biology, it's reshaping the global agricultural biotech industry and driving socio-economic development. China's agricultural biotechnology has seen rapid progress, marked by significant achievements in basic research, including comprehensive germplasm collection, systematic trait genetics analysis, and in-depth agricultural process evolution studies. Furthermore, the development and advancement of new breeding technologies like phenomics and intelligent design, along with cutting-edge biotechnologies such as transformation, genome editing, and synthetic biology, are revolutionizing the biological seed industry. The variety and maturity of biotech products are increasing, yet China still needs to deepen its basic agricultural research to reach international standards. Facing challenges in innovation and industrialization, China's agricultural biotechnology must prioritize strategic areas, focusing on cloning high-value genes and decoding complex traits. Developing genome editing tools and an intelligent breeding system using multidimensional data is key. Innovating seeds and fast-tracking transgenic and genome editing products are vital. This requires a strengthened strategic approach, technological innovation, and international cooperation to enhance China's global bio-agricultural impact and IP internationalization. Building national strategic technology forces and fostering cross-departmental collaboration is essential for the field's advancement.

15.6 Advances in Storage, Transportation and Processing of Agricultural Products

Storage, transportation and processing of agricultural products is a discipline that studies the physical, chemical and biological characteristics and changes of edible agricultural products such as animals, plants and microorganisms during the processes of storage, transportation and processing, as well as the scientific and technological issues related to the quality such as nutrition, safety and flavor of the processed products. It mainly includes key areas such as storage and processing technologies and equipment for grains and oils, fruits and vegetables, livestock products and aquatic products. The future development trends of this discipline are initial processing localization of agricultural products, tiered deep processing, intelligent moderate processing and standardized quality system of the whole chain.

Future trends in the storage, transportation, and processing of agricultural products include: ① fostering technological innovation through interdisciplinary integration, combining high-tech fields like biotechnology and nanotechnology with agricultural processing to create new healthy and functional industries, and utilizing big data and blockchain for intelligent processing and delivery; ② developing new theories and methods to overcome natural limitations in agricultural production, diversifying the food supply, and utilizing natural energy cycles; ③ strengthening cold chain logistics and storage, accelerating systems for fresh products, and developing green storage and pest control technologies for reduced losses and sustainable processing; ④ focusing on nutrition and function-oriented processing, meeting higher demands for nutritional health, exploring agricultural product components, and creating personalized health products aligned with “Healthy China 2030”.

16 Plant Protection

16.1 Introduction

Crop pests, diseases, weeds and rodents are important biological disasters in China, and they are important factors restricting food security, agricultural product quality and safety, and sustainable agricultural development. The subject of plant protection is to control crop biological disasters, protect agricultural ecosystems, control environmental pollution and foreign biological invasion, curb the continuous loss of biodiversity, and provide important scientific support and technical support for protecting national agricultural production safety, ensuring the quality and safety of agricultural products, reducing environmental pollution, safeguarding people's health, and promoting sustainable agricultural development. In recent years, the discipline of plant protection has made two-way efforts in the fields of basic science and applied science. Adhering to the four orientations, the discipline has continuously innovated and enriched the microanalytical methods and theories of major agricultural biological disasters, innovated and developed green prevention and control technologies and products of plant protection, and continuously developed and improved the prevention and control technology system of major agricultural biological disasters, making a practical contribution to promoting high-quality development of agriculture and rural areas and rural revitalization.

16.2 The Latest Research Progress of the Discipline in Recent Years

16.2.1 Proposing a number of new theories and new methods to lead the innovation and development of the discipline

Plant pathology: Pathogenicity of pathogens. It was found that the effector protein MoCDIP4 of

Magnapor the grisea targeted the mitochondrial division-related protein complex in rice, and regulating the mitochondrial division and immune response of rice. It was found that the effector protein Pst_A23 of stripe rust directly binds to the specific RNA motif of the alternative splicing site to regulate the alternative splicing of host disease resistance-related genes and inhibit the host immune response. The mechanism of action of multiple soybean Phytophthora effectors was identified and analyzed and why effectors could promote pathogen infection had been revealed.

Plant disease resistance　In terms of pathogen-associated molecular pattern-triggered immune response (PTI) pathway, E3 ligase PUB25/26 was found to mediate the degradation of non-activated BIK1, and CPK28 negatively regulated plant immune response by enhancing the above process. It was found that the NLR receptor protein of broad-spectrum resistance in rice protects the immune metabolic pathways from pathogen attack, synergistically integrates PTI and ETI, and thus endows a new mechanism of broad-spectrum resistance in rice. The main effect gene *Fhb7* of resistance to stripe rust from long-eared earwig was successfully cloned, and its molecular and genetic mechanism of resistance was revealed. Multiple disease-sensitive genes in wheat were identified, and the resistance of wheat to stripe rust was significantly improved by gene editing technology, realizing broad-spectrum resistance to stripe rust in wheat.

Agricultural entomology: Regulatory mechanisms of metamorphosis and diapause of insects. The promoting effect of epidermal growth factor receptor (EGFR) on JH synthesis and the mechanism of JH on insect egg formation and production were found. The mechanism of 20E signaling pathway mediated by dopamine receptor in regulating metamorphosis of insects was analyzed. Insect migration patterns and laws. The declining trend of the abundance of migratory phytophagous insects and natural enemies was found. Two key environmental factors were screened for the migration of *Vanessa cardui*, and a control model for the migration of *Nilaparvata lugens* was proposed.

Weed Science: Weed biology and ecology. The adaptation mechanism of weed rice and cultivated rice to low temperature was clarified. The tillering regulation mechanism of Jiujiu wheat under high distribution density stress was revealed, which provided a theoretical basis for effective control. The mechanism of herbicide safety. Herbicide could accelerate the degradation of appammonium in rice, reduce the lipid peroxidation and oxidative damage caused by appammonium in rice plants, revealing the mechanism of alleviating herbicide damage by selectively inducing the up-regulation of rice GSTs gene expression.

Rodent damage: Develop the ecological threshold research of rodent damage control, and objectively evaluate the function of rodents in different ecosystems. Important progress has been made in the invasion and outbreak mechanism of major rodent damage *R. norvegicus* in agricultural areas, the population outbreak mechanism of *R. orientalis* in Dongting Lake area, and the succession law of rodent community in this area.

Green pesticide innovation and application: Molecular targets and mechanism of action of crop pests and diseases. Original molecular targets and green pesticide varieties have been found. Chitin synthase PsChs1 in soybean Phytophthora was analyzed by using freeze–electron microscopy, laid a foundation for the precise design of new green pesticides with chitin synthase. Studies on pesticide resistance and control. It was found that the resistance mechanism of cotton aphid to fludioxonil and the potential adaptive costs such as feeding behavior and life history were reported.

Biological control science: It was found that miRNA of parasitic wasps could cross–regulate the expression of host ecdysone receptor to inhibit the growth and development of the host, obtain venom protein genes from bacteria through horizontal transfer, and regulate the host's immune response.

Invasion Biology: Disaster–inducing mechanisms of major agricultural invasive organisms. The omics database of alien invasive species, such as coding moth, tomato leaf miner, was constructed, revealing the invasion mechanism during the global invasion process.

Monitoring and early warning: Development of radar monitoring technology. The rotating polarization vertical insect radar, the distance measurement accuracy of the insect radar was improved a lot, and the radar blind area was largely reduced. The Ku band phase–coherent high–resolution full polarization insect radar with both scanning mode and beam vertical observation mode was developed.

16.2.2 Key technologies and products to promote industrial upgrading

Key control technologies: A high–resolution multi–dimensional cooperative radar measuring instrument consisting of a high–resolution phased array radar and three multi–frequency full polarization radars has been developed. The intelligent trap based on machine vision uses adhesive insect plate plus machine vision system to collect insect images, plus insect image automatic identification and counting method, making up for the photoelectric intelligent trap, and achieving the desired effect. The field application technology of a batch of new herbicides such as triazole sulfonazone, bicyclol sulfonazone on rice, wheat, corn and other crops has been established.

Key control products: The first new wheat variety Shannong 48 carrying the gene *Fhb7* was bred. The green prevention and control technology system was constructed, which has been applied to 116 million mu. The first genetically engineered microbial pesticide in China, *Bacillus thuringiensis* G033A, was developed. The first registered and commercialized nematode *Bacillus thuringiensis* preparation HAN055 was developed. A new type of mesionic insecticide, such as isozolidacil, epoxidine, pentapyridine, epoxoline, and so on, has been developed.

16.3 Development Trend and Prospect of this Discipline

Global climate change and the continuous adjustment of crop structure had promoted the development of plant protection and took new opportunities. In the next five years and longer, we will consolidate the theoretical innovation system of major plant protection, prioritize the development of forward-looking green plant protection technology, improve the promotion system of green plant protection technology, make up for the shortcomings of platform construction, further improve the capacity and level of major epidemic prevention and control, and make greater contributions to China's food security, biosecurity and ecological security.

16.3.1 Consolidating the theoretical innovation system of major plant protection

In the future, the research on the basic research will highlight the original scientific and technological innovation work from 0 to 1, and should promote the establishment of the theoretical innovation system of plant protection in China from many aspects. The first is to strengthen the development of interdisciplinary research. With the opening of the post–genomic era, the combined analysis of multi–omics big data provides the possibility for in–depth research. Second, more attentions will be paid to the expansion of emerging fields. The continuous emergence of new found will open up the study of the occurrence law and disaster mechanism of diseases and pests, but also promote the innovation of new plant protection technology. The third is to strengthen the orientation of industrial demand. Basic research of plant protection needs to be closely combined with the actual agricultural production, and promote more scientific and more accurate pest control technology innovation by analyzing the new laws and mechanisms of major pest occurrence.

16.3.2 Give priority to the development of forward-looking green plant protection technologies

We should make full use of monitoring and early warning, plant resistance to diseases and insects, biodiversity, biological pesticides, etc., to develop the most economical, effective and environmentally friendly damage control technology. Focusing on the research of forward–looking green plant protection technology, the basic and applied research of new green prevention and control technologies such as plant disease resistance utilization and induction technology, crop pest ecological prevention and control technology and biological control technology, RNAi and genetic precision control technology will be given priority. Support the use of artificial intelligence to develop modern pest control technologies, and integrate the construction of green plant pest prevention and control technologies to overcome obstacles to continuous cropping. Starting from the multi–species ecological food web, we should increase the investment in macro–control and macro–network research to break through the bottleneck of macro–ecology.

16.3.3 Improve the promotion system of green plant protection technology

Establish a public plant protection service system, and form a joint prevention and control mechanism supported by a monitoring and early warning system and a professional prevention and control team. Government, research organisms, enterprises and farmers should strengthen communication and establish feedback system.strengthen the publicity and guidance, responsibility fulfillment and social supervision of green plant protection. Continue to improve the new agricultural pest prevention and control theories with Chinese characteristics, including "public plant protection, green plant protection, scientific plant protection, and smart plant protection", make sure we could fully support the biosecurity, food security and ecological security.

16.3.4 Strengthen the construction of platform for plant protection resources and other conditions

Resource advantages will be the key to the identification of core pathogenic factors of pathogens and important crop disease resistance genes in the future. We did not have national natural enemies and biological control microorganisms resource banks so far. It is urgent to rely on these resource advantages to promote the development of research level. Crop pest monitoring and early warning has become an important means for the prevention and control of migratory pests, however, the current monitoring system is scattered and lacks of a nationwide monitoring and early warning platform. Existing resources need to be integrated to carry out long–term monitoring and accurate forecasting.

17 Crop Science

17.1 Introduction

This section offers a comprehensive analysis of recent insights, new viewpoints, emerging technologies, novel theories, achievements, and the forefront of developments in the field of crop science. It is based on the strategic needs of China's modernized agriculture development, national food security, food safety, ecological security, and increase in farmers' income. It delineates the research directions and key tasks for the future development of crop science. In this report, we aim to provide a window for various sectors of society to gain an accurate understanding of the developmental trends in the field of crop science; to serve as a scientific foundation for optimizing the structure of China's unique subject and professional system, rationally allocating innovative resources, and achieving independent control over the agricultural industry chain.

17.2 Recent Advances in Crop Science in China

Significant progress has been made in the collection, preservation, and innovative utilization of crop germplasm resources. The total number of crop germplasm resources has reached more than 540000, making China the second-largest holder of such resources in the world. Over 1000 new germplasm resources, including wild rice with warty grains and perennial wild soybeans, have been salvaged. The National Crop Gene Bank, a newly-constructed facility, started its trial operation in September 2021 and has the capacity to store 1.5 million samples of crop seeds and resources for long-term preservation.

Breakthroughs have been achieved in the genetic analysis of important crop traits. Convergence selection gene *KRN2/OsKRN2*, which simultaneously enhances maize and rice yields, and

OsDREB1C, the high-yield gene in rice, have been identified. The major effective resistance gene *Fhb7* against *Fusarium* head blight in wheat has been cloned. The first principal gene, *THP9*, controlling high protein content in maize, has been cloned from the wild maize. A transcription factor, *MYB61*, controlling cellulose and nitrogen utilization efficiency in rice, has been cloned. The *indica-japonica* hybrid sterility gene *RHS12* has been identified in rice. A new mechanism underlying broad-spectrum disease resistance in rice, involving NLR receptor proteins, has been uncovered.

Innovations in key breeding technologies for crops have yielded significant impacts. Simultaneously created mutations in four genes in rice has basically solved the technical bottleneck of unifying the production and breeding of "one-line method". Engineered double haploid technology for maize has been put into practical use. New wheat lines developed through chromosome engineering and mutagenesis have increased yield by more than 25% over control varieties, such as Zhoumai 18. Genetically modified insect-resistant maize, herbicide-tolerant maize, and herbicide-tolerant soybeans have been granted safety certificates for production and application. The foundational gene-editing tools Cas12i and Cas12j have obtained patents in Mainland China, the Hong Kong region of China, and Japan, marking China's entry into the forefront of the international gene editing technology.

Significant progress in creation of novel germplasm and breeding for new varieties. Remarkable advancements have been achieved in the creation of novel germplasm and the development of new crop varieties. A multitude of superior wheat varieties, each of them incorporating more than 10 disease resistance genes such as *Yr30*, *Lr27*, and *Sr2*, have been developed. Several maize varieties, such as Yufeng 303, Denghai 605, Dongdan 1331, MC121, Zhongyu 303, and Chuandan 99, have been chosen as the national leading varieties. The highly rust-resistant inbred line, Jing 2416K, has been used to develop more than 40 widely grown rust-resistant hybrid varieties (including Jingnongke 767). A new soybean variety, Heinong 84, exhibits resistance to the three major diseases, gray leaf spot, viral diseases, and cyst nematodes. A new high-yield rapeseed variety, Zhongyouza 501, is suitable for dense planting. Cotton variety Zhongmian 113 has expanded the north boundary of cotton cultivation in northern Xinjiang with superior quality over "Australian cotton".

Substantial breakthroughs have been made in unlocking the high-yield potential of crops and harmonizing the theory and technology of high-quality, high-yield coordinated cultivation. Crops like rice, wheat, and maize have continually achieved high-yield records. Wide refined sowing planting has become a leading technique in northern wheat producing regions. Compared to conventional row planting, wide-row planting has maintained stable quality of strong-gluten wheat while increasing yield and nitrogen utilization efficiency. The primary approaches to unlock the yield potential of maize have been elucidated, with innovations in practices such as dense planting, integrated water and fertilizer precision management, and mechanical grain harvesting.

Advancements in efficient utilization of crop resources and the development of "double reduction" green cultivation techniques. We have developed rice light and simplified nitrogen fertilizer management technology, wheat nutrient optimization management plan, rice field water-saving irrigation measures, wheat full film micro-ridge sowing technology, maize mulching drip irrigation technology, etc.

Crop production has seen a trend towards full mechanization, informatization, and smart cultivation technologies. The "high-yield full mechanization technology for densely planted maize" was included in China's top ten agricultural technologies for the years 2020–2022. Drone-based digital imaging and monitoring platforms have provided a technical support for dynamic assessments of crop leaf area index. The "unmanned green and high-yield cultivation technology for rice and wheat" was selected as a leading technology by the Ministry of Agriculture and Rural Affairs of the People's Republic of China in 2021.

17.3 Comparison of Domestic and International Research Progress in Crop Science

Research on crop germplasm resources: the National Crop Germplasm Preservation Center ranks the second in the world in terms of the number of preserved germplasm resources. The accuracy and scale of germplasm resource assessments are approaching international standards. Research on genome analysis, crop domestication and improvement, and germplasm formation rules of crops such as rice, potato, cotton, and edible beans is at the leading level. The following gaps from the international level are defined: the quantity and quality of our germplasm resources need to be improved simultaneously, the excavation and breeding application of excellent allelic variation lag behind, the innovative utilization platform of germplasm resources is not perfect.

Fundamental research in crop breeding: China is transitioning from "catching up" to "leading" in crop breeding research, particularly in fields such as rice and wheat genome researches, which is leading at the international forefront. However, China's foundational and theoretical researches in biological breeding is not in depth compared to developed countries. For crops like maize and soybeans, basic research is still catching up. The genetic basis and regulatory networks responsible for the formation of important traits have not been systematically studied.

Crop breeding methods and technological innovations: Compared to the developed countries, China's utilization of heterosis in rice, wheat, soybean and rapeseed are internationally leading, and breakthroughs have been made in molecular breeding, cell engineering, ploidy breeding, and other breeding technologies. However, the theoretical and technological innovation ability of crop genetics and breeding and the support ability for the seed industry still need to be improved. The original innovation ability of breeding technology is still weak. Some key technologies are subject to others. Breeding for new crop varieties and cultivars: the unit yield of maize and soybeans falls short of 60% than that of in the United States; the import of soybean is still on large scale; major varieties research and development lag behind; and there is a significant need to upgrade crop varieties. Intellectual property protection has led the seed enterprises in developed countries to become the main investors and innovators, whereas China has yet to establish a commercialized breeding system centered around enterprises.

Foundational research in crop cultivation and farming: Developed countries place greater emphasis on high-quality, green, low-carbon, ecological, and safe in crop cultivation and farming, and have strengthened the research and application of crop stress resistance cultivation mechanisms, efficient resource utilization mechanisms throughout the production process, and carbon footprint quantification assessment methods to address the frequent occurrence of non-biological or biological disasters such as high and low temperatures, droughts and floods, and diseases, pests, and weeds caused by climate change. China urgently needs to strengthen research in such areas, and form Chinese characteristics based on the research of cultivation and farming with high yield, quality, and efficiency of crops, accelerating the transformation from "following" to "leading".

Key cultivation and farming technologies: Compared with the developed countries, China has significant gaps in protective cultivation, efficient resource utilization, and large-scale production due to regional diversity and complexity of planting systems, which seriously restricts the improvement of comprehensive production capacity, efficiency of land.

Innovations in crop cultivation techniques: Developed countries such as European countries and the United States have achieved mechanized, intelligent, and standardized production, while Japan has a complete range of small-scale intelligent agricultural machinery equipment and intelligent control operations for crop production. China's research and development of high-throughput crop phenotyping and equipment development are relatively lagging behind, and there is a lack of accurate crop phenotype and growth regulation models.

17.4 Trends and Prospects in Crop Science

Original innovation in basic theory of biotech-breeding: Initiatives such as the Agricultural Biotechnology Innovation Action are being launched to focus on fundamental research in agricultural biotechnology and cutting-edge technology innovations, supporting the development of breakthrough varieties of major crops. This will involve the continuous use of multi–omics collaborative technologies to explore major–effect genes for important traits, as well as the genetic regulatory mechanisms underlying these traits.

Continual innovation in elite crop germplasm resources collection and utilization: Through implementation of initiatives for creating and applying excellent germplasm resources, a comprehensive innovation technology system is being developed. This system aims to create groundbreaking new germplasm resources at scale, promote significant research collaboration for superior varieties, and breed new germplasm with prominent target traits and excellent overall characteristics.

Breakthroughs in major key crop breeding technologies and management system: To advance seed industry innovation, essential and core technologies for seed sources are being pursued. Major biotech–breeding projects are being implemented, with a focus on promoting industrial applications. This involves combining resources, data, information, talent, and technology to facilitate breakthroughs in a series of new and major crop varieties.

Research on cultivation techniques for crop stress–resistance and disaster reduction: Comprehensive research on the mechanisms and regulatory networks of crop responses and adaptations to abiotic stresses is essential. Research will focus on uncovering the key cultivation and farming techniques that enhance crop stress resistance and disaster reduction. This includes understanding the collaborative relationship between plant stress responses and growth and development.

Establish new models for smart, high–yield, green and ecological crop development: The development of precision and efficient fertilization and pest control methods, as well as the reduction and control of pollution, is a priority to drive China's transition from resource–intensive to green and efficient crop production. Researchers are working on new theories and technologies for efficient crop resource utilization, reduced fertilizer and pesticide use, water conservation, and carbon sequestration.

Innovations in multi–cropping models and efficient complementary cultivation techniques: Researchers are combining multi–cropping systems with modern agricultural technologies to establish a wide range of high–yield and efficient planting models for grains, cash crops, and forage. This includes combining agriculture with animal husbandry, and agriculture with forestry, and introducing efficient integrated cultivation techniques.

18 Malignant Tumors

18.1 Introduction

Oncology is a discipline that studies the occurrence, development, prevention, diagnosis and treatment of tumors. In the past century, especially since the 1940s, with the increase in the incidence and mortality of malignant tumors, people have increased their awareness and attention of the harmfulness of tumors, increased efforts in tumor research. Both theoretical and clinical research have developed rapidly. It has not only become an independent discipline, but also formed many sub–disciplines. The scope of research involves multiple disciplines including basic medicine, clinical medicine, preventive medicine, physiology, biochemistry, psychology, sociology, economics and other areas related to tumors. In the past five years, the Chinese Anti–Cancer Association has actively implemented the Healthy China strategy, adhering to the core concept of "cancer prevention and treatment, winning through integration", and completed the task of "building a big army, holding conferences, writing big books, establishing big rules, creating big journals, and opening big journals". Guided by the two development goals of "expanding the team" and "improving academics" proposed by Academician Fan Daiming, in order to promote the construction of local guidelines in our country, the Chinese Anti–Cancer Association organized more than 13,000 experts to compile my country's first integrated tumor diagnosis and treatment guidelines, which has effectively promoted the progress of standardized treatment for malignant tumors. This section systematically summarizes the period from January 2018 to June 2023 in our country. An overview of the development of important disciplines in the discipline of malignant tumors in terms of technical methods, academic theories, research progress, talent training, academic structure, research teams, etc. At the same time, the research progress at domestic and abroad is also compared, and the future development trends and prospects of the subject are proposed.

18.2 The Latest Research Progress in Recent Years

18.2.1 Lung cancer

In recent years, with the development of targeted and immunotherapy and innovations in genetic testing and radiotherapy technology, the 5–year survival rate of lung cancer has improved, reaching 17%~32%. Progress in targeted therapy for lung cancer is mainly focused on the expansion of indications, treatment options after resistance to common mutation targets, and the development of new drugs for rare mutation targets. Progress in immunotherapy is mainly focused on the expansion of indications and multiple treatments that are organically combined with radiotherapy, etc.

18.2.2 Breast cancer

Breast cancer is the most common malignant tumor among Chinese women. Progress in the treatment of breast cancer is mainly focused on domestically produced targeted drug treatments such as DAWNA–1 research and PEONY research. Clinical trials have been widely carried out, and international guidelines are constantly being rewritten. In terms of basic research, the Fudan classification of triple–negative breast cancer provides a new research direction for the precise treatment of triple–negative breast cancer.

18.2.3 Gastric cancer

In recent years, basic research on the pathogenesis of gastric cancer is no longer limited to mutations or expression changes of a certain gene, but has shifted to the multi–gene/multi–locus or even the whole–genome level. Minimally invasive surgical treatment of gastric cancer has been explored for more than 30 years. Especially in the past decade, the Chinese Laparoscopic Gastrointestinal Surgery Study (CLASS) group has launched a series of high–level clinical studies on laparoscopic gastric cancer surgery, leading the paradigm innovation in this field. The discovery of combined treatment models and new therapeutic targets has further improved the efficacy of targeted therapy for gastric cancer. ADC drugs have become the standard treatment option for late–line treatment of HER2–positive advanced gastric cancer. Immune checkpoint inhibitors (ICIs) have shown better efficacy than traditional treatments in both first–line and late–line treatment of advanced gastric cancer.

18.2.4 Colorectal cancer

At present, the proportion of minimally invasive surgery in domestic hospitals at all levels is constantly increasing, and it has become the mainstream of colorectal cancer surgery. Optimizing the neoadjuvant chemo-radiotherapy model for rectal cancer, strengthening concurrent chemo-radiotherapy regimens, full neoadjuvant treatment, and short-course radiotherapy combined with chemotherapy can further improve the efficacy and give more patients the opportunity to preserve organs. Targeted therapy remains an important treatment for metastatic colorectal cancer. As the main signaling pathway in the occurrence and development of colorectal cancer, research on molecular targeted therapy of metastatic colorectal cancer targeting the RAS–RAF–MEK pathway continues to be in-depth.

18.2.5 Esophageal cancer

In recent years, the role of microorganisms in the occurrence and development of cancer has become a research hotspot. ERAS has obvious advantages in reducing postoperative complications, accelerating patient recovery, shortening hospitalization time, reducing medical expenses, and increasing patient satisfaction. Total thoracolaparoscopic radical esophagectomy is currently the most minimally invasive radical esophagectomy. A form of philosophy. Targeted therapy targeting EC genomic drivers has become a research hotspot. As a new treatment option in addition to surgery, radiotherapy and chemotherapy, immunotherapy has rewritten the treatment landscape of esophageal cancer. The immune combination therapy model is not only the main application model in current clinical practice, but also a direction that needs to be explored in depth in the future.

18.2.6 Primary liver cancer

In recent years, artificial intelligence (AI) –assisted pathological diagnosis has developed rapidly in liver cancer, including automatically extracting lesion areas, determining disease types, analyzing diseases more accurately, and processing some details and characteristic textures that are difficult to distinguish with the naked eye. At present, surgical treatment of liver cancer is still an important means for long-term survival of liver cancer patients, mainly including liver resection and liver transplantation. Molecular targeted drugs are the main means of treating advanced liver cancer. In recent years, research on molecular targeted drugs for liver cancer has made many new progress. With further exploration in clinical trials, immunotherapy has also become an important treatment method for neoadjuvant therapy and conversion therapy. In China, multiple immune checkpoint inhibitor single drugs have been approved for second-line treatment of liver cancer.

18.2.7 Biliary tract carcinomas

Currently, how to perform appropriate regional lymph node dissection and ensure negative corresponding margins for individual patients is an important part of precision biliary surgery. At the same time, with the development of new medical devices such as 3D laparoscopy and robots, BTC's minimally invasive treatment continues to achieve surgical innovation in this field. In the past decade, many studies have been carried out around the ABC–02 phase III trial, but the GC regimen is still the standard first–line treatment for advanced patients. As the understanding of BTC gene mutation spectrum, different subtype characteristics, immune escape, etc.gradually deepens, it is worth looking forward to the combination of chemotherapy with other treatments such as targeted immunity to improve the efficacy.

18.2.8 Urinary tract malignant tumors

In terms of prostate cancer, with the development of surgical technology and the popularization of robotic laparoscopy technology, robot–assisted laparoscopic radical prostatectomy can shorten the operation time and reduce intraoperative blood loss. RC48–ADC showed good efficacy and safety in HER2+ locally advanced/mUC patients. ICIs have been used as second–line treatment for patients with unresectable and metastatic MIBC, and as first–line treatment for patients who cannot tolerate platinum and are PD–L1 positive. In terms of renal cancer, 3D reconstruction combined with robot–assisted partial nephrectomy has been a hot topic in the past few years. Targeted therapy remains the first–line treatment strategy for advanced renal cancer.

18.2.9 Cervical cancer

Since the LACC trial, many studies have confirmed that open radical hysterectomy is safer than minimally invasive surgery in the treatment of early cervical cancer. Radiotherapy for cervical cancer includes external beam radiation and brachytherapy. The three–dimensional inter–tissue implant brachytherapy that emerged in recent years improved dose coverage of high–risk clinical target areas. Immunotherapy represents a new treatment option for cervical cancer with survival benefit in patients with recurrence.

18.2.10 Ovarian cancer

In order to delay the recurrence of epithelial ovarian cancer, a number of drugs are currently approved for first–line maintenance treatment of epithelial ovarian cancer. Currently, various clinical evidence and domestic and foreign guidelines show that the drugs recommended for first–line maintenance treatment of epithelial ovarian cancer mainly include PARP inhibitors and anti–angiogenic drugs. ICIs have been clinically studied in both first–line and late–line treatment of advanced ovarian cancer. Unfortunately, no breakthrough efficacy has been achieved in first–line treatment. Patients with platinum–resistant recurrent ovarian cancer and high expression of folate receptor α can consider using the FRα–targeting antibody conjugate drug Mirvetuximab Soravtansine.

18.2.11 Bone and soft tissue sarcoma

Uncovering the characteristic landscape of the immune microenvironment of osteosarcoma is crucial to improving the effectiveness of immunotherapy of osteosarcoma, and basic research continues to make progress. The development of existing medical 3D printing technology provides ideas for precise resection and personalized reconstruction of pelvic tumors. For advanced osteosarcoma, a consensus has been reached on the treatment of anti–angiogenic targeted drugs. The GALLANT study suggests metronomic chemotherapy combined with ICIs for the second/third–line treatment of advanced sarcoma. Cell therapy is expected to open a new journey in the treatment of advanced sarcoma.

18.2.12 Hematological malignancies

After decades of stagnation, several promising therapies for newly diagnosed and relapsed DLBCL have received FDA approval or are in the final stages of development, including enhanced monoclonal antibodies, ADCs, and bispecifics antibodies, etc. Currently, CAR–T cell therapy has received widespread attention as a second–line treatment for R/R DLBCL. ICIs have made significant progress in the field of lymphoma in recent years. Hodgkin lymphoma has mainly progressed in patients with advanced or refractory disease in recent years. In recent years, targeted therapy, immunotherapy and cell therapy have provided new treatment strategies for acute leukemia.

18.2.13 Head and neck malignancies

At present, there have been some studies on cetuximab combined with immunotherapy to achieve good efficacy in recurrent/metastatic head and neck squamous cell carcinoma, and have been included in guideline recommendations. Initial results of immunotherapy in perioperative treatment of HNSCC. As an emerging treatment model, neoadjuvant targeted therapy is expected to increase the R0/1 resection rate of locally advanced thyroid cancer and improve patient prognosis. In recent years, with the advancement of molecular detection technology and clinical breakthroughs in targeted drugs and ICIs, the management of anaplastic thyroid cancer has entered the era of precision diagnosis and treatment.

18.3 Comparison of Research Progress at Domestic and Abroad

Through the development of medicine in recent years, the treatment level of malignant tumors has been significantly improved. The improvement of the overall treatment level of malignant tumors is due to our deepening understanding of their biological behaviors and the progress of comprehensive treatments including surgery, chemotherapy, radiotherapy, endocrine therapy, targeted therapy, and immunotherapy. In terms of basic research, although there is still a certain gap with foreign countries, Chinese scholars have made significant progress in the pathogenesis and molecular classification of malignant tumors in recent years. Taking breast cancer as an example, Fudan University Cancer Hospital independently developed a multi-gene sequencing platform and drew China's first breast cancer gene mutation map for 1,000 people. It comprehensively analyzed the clinical characteristics and genomic characteristics of breast cancer in China and found that identified unique precise treatment targets for breast cancer in China. Professor Shao Zhimin's team has drawn the world's largest multidimensional omics map of triple-negative breast cancer and proposed the "Fudan Classification", which has pointed out a new direction for the precise diagnosis and treatment of triple-negative breast cancer. In terms of clinical research on new drugs, due to the long-term leading position of European and American countries in the field of new drug treatment research, related drugs have certain advantages in the quality of overseas clinical trials and the speed of new drug review. In recent years, domestic reforms in pharmaceutical administration, clinical trials, new drug approval and other aspects, especially since 2017, China has officially joined the International Council for Harmonisation of Technical Requirements for Pharmaceuticals for Human Use (ICH), providing better opportunities for Chinese researchers. With the joint efforts of a large number of Chinese researchers, and with the continuous accumulation of clinical experience and data in China, clinical trial data on Chinese patients are rapidly improving in terms of quality and influence. With the enhancement of domestic innovative drug research and development capabilities and the improvement of the international influence of local researchers, more and more "Chinese data" have appeared on the world stage and are constantly rewriting international guidelines.

18.4 Development Trends and Prospects

There are nearly 4 million new cases of malignant tumors in China each year, and 2.34 million cancer deaths. Overall, the incidence of malignant tumors is on the rise. Especially for tumors related to environmental lifestyle, the overall mortality rate is on the rise. The "Healthy China 2030" Planning Outline sets out the important goal of increasing the overall 5-year survival rate of malignant tumors by 15% by 2030. In order to achieve the goal of "Healthy China 2030", high-quality development of the discipline of malignant tumors is very necessary. The development trends and prospects of malignant tumors are mainly reflected in the following aspects: deepening the development of precision prevention; optimization of individualized and efficient screening strategies; precise exploration of adjuvant/neo-adjuvant treatment of early operable malignant tumors; molecular target exploration and genetic testing help achieve precise treatment of malignant tumors; surgical treatment of malignant tumors is developing towards precision, standardization, minimally invasive, appearance-focused, and function-preserving; immunotherapy is in the ascendant; domestic original drugs are constantly coming out and clinical research is booming; combination therapy based on different treatment strategies is expected to optimizing treatment options; artificial intelligence will have greater room for development in the subject of malignant tumors; and tumor science will continue to play an important role in the prevention and treatment of malignant tumors.

19 Obstetrics

19.1 Introduction

Obstetrics is a traditional medical discipline that is concerned with ensuring the health of pregnant women and facilitates in the birth of new life. Modern obstetrics has achieved a deep integration of clinical and preventive medicine concepts. With advancements in fundamental medical disciplines such as molecular biology, genetics, immunology, and clinical diagnostic and therapeutic technologies, obstetrics has expanded in its content and scope. With declining fertility and population growth rates, as well as increasing population ageing, planning and developing the obstetrics discipline is of great significance in ensuring the population security and promoting the sustained development of our country's economy and society. This report summarizes the major advancements in the field of obstetrics in China over the past five years (2018–2023) from three aspects: the physiological mechanisms of pregnancy, prevention and treatment of major obstetric diseases, and the evolution and innovation of obstetric medical service models.

19.2 Advances in Obstetrics

19.2.1 Research on physiological mechanisms of pregnancy

In the study of early embryonic development and embryo implantation, Chinese scholars have identified pathogenic genes that cause disorders in oocyte maturation and spermatogenesis from a genetic perspective. They have explored the molecular mechanisms of early embryonic development and embryo implantation across transcription, translation, and protein synthesis, and have mapped the spatiotemporal landscape of early embryonic development. Using multi–omics technologies, they have revealed potential causes of embryo implantation failure. Additionally, some teams have attempted to establish in vitro culture systems for non–human primate embryos, providing an crucial theoretical basis and technical support for understanding embryo implantation and development. In the field of placental development, researchers have studied the development of placental cell lineages and discovered various finely regulated mechanisms in the differentiation of trophoblast cells. They have innovatively elucidated mechanisms related to nutrient transport and metabolism, immune tolerance, and infection defense at the maternal–fetal interface. By harnessing human induced pluripotent stem cells, they have constructed placental/uterine–like organs, providing significant means to further explore key events in placental development. Furthermore, the "physiological clocks" of pregnancy, including the immune clock, endocrine clock, membrane clock, molting clock, and uterine muscle layer clock have been systematically described. The spatiotemporal landscape of physiological changes during pregnancy maintenance and labor initiation has been depicted.

19.2.2 Research on the mechanisms, prevention, diagnosis and treatment of major obstetric diseases

19.2.2.1 Prenatal screening, diagnosis, and prevention of birth defects

Chinese government and medical institutions have taken various measures to promote and standardize prenatal screening for pregnant women. Further development of the testing efficacy of cell–free fetal DNA in maternal peripheral blood has been carried out to achieve detection of common/sex chromosome aneuploidies, chromosomal microdeletion/microduplication syndromes. Various prenatal diagnostic methods such as CMA, CNV–Seq, WES, and even WGS are developed well and widely used. Breakthroughs have been made in some new prenatal diagnosis technologies, such as embryo diagnosis gradually shifting from monogenic diseases to polygenic diseases, providing a theoretical basis for the prevention and control of chronic diseases closely related to genetics.

19.2.2.2　Prevention and treatment of maternal severe and complicated illness

In the studies of postpartum hemorrhage, it was preliminarily found that abnormal immune inflammatory coagulation state during third trimester may be associated with atonic postpartum hemorrhage. With the use of randomized controlled trails and other high-evidence level studies, more progress has been made in the selection of preventive and therapeutic drugs, the development of hemostatic balloons, and hemostatic operation techniques. In terms of postpartum hemorrhage management, the National Obstetric Quality Control Center integrates domestic and international advancements, utilizes new strategies and tools to establish a comprehensive management strategy for postpartum hemorrhage that is multidimensional, well-organized, and combines prevention and treatment. This strategy is then promoted in various grassroots hospitals across the country. For placenta accreta spectrum, a placental implantation scoring system has been established in China to guide clinical prediction, typing, and perioperative management, and the blood transfusion rate and hysterectomy rate of patients with placental implantation have been significantly reduced through innovative surgical techniques such as the Nine-Step Surgery Method for invasive placenta accreta. For hypertensive disorders during pregnancy, Chinese scholars have taken the lead in analyzing the role of cell subtypes and trophoblast differentiation in the onset of preeclampsia at the single-cell level and revealed the role of the intestine-placenta axis. Preliminary development of a prediction model for early-and mid-pregnancy preeclampsia based on clinical high-risk factors and biological indicators and a perinatal management plan have been developed. In the field of gestational diabetes, a standardized diagnosis and treatment management model for pregnancy with diabetes has been established and formulated, and researchers have participated in the formulation of international GDM screening and diagnosis standards and clinical guidelines. In terms of obstetric critical illness, the multidisciplinary team model in China is becoming more mature and popular, and the application of some new diagnostic and treatment technologies such as extracorporeal membrane oxygenation (ECMO) has greatly improved the rescue success rate. The issuance of some representative guidelines or consensuses in recent years has provided a basis for the standardized diagnosis and treatment of critical maternal diseases. "Clinical management guidelines for acute fatty liver of pregnancy in China (2022)" was rated by the Guideline STAR as the best guideline in the field of obstetrics in 2022.

19.2.2.3　Prevention and treatment of fetal and neonatal illness

The preterm birth rate in China is rising annually. Domestic scholars have identified proteins and cell molecular signals related to the initiation of childbirth through multi-omics techniques, providing intervention targets for maintaining pregnancy and preventing spontaneous preterm birth. Many teams in China have reduced the occurrence of medically induced preterm birth to the greatest extent by developing predictive models for high-risk pregnancies. The treatment of very preterm infants have greatly improved, and the survival rate of extremely preterm infants of 26–28 weeks gestation is close to the level of developed countries. Progress in the field of stillbirth is focused on epidemiological research, pregnancy management, and placental pathology. Prospective cohort studies have found that stillbirth is related to ethnicity, age, education, parity, and pre-pregnancy body mass index. However, the current situation is that the low rate of autopsy of stillbirths and the lack of attention to the results of pathological examinations of tissues such as the placenta and umbilical cord limit the exploration of the causes of stillbirth. In the research on fetal growth restriction (FGR), the largest twin FGR birth cohort in the world has been established by Peking University Third Hospital. For the first time, it was revealed that genetic factors and adverse intrauterine environment shape the early life microbiota, and key gut microbes and their metabolic products are related to the physiological and neurobehavioral development of growth-restricted newborns. For the first time, it was found through animal experiments that platelet-derived growth factor nanoparticles can strengthen placental blood supply and promote nutrient uptake ability in fetal skeletal muscles, thereby achieving fetal development intervention during the intrauterine period. In terms of twins pregnancy, a standardized diagnosis and treatment and referral process for twin pregnancy has been established in China. Placental perfusion, placental blood vessel anastomosis, and Doppler blood flow characteristics are used for etiological exploration. The focus is on multi-omics analysis such as placental omics, twin newborn gut microbiome metagenomics, and metabolomics, exploring the occurrence mechanism of complex twins from different levels.

19.2.2.4 Obstetric pharmacology research

China has formulated a number of norms and standards, such as the "Technical Specifications for Pregnancy Drug Teratogenic Risk Consultation" and the "Expert Consensus on Pregnancy Drug Registration in China", to establish a unified domestic standard for consultation and pregnancy drug registration. For the first time, a nationwide registration of epilepsy medication use during pregnancy has been carried out, laying a foundation for the future establishment of a pregnancy medication data registration system guided by specialty disease medication. Accurate medication research for pregnant women based on the PBPK model is being conducted, and an in vitro model of drug transport through the placental barrier has been established to evaluate the safety of drugs for the fetus.a placental barrier microcurrent biomimetic chip and a "placenta organoid" model were expected to be used to evaluate the safety of drugs during pregnancy. Currently, the use of drugs for pregnancy complicated by immune system diseases lacks high-level evidence-based medical evidence as guidance, which requires multi-disciplinary teams to jointly carry out multi-center clinical research in the field of autoimmune diseases during pregnancy, and to evaluate the safety of standardized treatment options.

19.2.3 Innovation and evolution of obstetric service content and model

In the past 5 years, there has been a significant change in the content and model of obstetric services in China. The connotation of obstetric services is constantly being extended and expanded, from solely obstetric treatment to comprehensive support for physiological, psychological, and social adaptation, from focusing on individual medical care to emphasizing health education and group health care, and improving the self-care ability of pregnant women; the service targets have also expanded from pregnant women and infants to all family members. Evidence-based practices are gradually being applied to obstetrics-related fields, and the development and application of new scientific technologies such as internet medicine and artificial intelligence have brought obstetric medical services to a new stage of development. At the national policy level, China is promoting the construction of national medical centers on the one hand, allowing high-level hospitals to play a leading role in medical research and practices nationwide and even globally; on the other hand, it is building national regional health centers to promote the vertical and horizontal flow of high-quality medical resources, improve the technical level of medical services in the country and various regions, and narrow the gap in medical technology and service levels between regions. At the same time, the implementation of the Five Strategies for Maternal and Newborn Safety has built regional cooperative networks of maternal and newborn care centered on the referral and treatment center for critically ill pregnant women and newborns. The concepts, systems and framework of obstetric quality management are continuously improving, and the evidence-driven and goal-oriented obstetric quality improvement work is being promoted nationwide.

19.3 Comparison of Research Progress in Obstetrics

19.3.1 Comparison of research quantity and quality

From the National Natural Science Foundation of China and the National Key Research and Development Program, funding for research projects in the field of obstetrics has been increasing year by year. The number of articles from China is relatively leading internationally, but the number of high-quality articles still needs improvement, especially in clinical research such as randomized controlled studies. Compared to European and American countries, the percentage of international collaborative research is less than one-fourth, indicating the need to strengthen international collaboration and communication.

19.3.2 Comparison of research methods and techniques

In recent years, several domestic interdisciplinary teams combining basic science and obstetric clinical research have been at the forefront of using various emerging cutting-edge technologies, including single-cell multi-omics techniques and organoid technologies, to explore the mechanisms of obstetric-related diseases in-depth. However, the clinical translation of existing research achievements remains challenging. In contrast, foreign research places more emphasis on addressing clinical problems and conducting high-quality studies on the mechanisms and therapeutic strategies of maternal and fetal diseases.

19.3.3 Research hotspot areas and competitive situation analysis

Research in the field of obstetrics still focuses on the mechanisms and prevention of high-risk pregnancies. China has achieved international prominence in terms of research on early embryonic development and cell differentiation in humans and non-human primates. We have caught up with the international community in areas such as the mechanism of metabolism and nutrient transfer at the maternal-fetal interface, and immune tolerance regulation. Prenatal diagnostic technology, placental implantation surgery, fetal surgery and other technologies have reached an internationally advanced level. However, overall, our country's research on the pathogenesis of many complications and comorbidities is not deep enough. The number of national multi-center large-sample cohort studies and randomized controlled clinical trials is limited. The translation and application of scientific research results are not enough. The quantity of high-level research teams and high-quality research results lags behind developed countries like the United States and the United Kingdom, resulting in limited global influence.

19.4 Development Trends and Suggestions

Faced with new challenges and opportunities under the current population development trend, obstetrics needs to innovate across various aspects, such as talent training, discipline planning, research transformation, service models, and quality management, in order to achieve new breakthroughs in discipline development.

The development of obstetrics in China faces prominent challenges brought about by the complex changes in the characteristics of the pregnant population. Future research should be driven by major scientific issues and clinical needs in obstetrics, achieving a bidirectional development driven by application and basic research, and performing clinical studies such as multicenter cohort studies and randomized controlled studies. Actively exploring new diagnostic and treatment technologies and management models such as multidisciplinary cooperation in treatment, and improving obstetric clinical diagnosis and treatment strategies in our country with high-quality evidence-based medical evidence.

Current major challenges, including the long-term health impact of advanced maternal age, management of major obstetric diseases, integration of large obstetric data information resources, development of sub-specialties such as fetal medicine, etc., urgently require an overall planning, and top-level design for disciplinary development and industrial layout. A nationwide or regional multi-center interdisciplinary joint research network needs to be established to provide a powerful support platform for original innovation. Using national key projects or plans to promote collaborative innovation, supporting multi-disciplinary, multi-center aggregation of scientific research resources to overcome difficulties, and promote clinical practice and health decision-making.

The formulation and implementation of standardized guidelines/consensus, strengthening of medical service quality monitoring, optimization of the quality assessment system and performance management mechanisms, establishment of a scientific, efficient, and fair obstetric quality management model, and exploration of innovative models for quality improvement should be emphasized. Through optimizing medical resources, integrating service content, promoting the improvement and quality enhancement of medical services, a high-quality service chain for the entire childbirth process can be created, promoting maternal and infant health.

20 Urban Science

20.1 Introduction

As component of science and technology, Urban Science has been researched with the goal of building socialism with Chinese characteristics and trying to government policies and discipline, which focuses on important and difficult issues in the progress of urbanization. With the development of our world, the urbanization rate has exceeded 50%, meanwhile Chinese urbanization rate is 66.16% at the end of 2023. As it is a crucial period that urbanization rate is approaching 70%, China would improve developing quality, and urban science is more and more important for the progress.

In the information age, everything in a city is subtly interconnected in networks, and data has become more and more significant. Cities are no longer just containing information of place or space, but a complex system with various elements of urban, which are connected with each other as networks and flows. Urban science has become synthetic discipline supported by complex adaptive system theory, information technology, social physics, urban economics, transportation theory, regional science, and urban geography to study structure and dynamic of urban development and urbanization.

20.2 Latest Research Progress in this Discipline

20.2.1 Theory of urban science

Urbanization is the trend of human social development and the symbol of modernization. According to human settlement theory, following the laws of development, adapting to the situation, seeking benefits and avoiding disadvantages are the ways to promoting urbanization and our living environment. Complex adaptive systems emphasize on agent's adaptive ability which is main difference compared with traditional complex system. Cities as multi-agent system, is complex adaptive system exhibiting various properties: adaption, interconnectivity, scaling and so on. It is necessary to use the theory of complex adaptive system to study urban science in urbanization process. Urban science also emphasizes on inheritance and innovation, that not only continues urban history but oriented to the future of city.

20.2.2 Focused areas

20.2.2.1 Heritage conservation with new technologies and methods

Nowadays, the research and practice of historical and cultural heritage conservation should be carried out with new ideas, new methods and new technologies. Computer, informatics, internet technology, digital twinning, urban information systems and other technology methods are used in the process of historical and cultural heritage conservation. The data and information of historical and cultural resources have been gathered on a three-dimensional information platform by some cities. Combined with digital technology, twinning technology, protections and application of historical and cultural heritage could be performed effectively, which is useful for urban tourism and development.

20.2.2.2 Intelligent city converting to smart city

The design and construction of smart cities should be based on the third generation of systems theory, Complex Adaptive Systems (CAS) theory, which organically combines "top to down composition" with "bottom to up generation". In the process, the government should pay more attention to smart cities' public goods. Building the "multiple pillars' of smart cities" public goods and guiding market agents to use CAS's three — generation mechanisms (building blocks, internal model and tagging) to generate a bottom — up smart city that can evolve autonomously according to the needs of the public. This will further enhance the ability of urban governance and help city governments to become a qualified service oriented government.

20.2.2.3 Low carbon city, healthy city and resilient city

The development of low–carbon city theory has four stages such as concept proposal, connotation definition, evaluation indicators, and practical exploration, Low–carbon city, is still improving and innovating. The core issue of low–carbon city theory is how to construct a low–carbon city development model, which is how to optimize the urban industrial structure, spatial structure, transportation structure, and energy structure for maintaining economic growth while urban carbon emissions is decreasing.

Entering the 21st century, with the rapid advancement of industrialization and urbanization, human society is facing new challenges brought about by population, aging, new diseases, and ecological environment changes. Many countries have proposed goals and strategies of healthy cities at the national level.

Resilient city is the inevitable choice to deal with Black Swan. Its resilience is reflected in three aspects: structural resilience, process resilience and system resilience. The new sources of uncertainty in cities can be summarized as extreme weather, scientific and technological revolution, corona virus and sudden attacks. Ten steps are needed to build a resilient city: first, change the ideology; second, innovate design research institutions; third, make management plan, included in 5–year–plan; fourth, work out the transformation plan of lifeline engineering group; fifth, in every community, every group to supplement the original microcirculation; sixth, use information technology to coordinate each cluster and the micro–circulation facilities within the cluster; seven, transform the old community, make up for the shortcomings of the community unit; eighth, the combination of epidemic disease and disaster should be considered in the construction and reconstruction of public buildings; ninth, build the "micro–center" of "anti–magnetic force" in the surrounding area of the megalopolis; tenth, deepen network format management and build smart cities.

20.2.2.4 Sponge city, eco city and livable city

Sponge city refers to a city that, like a sponge, has good "resilience" in adapting to environmental changes and responding to natural disasters. When it rains, it absorbs, stores, seeps, and purifies water. When it is needed, the stored water is "released" and utilized to enhance the functionality of the urban ecosystem and reduce the occurrence of urban floods.

Ecological city, as a widely recognized urban development model, has been first proposed by UNESCO in the Man and the Biosphere Program in the 1970s. After 50 years of development, its theory has continuously evolved and deepened, and practices have been widely carried out in many cities around the world, covering all continents.

A livable city refers to a city where the economy, society, culture, and environment develop in a coordinated manner, with a good living environment that can meet the material and spiritual needs of residents, and is suitable for human work, life, and residence.

20.2.2.5 Urban renewal and new urban science

The term "urban renewal" comes from the West, and the definition of urban renewal varies internationally. Broadly, urban renewal refers to the urban reconstruction of European cities since the end of World War II, while narrowly, it refers to the urban construction methods to solve urban problems. Urban renewal in China can be divided into four parts, urban old blocks, old factory areas, old residential areas and urban village renovation.

New urban science, which relies on quantitative analysis and data calculation to study cities brings innovation to urban design, with the support of various new technologies such as urban computing, virtual reality, human–computer interaction, and so on. New urban science also proposes that city research should not only focus on urban space, but also learn about urban networks and flows for shaping cities. New urban science can improve understanding of urban systems and structures.

20.3 Comparison of Research Progress in this Discipline at Home and Abroad

Internationally, sustainable development has become consensus, the development of information and digital technology, prompting cities to be smart. In China, new urbanization and carbon peaking and carbon neutrality goals are carried out to promote the high–quality development. Urban science, combined with complex adaptive system and promoted by smart city, which is supported by internet, big data, artificial intelligence, information systems, digital twins and other technologies.

20.4 Development Trend and Prospect of this Subject

20.4.1 Urban science theory system is forming

Urban science relies on traditional economics, geography, sociology and complex science, informatics, artificial intelligence, etc. Combined with complex adaptive system, urbanization and human settlement theory, based on multidisciplinary theoretical research, urban science is gradually formed a theoretical system, which has city as its research object and takes complexity, diversity and mobility as its main characteristics.

20.4.2 Urban science is compatible with disciplines, and becomes more scientific

Urban geography, urban sociology, urban economics, urban management, urban history, urban ecology, urban geographic information systems, and complex science are applied to study city, urban science is supported by kinds of theories. Urban planning and construction change from practice with engineer to practice with research. In recent years, the development of urban science in our country has changed from urban and rural planning to regional spatial planning, which is the same as the trend of international urban planning. Urban planning is more and more concerned with urban economic, social, environmental and so on.

This trend of international and domestic disciplinary transformation, interdisciplinary research on cities, enriches the content of urban science and promotes its development.

20.4.3 Urban science is multidiscipline

Urban science has formed special characteristics through the intersection and recombination of corresponding disciplines. Urban science research is based on the requirements of the country's urban strategy, focuses on the ten major areas, such as history and cultural city, smart city, low-carbon city, healthy city, resilient city, sponge city, ecological city, livable city, urban renewal, and new urban science, and is forming multidiscipline and interdisciplinary subject group, which includes sub-disciplines.

附　件

附件 1 2023 年政府间国际科技创新合作重点专项情况

序号	指南方向	领域方向	拟支持项目数	拟支持经费	合作国家 / 地区	项目实施周期	批次
1	中国和欧洲国家联合实验室合作项目	环境（包括气候变化和碳中和）；卫生健康；农业、食品和生物技术；基础科学；航空航天；人工智能；先进材料；智慧城市；生产技术（包括智能制造）；清洁技术；信息通信技术等	10 个	3000 万元	英国、法国、意大利、西班牙、比利时	3 年	第一批
2	中国和意大利政府间科技合作项目	农业与食品科学；应用于文化遗产的人工智能；天体物理学与物理学；绿色能源相关研究；生物医学	10 个	2000 万元	意大利	2 年	第一批
3	中国和西班牙政府间科技合作项目	智慧城市；生产制造技术，包括智能制造；生物医学与医疗技术，包括可应对全球流行疾病的医疗设备、生物技术应用或制药技术；清洁技术，包括环境、可再生能源或水处理技术；现代农业，包括渔业技术、食品加工和食品安全；先进材料，不包括从原材料到其循环使用阶段会造成环境破坏的材料	不超过 20 个	6000 万元	西班牙	2~3 年	第一批
4	中国和欧盟政府间科技合作项目（一）中国－欧盟研究创新旗舰合作计划项目	食品、农业和生物技术（FAB）	不超过 2 个	7000 万元	欧盟	不超过 4 年	第一批
		气候变化与生物多样性（CCB）	不超过 4 个	6000 万元			
5	中国和欧盟政府间科技合作项目（二）中国－欧盟科技创—15—新合作联合资助机制一般类研究创新合作项目	食品、农业和生物技术及与此相关的领域方向；气候变化和生物多样性及与此相关的领域方向	10 个项目左右	3000 万元	欧盟	不超过 4 年	第一批
6	中国和芬兰政府间科技合作项目	智能绿色能源领域； 智能绿色出行； 智能绿色产业领域； 健康和老龄科技领域	不超过 15 个	6000 万元	芬兰	不超过 3 年	第一批
7	中国和奥地利政府间产学研合作项目	绿色建筑和建筑能效；城市环境中关于碳中和、循环经济（包括制造）和循环资源管理的可持续城市技术；城市环境背景下的可持续交通	4 个左右	1000 万元	奥地利	不超过 3 年	第一批

续表

序号	指南方向	领域方向	拟支持项目数	拟支持经费	合作国家 / 地区	项目实施周期	批次
8	中国和德国政府间科技合作项目	氢能及燃料电池汽车领域	5 个	2000 万元	德国	2~3 年	第一批
9	中国和以色列政府间联合研究项目	医疗研究，包括通过疫苗、药物开发预防和治疗传染性疾病等；水技术，包括海水淡化、饮用水净化等	10 个	2000 万元	以色列	2 年	第一批
10	中国和韩国政府间联合研究项目	生物科技；信息通信；可再生能源；医疗医学；航空航天和气候变化（适应）	6 个	600 万元	韩国	不超过 3 年	第一批
11	中国和韩国政府间能源技术联合研究项目	基于细颗粒物（颗粒物）治理的清洁热电技术（包括碳捕集、利用与封存技术，发电设施安全和燃气安全）；可再生能源（包括氢能和燃料电池技术）	2 个	1500 万元	韩国	不超过 3 年	第一批
12	中国和南非政府间联合研究项目	信息通信（人工智能、大数据、云计算，先进制造、现代矿业和单位管理能力提升相关的信息通信技术）；生物技术（生物技术在现代农业和医疗健康方面的应用）；能源创新（聚焦新能源和可再生能源）；空间科技（导航卫星、射电天文、遥感）；海洋科学	15 个	3000 万元	南非	不超过 2 年	第一批
13	中国和南非政府间联合研究旗舰项目	氢能	1 个	300 万元	南非	不超过 2 年	第一批
14	中国与阿根廷政府间联合研究项目	植物蛋白绿色高效改性关键技术研究及功能产品开发；食物（奶酪、肉类、水果等）加工及产后减损技术创新与新产品开发；纳米微胶囊技术和生物合成技术在食品领域中的应用；食品安全与品质评价：高附加值特色农产品的真实性评价鉴别技术研究与品质评价	4 个	200 万元	阿根廷	2 年	第一批
15	中国与俄罗斯政府间科技合作项目	信息通信技术（人工智能、物联网、数字技术等）；低碳技术（碳捕获、碳封存、节能技术等）；医学与生物技术（新冠病毒疫苗及药物研究等新发传染病防控技术）；新材料（纳米技术等）；合理利用自然资源和环境保护、农业技术；食品科学；新型交通系统；现代机械制造；海洋研究	不超过 20 个	6000 万元	俄罗斯	不超过 2 年	第一批

续表

序号	指南方向	领域方向	拟支持项目数	拟支持经费	合作国家 / 地区	项目实施周期	批次
16	中国和新西兰政府间科技合作项目	食品科学；健康和生物医学；环境科学	6 个	1080 万元	新西兰	—	第二批
17	中国和美国政府间科技合作项目	医药卫生；能源；环境；农业技术；基础科学	70 个	1.05 亿元	美国	—	第二批
18	中国和日本理化学研究所（RIKEN）联合资助项目	医药卫生；环境；基础科学等	10 个	3000 万元	日本	—	第二批
19	中国和埃及政府间联合研究项目	缓解气候变化影响；食品与农业；先进制造；可再生能源；信息通信技术；卫生；液体和固体废物管理；水	10 个	2000 万元	埃及	—	第二批
20	中国和以色列产业技术研发合作项目	不限领域	不超过 10 个	2000 万元	以色列	—	第二批
21	中国和阿拉伯国家联合实验室项目	生命健康；人工智能；绿色低碳；信息通信；空间信息	不超过 5 个	1000 万元	阿拉伯国家（包括阿尔及利亚、阿联酋、阿曼、埃及、巴勒斯坦、巴林、吉布提、卡塔尔、科威特、黎巴嫩、利比亚、毛里塔尼亚、摩洛哥、沙特阿拉伯、苏丹、索马里、突尼斯、叙利亚、也门、伊拉克、约旦、科摩罗）	一般为 3 年	第二批
22	中国和非洲国家联合实验室项目	不限领域	不超过 10 个	2000 万元	非洲国家（阿尔及利亚、埃及、吉布提、利比亚、毛里塔尼亚、摩洛哥、苏丹、索马里、突尼斯、科摩罗等阿拉伯国家也可申报中阿联合实验室，但只能通过一个途径申报）	一般为 3 年	第二批

续表

序号	指南方向	领域方向	拟支持项目数	拟支持经费	合作国家 / 地区	项目实施周期	批次
23	中国和英国政府间流行病预防与抗微生物耐药性“大健康”旗舰挑战计划项目	潜在传染病病原体研究	3 个左右	1300 万元	英国	一般为 3 年	第二批
		从产生到演进的病原体耐药性机制	3 个左右	1300 万元			
		确定多重耐药病原体的驱动因素	2 个左右	2500 万元			
24	中国和马耳他政府间联合研究项目	健康；绿色和蓝色经济转型；数字技术	4 个	640 万元	马耳他	不超过 2 年	第二批
25	中国和欧洲国家联合实验室合作项目	环境（包括气候变化和碳中和）；卫生健康；农业、食品和生物技术；基础科学；文化遗产；航空航天；人工智能；先进材料；信息通信技术；海洋和海事技术；智慧城市；生产技术（包括智能制造）；清洁技术（包括新能源、可再生能源、节能减排）等领域	15 个	3000 万元	英国、法国、意大利、西班牙、葡萄牙、希腊、比利时（仅限瓦隆和布鲁塞尔大区）	一般为 3 年	第二批
26	中国意大利政府间联合研究项目	生物多样性；极地科学；信息通信技术——面向智能制造的人工智能；个性化医疗和用于医疗器械的创新型生物材料	不超过 4 个	160 万元	意大利	2 年	第二批
27	中国—东盟联合研究项目	工业 4.0（如利用信息技术提升先进制造价值链）；数字基础设施（如 5G/6G 组件）；人工智能与先进数据分析；清洁能源（如太阳能、风能、氢能）；医疗技术（如疫苗、诊断方式、药物、医疗器械）	不超过 5 个	1000 万元	合作方必须来自 2 个及以上东盟成员国	3 年	第二批
28	中国和比利时政府间科技合作项目	生物多样性；气候变化	不超过 2 个	300 万元	比利时	2 年半到 3 年	第三批
29	中国和新加坡科技创新合作旗舰项目	清洁能源（海洋能、氢能、储能系统、地热能）	不超过 2 个	2800 万元	新加坡	3 年	第三批

附件 2　2022 年分行业规模以上工业企业研究与试验发展（R&D）经费情况

行业	R&D 经费（亿元）	R&D 经费投入强度（%）
合 计	19361.8	1.39
采矿业	466	0.67
煤炭开采和洗选业	182.6	0.44
石油和天然气开采业	121.8	0.96
黑色金属矿采选业	44.1	0.88
有色金属矿采选业	35.6	0.96
非金属矿采选业	32.2	0.7
开采专业及辅助性活动	49.5	2.03
其他采矿业	0.2	0.97
制造业	18619.6	1.55
农副食品加工业	346	0.58
食品制造业	164.8	0.72
酒、饮料和精制茶制造业	67.7	0.4
烟草制品业	25.8	0.2
纺织业	246.3	0.93
纺织服装、服饰业	117.8	0.79
皮革、毛皮、羽毛及其制品和制鞋业	117	1.03
木材加工和木、竹、藤、棕、草制品业	96	0.91
家具制造业	101.8	1.32
造纸和纸制品业	138.4	0.91
印刷和记录媒介复制业	111.7	1.44
文教、工美、体育和娱乐用品制造业	105.9	0.72
石油、煤炭及其他燃料加工业	170.6	0.27
化学原料和化学制品制造业	1004.9	1.06
医药制造业	1048.9	3.57
化学纤维制造业	171	1.56
橡胶和塑料制品业	535.5	1.76
非金属矿物制品业	628.7	0.92
黑色金属冶炼和压延加工业	816.4	0.94
有色金属冶炼和压延加工业	505.1	0.67
金属制品业	757.5	1.53
通用设备制造业	1190.6	2.46

续表

行业	R&D 经费（亿元）	R&D 经费投入强度（%）
专用设备制造业	1150.1	2.96
汽车制造业	1651.7	1.83
铁路、船舶、航空航天和其他运输设备制造业	633.2	4.64
电气机械和器材制造业	2098.5	2.02
计算机、通信和其他电子设备制造业	4099.9	2.63
仪器仪表制造业	354.1	3.53
其他制造业	70.5	3.18
废弃资源综合利用业	70.2	0.61
金属制品、机械和设备修理业	22.9	1.32
电力、热力、燃气及水生产和供应业	276.2	0.24
电力、热力生产和供应业	217.9	0.23
燃气生产和供应业	37.6	0.24
水的生产和供应业	20.7	0.45

附件 3　2022—2023 香山科学会议学术讨论会一览

序号	会议号	会议主题	召开日期
2022 年香山科学会议学术讨论会一览			
1	715	低碳生物合成	2022/7/21
2	716	单分子科学与技术	2022/8/2
3	717	远距离大容量连续无线功率传输的科学与技术问题	2022/8/15
4	718	中国居民营养素摄入量和慢性病防控	2022/8/18
5	S66	“宁静中国”与噪声治理关键技术	2022/8/22
6	719	食物系统转型的科学与创新	2022/8/23
7	S68	地球系统与全球变化	2022/8/24
8	720	数字眼科与全身疾病认知方法与关键技术	2022/8/25
9	722	黑土地保护与利用	2022/8/30
10	723	面向人体系统调控的生命间质结构、功能与行为	2022/8/30
11	724	大数据驱动中西医结合精准诊疗	2022/9/1
12	726	细胞治疗临床研究和监管	2022/9/3
13	725	超快光电子成像技术及应用	2022/9/3
14	727	健康医学的理论体系和技术方法	2022/9/5
15	728	中药资源的可持续开发与利用	2022/9/8
16	729	长寿命道路材料本构与结构力学行为理论	2022/9/15

续表

序号	会议号	会议主题	召开日期
17	730	激光加速器和深空探测：机遇和挑战	2022/9/20
18	732	火星探测中重要科学与技术问题	2022/9/22
19	733	基于多学科视角的“脑－肠－微生态”前沿与技术	2022/9/24
20	Y6	智能医学的基础理论与关键技术	2022/9/26
21	734	耕地内涵外延及其调查监测评价	2022/9/27
22	735	新型精神健康诊疗技术的挑战与机遇	2022/11/9
23	736	实现建筑碳中和的挑战与应对技术路径	2022/11/14
24	S69	科学数据治理与利用的前沿和热点	2022/11/19
2023 年香山科学会议学术讨论会一览			
1	737	如何加速解决我国粮油危机和多元化问题	2023/2/9
2	738	多学科交叉融合与整合药理学发展	2023/2/17
3	739	本征柔性电子学的科学问题和技术瓶颈前沿	2023/2/20
4	740	超常环境介质力学与工程中的关键科学问题	2023/3/30
5	741	卫生健康高质量发展趋势——“数字卫生”	2023/4/12
6	721	脊柱畸形预防与诊治关键科学技术	2023/4/15
7	743	多场条件超高分辨近场光学技术	2023/4/18
8	744	主动健康理论、方法与技术	2023/4/20
9	Y7	集成微腔光梳物理基础与重大应用	2023/4/25
10	745	无膜细胞器	2023/4/27
11	747	悬浮偶极场磁约束概念前沿科学和关键技术问题	2023/5/16
12	748	地下储能	2023/5/18
13	750	智能纳米药物——智慧医疗时代的药物新样态	2023/5/30
14	751	极端磁场下核磁共振的科学机遇与技术挑战	2023/7/11

附件 4　2022 年度“中国科学十大进展”

序号	进展名称
1	祝融号巡视雷达揭秘火星乌托邦平原浅表分层结构
2	FAST 精细刻画活跃重复快速射电暴
3	全新原理实现海水直接电解制氢
4	揭示新冠病毒突变特征与免疫逃逸机制
5	实现高效率的全钙钛矿叠层太阳能电池和组件
6	新原理开关器件为高性能海量存储提供新方案
7	实现超冷三原子分子的量子相干合成

续表

序号	进展名称
8	温和压力条件下实现乙二醇合成
9	发现飞秒激光诱导复杂体系微纳结构新机制
10	实验证实超导态“分段费米面”

附件 5　2023 年度“中国科学十大进展”

序号	进展名称
1	人工智能大模型为精准天气预报带来新突破
2	揭示人类基因组暗物质驱动衰老的机制
3	发现大脑“有形”生物钟的存在及其节律调控机制
4	农作物耐盐碱机制解析及应用
5	新方法实现单碱基到超大片段 DNA 精准操纵
6	揭示人类细胞 DNA 复制起始新机制
7	“拉索”发现史上最亮伽马暴的极窄喷流和十万亿电子伏特光子
8	玻色编码纠错延长量子比特寿命
9	揭示光感受调节血糖代谢机制
10	发现锂硫电池界面电荷存储聚集反应新机制

附件 6　2022 年“中国十大科技进展新闻”

序号	科技进展
1	中国天眼 FAST 取得系列重要进展
2	中国空间站完成在轨建造并取得一系列重大进展
3	我国科学家发现玉米和水稻增产关键基因
4	科学家首次发现并证实玻色子奇异金属
5	我国科学家将二氧化碳人工合成葡萄糖和脂肪酸
6	我国迄今运载能力最大固体运载火箭“力箭一号”首飞成功
7	“夸父一号”发射成功，并发布首批科学图像
8	新技术可在海水里原位直接电解制氢
9	国家重大科技基础设施“稳态强磁场实验装置”实现重大突破
10	“巅峰使命”珠峰科考创造多项新纪录

附件 7 2023 年“中国十大科技进展新闻”

序号	科技进展
1	全球首座第四代核电站商运投产
2	“神舟十六号”返回 空间站应用与发展阶段首次载人飞行任务圆满完成
3	超越硅基极限的二维晶体管问世
4	我国科学家发现耐碱基因可使作物增产
5	天问一号研究成果揭示火星气候转变
6	我国首个万米深地科探井开钻
7	液氮温区镍氧化物超导体首次发现
8	FAST 探测到纳赫兹引力波存在证据
9	世界首个全链路全系统空间太阳能电站地面验证系统落成启用
10	科学家阐明嗅觉感知分子机制

附件 8 2023 年中国科学院值得回顾的科技成果

序号	科技成果
1	“拉索”完整记录“宇宙烟花”爆发全过程 精确测量迄今最亮伽马暴的高能辐射能谱
2	中国国家太空实验室正式运行 首次公布中国空间站全貌高清图像
3	构建量子计算原型机“九章三号”刷新光量子信息技术世界纪录
4	全球光学时域巡天能力最强设备 墨子巡天望远镜正式投入观测
5	中国天眼看到“时空的涟漪”发现纳赫兹引力波存在的关键性证据
6	我国科学家揭示衰老新机制 为衰老相关疾病防治提供新的策略
7	作物主效耐碱基因及其作用机制首次揭示 在盐碱地上能够促进粮食增产
8	圆环阵太阳射电成像望远镜正式建成 可监测太阳爆发活动
9	发现 M87 星系黑洞喷流周期性进动 符合爱因斯坦广义相对论预测
10	实现世界最大规模 51 比特量子纠缠态制备 刷新世界纪录
11	人体免疫系统发育图谱绘制 有助于深入理解免疫系统功能和调控机制
12	中国“人造太阳”EAST 获重大突破 创造高约束模式等离子体运行时长新纪录
13	中国首次火星探测火星全球影像图发布 为人类深入认知火星作出中国贡献
14	我国科考队首次登顶卓奥友峰 首次在该区域开展综合科学考察
15	300MW 先进压缩空气储能系统膨胀机 我国先进压缩空气储能技术迈上新台阶
16	新一代国产 CPU 龙芯 3A6000 发布 自主研发、自主可控

附件 9　2022—2023 年未来科学大奖获奖者

年度	奖项	获奖者	获奖理由
2022	生命科学奖	李文辉	奖励他发现了乙型和丁型肝炎病毒感染人的受体为钠离子－牛磺胆酸共转运蛋白（NTCP），有助于开发更有效的治疗乙型和丁型肝炎的药物
2022	物质科学奖	杨学明	奖励他研发新一代高分辨率和高灵敏度量子态分辨的交叉分子束科学仪器，揭示了化学反应中的量子共振现象和几何相位效应
2022	数学与计算机科学奖	莫毅明	奖励他创立了极小有理切线簇 (VMRT) 理论并用以解决代数几何领域的一系列猜想，以及对志村簇上的 Ax–Schanuel 猜想的证明
2023	生命科学奖	柴继杰	奖励他们为发现抗病小体并阐明其结构和在抗植物病虫害中的功能作出的开创性工作
2023	生命科学奖	周俭民	奖励他们为发现抗病小体并阐明其结构和在抗植物病虫害中的功能作出的开创性工作
2023	物质科学奖	赵忠贤	表彰他们对高温超导材料的突破性发现和对转变温度的系统性提升所作出的开创性贡献
2023	物质科学奖	陈仙辉	表彰他们对高温超导材料的突破性发现和对转变温度的系统性提升所作出的开创性贡献
2023	数学与计算机科学奖	何恺明	奖励他们提出深度残差学习，为人工智能作出了基础性贡献
2023	数学与计算机科学奖	孙剑	奖励他们提出深度残差学习，为人工智能作出了基础性贡献
2023	数学与计算机科学奖	任少卿	奖励他们提出深度残差学习，为人工智能作出了基础性贡献
2023	数学与计算机科学奖	张祥雨	奖励他们提出深度残差学习，为人工智能作出了基础性贡献

附件 10　第十七届中国青年科技奖

中国青年科技奖特别奖		
序号	姓名	工作单位
1	王磊	复旦大学
2	王志鹏	北京航空航天大学
3	朱美萍（女）	中国科学院上海光学精密机械研究所
4	刘洪涛	中国石油天然气股份有限公司塔里木油田分公司
5	杨丽（女）	西安电子科技大学
6	杨健	系统工程研究院
7	宋海军	中国地质大学（武汉）

续表

序号	姓名	工作单位
8	范大明	江南大学
9	周颖（女）	同济大学
10	黄佳琦	北京理工大学

中国青年科技奖

序号	姓名	工作单位
1	万蕊雪（女）	西湖大学
2	王伟	南京大学
3	王拓	天津大学
4	王芳（女）	中国汽车技术研究中心有限公司
5	王艳（女）	中国医学科学院肿瘤医院
6	王博	中国科学院云南天文台
7	王琦	中国矿业大学（北京）
8	王磊	复旦大学
9	王双印	湖南大学
10	王志鹏	北京航空航天大学
11	王宏伟	山东农业大学
12	王祥喜	中国科学院生物物理研究所
13	尤延铖	厦门大学
14	毛相朝	中国海洋大学
15	方璐（女）	清华大学
16	方博汉	北京大学
17	尹升华	北京科技大学
18	邓方	北京理工大学
19	邓海啸	中国科学院上海高等研究院
20	龙笛	清华大学
21	田晖	北京大学
22	田怀玉	北京师范大学
23	田贵华（女）	北京中医药大学东直门医院
24	史浩飞	中国科学院重庆绿色智能技术研究院
25	付长庚	中国中医科学院西苑医院
26	冯旭	北京大学
27	朱美萍（女）	中国科学院上海光学精密机械研究所
28	乔庆庆（女）	中国科学院新疆生态与地理研究所
29	乔英云（女）	中国石油大学（华东）

续表

序号	姓名	工作单位
30	任玉龙	中国农业科学院作物科学研究所
31	刘明	北京农学院
32	刘英（女）	西安电子科技大学
33	刘斌	山东大学
34	刘瑜	海军航空大学
35	刘奕群	清华大学
36	刘洪涛	中国石油天然气股份有限公司塔里木油田分公司
37	许操	中国科学院遗传与发育生物学研究所
38	孙涛	中国科学院沈阳应用生态研究所
39	孙庆丰	浙江农林大学
40	李昂	中国科学院上海有机化学研究所
41	李昺	中国科学院金属研究所
42	李敏	武汉大学
43	李帝铨	中南大学
44	李隆球	哈尔滨工业大学
45	杨丽（女）	西安电子科技大学
46	杨健	系统工程研究院
47	杨元合	中国科学院植物研究所
48	肖振宇	北京航空航天大学
49	吴云	空军工程大学
50	吴巨友	南京农业大学
51	吴志勇	61001 部队
52	吴富梅（女）	61540 部队
53	邱雷	南京航空航天大学
54	何耀	苏州大学
55	何蓉蓉（女）	暨南大学
56	余倩（女）	浙江大学
57	余碧莹（女）	北京理工大学
58	汪萌	合肥工业大学
59	沈少华	西安交通大学
60	宋勇峰	山东第一医科大学附属省立医院
61	宋海军	中国地质大学（武汉）
62	张丰收	同济大学
63	张东菊（女）	兰州大学
64	张冀聪	北京航空航天大学
65	陈娟（女）	重庆医科大学

续表

序号	姓名	工作单位
66	陈蓉（女）	华中科技大学
67	陈韬	中国电子科技集团公司第五十五研究所
68	陈蕾（女）	四川大学华西医院
69	陈浩森	北京理工大学
70	范大明	江南大学
71	郁昱	上海交通大学
72	欧欣	中国科学院上海微系统与信息技术研究所
73	欧阳斌	海军工程大学
74	周颖（女）	同济大学
75	周欢萍（女）	北京大学
76	郑翠娥（女）	哈尔滨工程大学
77	赵远锦	南京大学医学院附属鼓楼医院
78	赵晓丽（女）	中国环境科学研究院
79	郝格非	贵州大学
80	胡殿印（女）	北京航空航天大学
81	钟武律	核工业西南物理研究院
82	胥蕊娜（女）	清华大学
83	袁星	南京信息工程大学
84	袁荃（女）	湖南大学
85	耿华	清华大学
86	柴人杰	东南大学
87	徐峰	山东大学齐鲁医院
88	徐通达	福建农林大学
89	高波（女）	中国科学院理化技术研究所
90	郭兆将	中国农业科学院蔬菜花卉研究所
91	黄火清	中国农业科学院北京畜牧兽医研究所
92	黄佳琦	北京理工大学
93	崔宁博	四川大学
94	符利勇	中国林业科学研究院资源信息研究所
95	董捷	北京空间飞行器总体设计部
96	董焕丽（女）	中国科学院化学研究所
97	程鹏	浙江大学
98	程方益	南开大学
99	谭韬	昆明理工大学
100	谭敏佳	中国科学院上海药物研究所

附件 11　2023 年国家工程师奖

国家卓越工程师拟表彰对象（按姓氏笔画排序）

序号	姓名	工作单位
1	丁文红（女）	武汉科技大学
2	万步炎	湖南科技大学
3	王军	中国中车集团有限公司
4	王珏	中国运载火箭技术研究院
5	王大轶	北京空间飞行器总体设计部
6	王仁坤	中国电建集团成都勘测设计研究院有限公司
7	王过中	中国人民解放军 61886 部队
8	王建华	中国人民解放军 93114 部队
9	王海峰	北京百度网讯科技有限公司
10	王维庆	新疆大学
11	王增全	中国北方发动机研究所
12	方向晨	中石化（大连）石油化工研究院有限公司
13	叶浩文	中建科技集团有限公司
14	史聪灵	中国安全生产科学研究院
15	朱衍波	民航数据通信有限责任公司
16	任国春	中国人民解放军陆军工程大学
17	刘书杰	中海石油（中国）有限公司海南分公司
18	刘继忠	探月与航天工程中心
19	刘清宇	中国人民解放军海军研究院
20	刘增宏	自然资源部第二海洋研究所
21	闫大鹏	武汉锐科光纤激光技术股份有限公司
22	严卫钢	中国航天科工集团有限公司
23	苏权科	香港科技大学（广州）
24	杜选民	汉江实验室
25	李平（女）	中国铁道科学研究院集团有限公司
26	李久林	北京城建集团有限责任公司
27	李少平	湖北兴发化工集团股份有限公司
28	李永胜	山东天瑞重工有限公司
29	李先广	重庆机电智能制造有限公司
30	李红霞（女）	中钢集团洛阳耐火材料研究院有限公司
31	李恒年	中国西安卫星测控中心
32	吴凯	宁德时代新能源科技股份有限公司

续表

序号	姓名	工作单位
33	吴晓光	中国船舶集团有限公司
34	邱旭华	公安部第一研究所
35	汪小刚	中国水利水电科学研究院
36	宋神友	深中通道管理中心
37	张弘	江西洪都航空工业集团有限责任公司
38	张军	中国科学院微小卫星创新研究院
39	张勇	沈鼓集团股份有限公司
40	张志清	国家卫星气象中心
41	张来勇	中国寰球工程有限公司
42	张利民	香港科技大学
43	张金涛	中国计量科学研究院
44	张春生	中国电建集团华东勘测设计研究院有限公司
45	张春江	中国农业科学院农产品加工研究所
46	张修社	中国电子科技集团有限公司
47	陆铭华	中国人民解放军海军潜艇学院
48	陈勇	中国商用飞机有限责任公司
49	林明智	广西柳工机械股份有限公司
50	林铁坚	广西玉柴机器集团有限公司
51	林毅峰	上海勘测设计研究院有限公司
52	易小刚	三一集团有限公司
53	周琦	贵州省地质调查院
54	周常河	暨南大学
55	周福建	中国石油大学（北京）
56	单增海	徐工集团工程机械股份有限公司
57	房子河	公安部大数据中心
58	赵斗	中国铁路设计集团有限公司
59	胡建华	湖南轨道交通控股集团有限公司
60	洪开荣	中铁隧道局集团有限公司
61	洪家光	中国航发沈阳黎明航空发动机有限责任公司
62	贺建华	东方电气风电股份有限公司
63	顾明	中交天津航道局有限公司
64	钱林方	中国兵器工业集团有限公司
65	徐先英	甘肃省治沙研究所
66	高成臣	北京大学
67	曹堪宇	长鑫存储技术有限公司

续表

序号	姓名	工作单位
68	崔鹤	青岛海关技术中心
69	梁建英（女）	国家高速列车青岛技术创新中心
70	彭云彪	核工业二〇八大队
71	蒋开喜	紫金矿业集团股份有限公司
72	韩佳彤	呼和浩特市现代信息技术学校
73	覃大清	哈尔滨电气集团有限公司
74	景来红	黄河勘测规划设计研究院有限公司
75	程芳琴（女）	山西大学
76	廉玉波	比亚迪股份有限公司
77	窦强	飞腾信息技术有限公司
78	蔡蔚	哈尔滨理工大学
79	蔡树军	中国电子科技集团公司第五十八研究所
80	谭旭光	山东重工集团有限公司
81	熊大和	赣州金环磁选科技装备股份有限公司
82	熊盛青	中国自然资源航空物探遥感中心
83	薛峰	中国中建设计研究院有限公司

国家卓越工程师团队拟表彰对象（按团队名称首字笔画排序）

序号	团队名称	依托单位
1	5G 标准与产业创新团队	中国信息通信研究院
2	12 英寸减压外延团队	北京北方华创微电子装备有限公司
3	400 万吨 / 年煤间接液化成套技术创新开发及产业化团队	国家能源集团宁夏煤业有限责任公司
4	工业废水治理技术与装备团队	南京大学
5	工业透平研发创新团队	杭州汽轮动力集团股份有限公司
6	大气污染物与温室气体协同控制团队	清华大学
7	大庆油田化学驱油技术研发团队	大庆油田有限责任公司
8	大型水轮发电机组安装与调试团队	中国水利水电第四工程局有限公司
9	广汽动力总成自主研发团队	广州汽车集团股份有限公司
10	天河超级计算创新应用团队	国家超级计算天津中心
11	云南省三江成矿系统与评价创新团队	昆明理工大学
12	中国天眼工程团队	中国科学院国家天文台
13	中核集团“华龙一号”创新团队	中国核电工程有限公司
14	水库大坝安全与管理创新团队	水利部交通运输部国家能源局南京水利科学研究院
15	化合物芯片技术团队	中国电科产业基础研究院

续表

序号	团队名称	依托单位
16	网络信息系统科研创新团队	中国人民解放军军事科学院
17	先进飞行器技术研发团队	中国航天科技集团有限公司第一研究院
18	先进发动机研制团队	中国航空发动机集团有限公司沈阳发动机研究所
19	先进核电系统堆芯支撑及堆内装置高端制造研究团队	上海第一机床厂有限公司
20	全球数值天气预报系统工程技术团队	中国气象局地球系统数值预报中心
21	军委联合参谋部某研究团队	军委联合参谋部
22	“两观三性”建筑创新实践与研究团队	华南理工大学
23	苏博特重大基础设施工程材料创新团队	江苏苏博特新材料股份有限公司
24	歼 -20 飞机研制团队	中国航空工业集团有限公司成都飞机设计研究所
25	青藏高原地质资源工程团队	西藏大学
26	城市轨道交通系统安全与运维保障国家工程研究中心	广州地铁集团有限公司
27	贵州交通山区峡谷桥梁建造技术团队	贵州交通建设集团有限公司
28	重型高端复杂锻件制造技术变革性创新研究团队	中国一重集团有限公司
29	复兴号高速列车研发创新团队	中国国家铁路集团有限公司
30	信息显示玻璃研发和产业化团队	中建材玻璃新材料研究院集团有限公司
31	盾构创新研发团队	中铁工程装备集团有限公司
32	起重机械技术创新团队	中联重科股份有限公司
33	核燃料专用装备研发创新团队	核工业理化工程研究院
34	特高压直流与柔性输电高端装备攻关团队	南京南瑞继保电气有限公司
35	特高压柔性直流输电技术研发团队	中国南方电网有限责任公司
36	高性能大跨度空间结构工作室	中国建筑西南设计研究院有限公司
37	高速铁路大跨度桥梁创新团队	中铁大桥勘测设计院集团有限公司
38	高端装备轻合金铸造技术科技创新团队	中国机械总院集团沈阳铸造研究所有限公司
39	高端聚氨酯原料 ADI 全产业链技术攻关团队	万华化学集团股份有限公司
40	海康威视创新团队	杭州海康威视数字技术股份有限公司
41	救捞工程关键技术攻关团队	交通运输部上海打捞局
42	眼科诊疗技术研发团队	中国医学科学院生物医学工程研究所
43	液氧煤油发动机工程师团队	中国航天科技集团有限公司第六研究院
44	超导材料制备及应用技术创新团队	西部超导材料科技股份有限公司
45	超级建筑工程设计创新团队	北京市建筑设计研究院有限公司
46	智能微系统团队	启元实验室
47	敦煌研究院文物保护团队	敦煌研究院
48	新型水下装备研制团队	中国船舶集团有限公司第七一九研究所
49	煤矿瓦斯防治与智能绿色开采团队	中国平煤神马控股集团有限公司
50	煤矿安全开采地质保障与生态修复团队	中国矿业大学（北京）

附件 12 新基石研究员项目首期资助名单

领域	学科	姓名	性别	工作单位
数学与物质科学	数学	何旭华	男	香港中文大学
		林华珍	女	西南财经大学
		刘钢	男	华东师范大学
		刘若川	男	北京大学
		沈维孝	男	复旦大学
		孙斌勇	男	浙江大学
		张旭	男	四川大学
	物理	曹俊	男	中国科学院高能物理研究所
		丁洪	男	上海交通大学
		段路明	男	清华大学
		封东来	男	中国科学技术大学
		胡江平	男	中国科学院物理研究所
		刘继峰	男	中国科学院国家天文台
		刘仁保	男	香港中文大学
		陆朝阳	男	中国科学技术大学
		童利民	男	浙江大学
		王亚愚	男	清华大学
		吴从军	男	西湖大学
		张霜	男	香港大学
		张远波	男	复旦大学
	化学	陈鹏	男	北京大学
		樊春海	男	上海交通大学
		黎书华	男	南京大学
		李景虹	男	清华大学
		马丁	男	北京大学
		王兵	男	中国科学技术大学
		吴骊珠	女	中国科学院理化技术研究所
		杨黄浩	男	福州大学
		游书力	男	中国科学院上海有机化学研究所
		俞书宏	男	中国科学技术大学

续表

领域	学科	姓名	性别	工作单位
生物与医学科学		曹彬	男	中国医学科学院北京协和医学院
		陈玲玲	女	中国科学院分子细胞科学卓越创新中心
		董晨	男	上海交通大学
		傅向东	男	中国科学院遗传与发育生物学研究所
		郭红卫	男	南方科技大学
		胡海岚	女	浙江大学
		黄志伟	男	哈尔滨工业大学
		赖仞	男	中国科学院昆明动物研究所
		李栋	男	中国科学院生物物理研究所
		李毓龙	男	北京大学
		刘颖	女	北京大学
		鲁伯埙	男	复旦大学
		罗敏敏	男	北京脑科学与类脑研究中心
		祁海	男	清华大学
		瞿礼嘉	男	北京大学
		邵峰	男	北京生命科学研究所
		时松海	男	清华大学
		王二涛	男	中国科学院分子植物科学卓越创新中心
		王文	男	西北工业大学
		王晓群	男	北京师范大学
		颉伟	男	清华大学
		徐彦辉	男	复旦大学
		于洪涛	男	西湖大学
		曾艺	女	中国科学院分子细胞科学卓越创新中心
		张宏	男	中国科学院生物物理研究所
		周斌	男	中国科学院分子细胞科学卓越创新中心
		朱冰	男	中国科学院生物物理研究所
		朱听	男	西湖大学

附件 13　新基石研究员项目第二期资助名单

领域	学科	姓名	性别	工作单位
数学与物质科学	数学与理论计算机	董彬	男	北京大学
		肖梁	男	北京大学
		薛金鑫	男	清华大学
		尹一通	男	南京大学
		于品	男	清华大学
	物理学	陈宇翱	男	中国科学技术大学
		戴希	男	香港科技大学
		江颖	男	北京大学
		李菂	男	中国科学院国家天文台
		刘江来	男	上海交通大学
		彭承志	男	中国科学技术大学
		孙超	男	清华大学
		肖云峰	男	北京大学
		姚望	男	香港大学
	化学	陈春英	女	国家纳米科学中心
		程建军	男	西湖大学
		高毅勤	男	北京大学
		焦宁	男	北京大学
		刘国生	男	中国科学院上海有机化学研究所
		刘磊	男	清华大学
		郑南峰	男	厦门大学
生物与医学科学	生物学	陈良怡	男	北京大学
		陈学伟	男	四川农业大学
		丁胜	男	清华大学
		高彩霞	女	中国科学院遗传与发育生物学研究所
		胡凤益	男	云南大学
		金鑫	男	华东师范大学
		李国红	男	武汉大学
		刘默芳	女	中国科学院分子细胞科学卓越创新中心
		刘清华	男	北京生命科学研究所
		彭汉川	男	东南大学
		戚益军	男	清华大学

续表

领域	学科	姓名	性别	工作单位
生物与医学科学	生物学	汤富酬	男	北京大学
		王佳伟	男	中国科学院分子植物科学卓越创新中心
		王四宝	男	中国科学院分子植物科学卓越创新中心
		肖百龙	男	清华大学
		徐浩新	男	良渚实验室
		薛天	男	中国科学技术大学
	医学科学	程功	男	清华大学
		雷群英	女	复旦大学
		黄秀娟	女	香港中文大学医学院
		苏士成	男	中山大学孙逸仙纪念医院
		孙金鹏	男	山东大学
		徐文东	男	复旦大学附属华山医院
		杨胜勇	男	四川大学
		张雁	男	天津大学

注　释

研究与试验发展（R&D）： 指在科学技术领域，为增加知识总量以及运用这些知识去创造新的应用而进行的系统的、创造性的活动，包括基础研究、应用研究、试验发展三类活动。

基础研究： 指为了获得关于现象和可观察事实的基本原理的新知识（揭示客观事物的本质、运动规律，获得新发展、新学说）而进行的实验性或理论性研究，它不以任何专门或特定的应用或使用为目的。

应用研究： 指为获得新知识而进行的创造性研究，主要针对某一特定的目的或目标。应用研究是为了确定基础研究成果可能的用途，或是为达到预定的目标探索应采取的新方法（原理性）或新途径。

试验发展： 指利用从基础研究、应用研究和实际经验所获得的现有知识，为产生新的产品、材料和装置，建立新的工艺、系统和服务，以及对已产生和建立的上述各项作实质性的改进而进行的系统性工作。

R&D 经费： 统计年度内全社会实际用于基础研究、应用研究和试验发展的经费支出。包括实际用于研究与试验发展活动的人员劳务费、原材料费、固定资产购建费、管理费及其他费用支出。

R&D 经费投入强度： 全社会 R&D 经费支出与国内生产总值（GDP）之比。

研究人员： 指 R&D 人员中具备中级以上职称或博士学历（学位）的人员。

R&D 人员全时当量： 是国际上通用的、用于比较科技人力投入的指标。指 R&D 全时人员（全年从事 R&D 活动累积工作时间占全部工作时间的 90% 及以上人员）工作量与非全时人员按实际工作时间折算的工作量之和。例如：有 2 个 R&D 全时人员（工作时间分别为 0.9 年和 1 年）和 3 个 R&D 非全时人员（工作时间分别为 0.2 年、0.3 年和 0.7 年），则 R&D 人员全时当量 =1+1+0.2+0.3+0.7=3.2（人年）。

规模以上工业企业： 年主营业务收入 2000 万元及以上的工业法人单位。

发文量： 基于各数据库统计的论文数量，其中 WOS 论文量是指被 WOS 核心合集中的四大期刊数据库收录的论文数量。

国际被引频次： 指论文被来自 WOS 核心合集论文引用的次数。

国内被引频次： 指论文被来自《中国学术期刊（网络版）》《中国博士学位论文全文数据库》《中国优秀硕士学位论文全文数据库》《国内外重要会议论文全文数据库》文献引用的总次数。

影响因子： 影响因子（Impart Factor，IF）是科睿唯安（Clarivate Analytics）出品的期刊引证报告（Journal Citation Reports，JCR）中的一项数据。即某期刊前两年发表的论文在该报告年份（JCR year）中被引用总次数除以该期刊在这两年内发表的论文总数。这是一个国际上通行的期刊评价指标。

国际合作论文： 指由两个或两个以上国家和 / 或地区作者合作发表的被 WOS 收录的论文。本报告中，中国的国际合作论文特指中国大陆学者与国外学者合作发表的论文。合作论文的计数方式为，每一篇合作论文在每个参与国家和（或）地区中均计作一篇发文。

高被引论文占比：指基于合作论文总量的高被引论文占比。若国际合作论文中高被引论文总数为 A，国际合作论文总量为 B，则高被引论文占比为 A/B。

高水平论文：指 ESI 数据库中遴选的高被引论文和热点论文之和。

ESI 22 学科：ESI 设置的 22 个学科为生物学与生物化学、化学、计算机科学、经济与商业、工程学、地球科学、材料科学、数学、综合交叉学科、物理学、社会科学总论、空间科学、农业科学、临床医学、分子生物学与遗传学、神经系统学与行为学、免疫学、精神病学与心理学、微生物学、环境科学与生态学、植物学与动物学、药理学和毒理学。

新基石研究员项目：是一项聚焦原始创新、鼓励自由探索、公益属性的新型基础研究资助项目。在中国科学技术协会的指导下，由科学家主导，腾讯公司出资，独立运营。该项目旨在长期稳定地支持一批杰出科学家潜心基础研究、实现“从 0 到 1”的原始创新，推动科学发展，增进人类福祉。腾讯公司将在 10 年内投入 100 亿元人民币，支持在中国内地及港澳地区全职工作的科学家。